**Stability, Instability and Chaos:**
an introduction to the theory of
nonlinear differential equations

**Cambridge Texts in Applied Mathematics**

Maximum and Minimum Principles
M.J. SEWELL

Solitons
P.G. DRAZIN AND R.S. JOHNSON

The Kinematics of Mixing
J.M. OTTINO

Introduction to Numerical Linear Algebra and Optimisation
PHILIPPE G. CIARLET

Integral Equations
DAVID PORTER AND DAVID S.G. STIRLING

Perturbation Methods
E.J. HINCH

The Thermomechanics of Plasticity and Fracture
GERARD A. MAUGIN

Boundary Integral and Singularity Methods for Linearized Viscous Flow
C. POZRIKIDIS

Nonlinear Systems
P.G. DRAZIN

Stability, Instability and Chaos
PAUL GLENDINNING

Applied Analysis of the Navier-Stokes Equations
C.R. DOERING AND J.D. GIBBON

Viscous Flow
H. OCKENDON AND J.R. OCKENDON

# *Stability, Instability and Chaos:*

## *an introduction to the theory of nonlinear differential equations*

PAUL GLENDINNING
*Department of Applied Mathematics and Theoretical Physics*
*University of Cambridge*

CAMBRIDGE UNIVERSITY PRESS
Cambridge, New York, Melbourne, Madrid, Cape Town, Singapore,
São Paulo, Delhi, Dubai, Tokyo, Mexico City

Cambridge University Press
The Edinburgh Building, Cambridge CB2 8RU, UK

Published in the United States of America by
Cambridge University Press, New York

www.cambridge.org
Information on this title: www.cambridge.org/9780521425667



First published 1994
Reprinted 1995, 1996, 1999

*A catalogue record for this publication is available from the British Library*

*Library of Congress Cataloguing in Publication Data*

ISBN 978-0-521-41553-8 Hardback
ISBN 978-0-521-42566-7 Paperback

For Fiona

# *Contents*

# *Preface*

As the theory of dynamical systems and chaos develops, more and more recent results are filtering through to undergraduate courses. The aim of this book is to provide a coherent account of the qualitative theory of ordinary differential equations which deals in an even handed and consistent way with both the 'classical' results of Poincaré and Liapounov and the more recent advances in bifurcation theory and chaos. The book covers two undergraduate courses: a first course in nonlinear differential equations and an introduction to bifurcation theory.

Throughout, the emphasis is on understanding and the ability to apply theory to examples rather than on rigorous mathematical developments. Although there are theorems, the level of rigour is not that of a pure mathematical text. None the less, it is vital to appreciate the restrictions and limitations of any method and so wherever possible I have stated results in a precise form. The choice of topics has also been influenced by a desire to cover material which can be examined sensibly.

This book has developed out of courses given to third year undergraduates at the University of Warwick and the University of Cambridge. In both places I have been fortunate to inherit the notes of previous lecturers: Tony Pritchard at Warwick, Peter Swinnerton-Dyer, John Hinch and others at Cambridge. The first seven chapters owe an enormous debt to these people and in some places it is hard for me to see where they end and I begin (although I retain full responsibility for any errors). There are a number of exercises, some in the main text and some at the end of chapters. Those in the main text are intended to help the reader follow some argument. There is some repetition of important exercises and some exercises are answered later in the text. Many of the exercises at the end of the chapters are taken from example sheets and examination questions from the third year courses on *Nonlinear Differential Equations* and *Dynamical Systems* at the University of Cambridge.

James Glover, Alistair Mees and a class at the University of Western Australia in Perth acted as guinea pigs for an early version of the book. Their comments and corrections were immensely useful. An anonymous reader for CUP also made a number of good suggestions, and I would like to thank Alan Harvey, Roger Astley and David Tranah at CUP for their support. The diagrams were expertly produced by Margaret Downing (by hand) and Alastair Rucklidge (by computer). I have received help and encouragement from many people whilst writing this book. In particular, Bob Devaney, Toby Hall, Guilermo Procida, Mike Proctor, Colin Sparrow, Ian Stewart, Sebastian van Strien and Nigel Weiss all made helpful remarks or pointed out painful mistakes, and Fiona Russell helped me keep a sense of perspective and focus throughout the period of writing.

Paul Glendinning
Cambridge, January 1994.

# *Notation*

Most of the notation used in this book is standard and either does not need explanation or is explained in the text. Vectors and vector valued functions (vector fields) are neither underlined nor in bold. Derivatives with respect to time, $t$, which is generally the independent variable, are denoted by dots (as in $\dot{x}$) whilst derivatives of real valued functions of a single variable which is not time are denoted by primes (as in $f'(x)$). Partial derivatives are sometimes indicated by subscripts, so $f_{xy}$ would denote $\frac{\partial^2 f}{\partial x \partial y}$. Derivatives of vector fields are viewed as matrices, so if $f : \mathbf{R}^n \to \mathbf{R}^n$ is a smooth vector field which is a function of $x \in \mathbf{R}^n$, then the derivative (or Jacobian matrix) is the $n \times n$ matrix $Df(x)$ defined in Cartesian coordinates, $f(x) = (f_1(x), \ldots, f_n(x))$, $x = (x_1, \ldots, x_n)$, by

$$[Df(x)]_{ij} = \frac{\partial f_i}{\partial x_j}(x),$$

$0 \leq i, j \leq n$. If $A$ is an open set then the closure of $A$ is denoted by $cl(A)$, whilst if $B$ is a closed set then the interior of $B$ is denoted by $int(B)$. The function $sign(x)$ is simply the sign of $x$, i.e. $+1$ if $x$ is positive, $-1$ if $x$ is negative and 0 if $x$ equals zero.

# 1

# *Introduction*

Differential equations are used throughout the sciences to model dynamic processes. They provide the most simple models of any phenomenon in which one or more variables depend continuously on time without any random influences. They are also fascinating mathematical objects in their own right. If a differential equation is derived from some physical situation it is clearly desirable to know something about solutions to the equation. Indeed, there is little point deriving a model if it is then impossible to gain any information from it! This poses a big problem. Whilst most differential equations in university courses have closed form solutions, typical nonlinear differential equations do not have solutions which can be written down in terms of familiar special functions such as sines and cosines. This means that when faced with general (nonlinear) differential equations we need to change our approach. We will rarely solve differential equations, instead we will try to obtain qualitative information about the long term, or *asymptotic*, behaviour of solutions: are they periodic? eventually periodic? attracting? and so on. This shift from the quantitative to the qualitative is reflected in a shift in the mathematical techniques which are used to analyse equations: much of the analysis will be geometric rather than analytic.

Initially we will concentrate on *hyperbolic* solutions. Roughly speaking a solution is hyperbolic if all sufficiently small perturbations of the defining differential equation have similar behaviour close to that solution (this is not simply a statement about continuity). This leads on to the idea of differential equations which depend on a parameter. For example, if the differential equation models some physical situation then a coefficient in the equation may depend upon temperature. In such a case it may be useful to know the dependence of solutions on the ambient temperature, i.e. to analyse the differential equation for several different values of the coefficient. If at some value of this coefficient a solution is hyperbolic, then a small change in the coefficient, and hence a small change in the ambient temperature, does not alter the qualitative behaviour of the system *near that solution.* The second half of this book introduces ideas from bifurcation theory, which describes the qualitative changes that can occur near non-hyperbolic solutions. This corresponds to situations in which small changes in the coefficients of the defining equations can lead to qualitatively different behaviour of solutions.

Before these terms are given more precise definitions it is worth thinking about the possible behaviour we might expect to meet. A standard introductory example in physics is the model of the ideal pendulum (Fig. 1.1). A simple application of Newton's laws of motion shows that the angle $\theta$ of the pendulum changes with time satisfying an equation of the form

$$\frac{d^2\theta}{dt^2} + \sin\theta = 0 \tag{1.1}$$

after rescaling time so as to make the various physical constants that appear equal to unity.

If we consider only small amplitude oscillations then $\sin\theta \approx \theta$ and so we obtain the simplified equation

$$\frac{d^2\theta}{dt^2} + \theta = 0 \tag{1.2}$$

which has solutions $\theta = A\sin t + B\cos t$, where $A$ and $B$ are constants determined by the initial position and angular velocity of the pendulum. If $A = B = 0$ then $\theta = 0$ is a solution. This solution corresponds to the

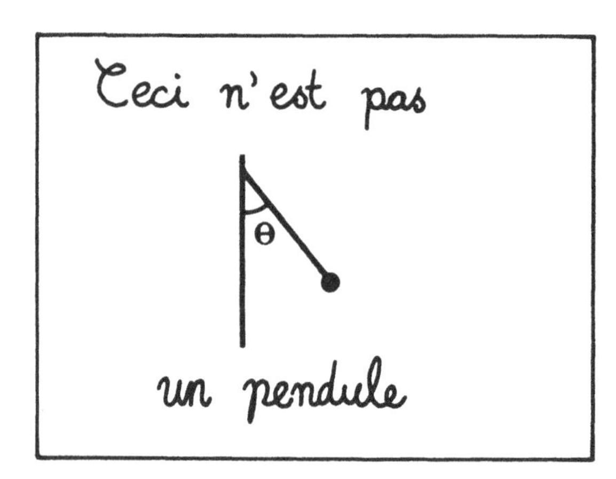


Fig. 1.1 The ideal pendulum (with apologies to Magritte).

stationary pendulum, where it simply hangs directly downwards with no oscillation. There is no motion. So, the first sort of dynamics that we can identify is trivial: no motion. However, for more general choices of initial conditions the solution $A \sin t + B \cos t$ is periodic: the position and angular velocity of the solution is the same at time $t$ and time $t + 2\pi$. This is called periodic motion with period $2\pi$ and corresponds to the simple periodic oscillations of the pendulum. By complicating the equation a little we can get examples of more complicated dynamics. For example, if

$$\frac{d^2\theta}{dt^2} + \theta = a(1 - \omega^2) \cos \omega t, \tag{1.3}$$

with $\omega \neq \pm 1$, then solutions are

$$\theta(t) = A \sin t + B \cos t + a \cos \omega t. \tag{1.4}$$

Is this solution periodic? The first two terms are periodic with period $2\pi$ and the third term is periodic with period $\frac{2\pi}{\omega}$. The solution is periodic if there exists a time $T$ such that both $\theta(0) = \theta(T)$ and $\frac{d\theta}{dt}(0) = \frac{d\theta}{dt}(T)$, i.e. if $T$ is a multiple of both $2\pi$ and $\frac{2\pi}{\omega}$. So solutions are periodic if there exist integers $p$ and $q$ such that $2\pi q = \frac{2\pi p}{\omega}$ or $\omega = \frac{p}{q}$, a rational number. If $\omega$ is irrational then we say that the solution is *quasi-periodic* with two independent frequencies. Although the solution is not periodic it does have a regular structure (see Fig. 1.2).

One of the most exciting developments in the recent theory of differential equations is the discovery that relatively simple differential equations can have solutions which are much more complicated than these periodic and quasi-periodic solutions. Very roughly, a differential equation is said to be *chaotic* if there are bounded solutions which are neither periodic nor quasi-periodic and which diverge from each other locally. The existence of chaotic solutions has had a profound effect on thinking in many disciplines. One immediate corollary of the local divergence of nearby solutions is that one loses predictive power in practical situations. The solutions of differential equations are deterministic in the sense that if the initial conditions are precisely specified then the solution is completely determined and so, in principle, we should be able to predict the value of the solution at some later time. Of course, in practice the initial condition can only be known to some finite precision and so if the equation is chaotic we rapidly lose information about the system since our solution through the approximate initial condition does not stay close to the desired solution. In experiments this can manifest itself in an apparent unrepeatability of the results: many physicists have

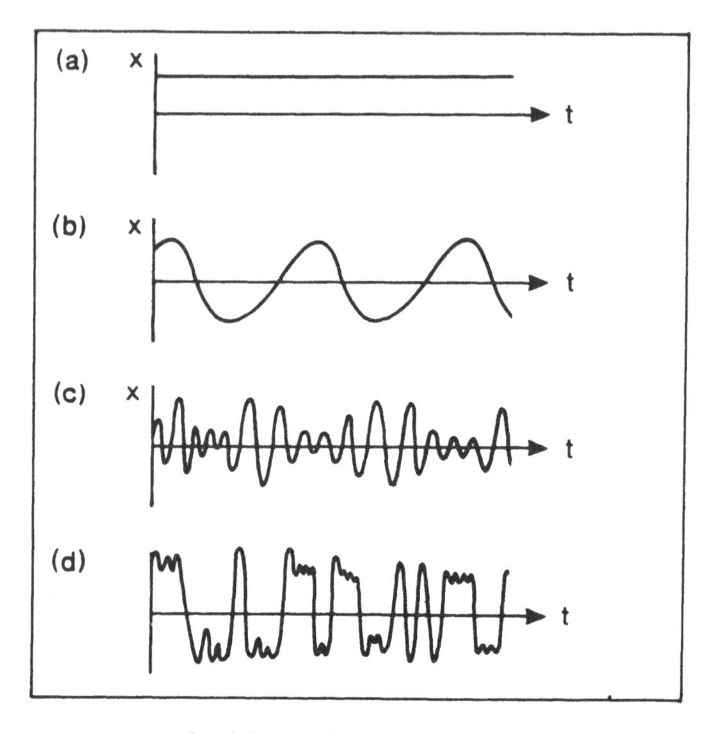


Fig. 1.2 Time series for differential equations. (a) No motion; (b) periodic motion; (c) quasi-periodic motion; (d) possibly chaotic motion.

dusted off experiments which were rejected in the 1960s on the grounds that results were not repeatable. At the time it was assumed that there was some sort of background noise or random fluctuation which had not been eliminated, but these are now recognised as being examples of chaotic behaviour. The results *are* repeatable, but only if looked at from the right point of view. A useful example is the pinball machine: imagine trying to reproduce a sequence of scores! Yet there is no random element, it is all just Newton's laws in action (assuming that the table is not being jiggled overvigorously). The source of this complexity lies in the rounded buffers: small differences in the trajectory of the ball are magnified each time the ball strikes a buffer. Another, more mathematical, example is the difference equation

$$x_{n+1} = 10x_n \pmod 1. \tag{1.5}$$

Given an initial number, $x_0$, with decimal expansion $0.a_0a_1a_2\ldots$, this difference equation generates a new number $x_1 = 0.a_1a_2a_3\ldots$ which generates a new number $x_2$ and so on. Now consider the sequence $(x_i)_{i\geq 0}$.

We say the sequence is eventually periodic of period $p$ if $x_n = x_{n+p}$ for all $n \geq N$. Since $x_n$ is obtained from $x_{n-1}$ by simply deleting the first term in the decimal expansion of the number we can see that every rational number is eventually periodic, since the decimal expansion of a rational number is eventually periodic, and every irrational number is aperiodic. Furthermore, suppose you wanted to predict the motion of a given point on the interval but the number is only known to a finite precision (7 decimal places, say). Then you would know $x_1$ to six decimal places, $x_2$ to five decimal places, $x_3$ to four decimal places, $x_4$ to three decimal places ... and $x_7$ could be anywhere! This illustrates the loss in predictive power which also seems to be at work in weather forecasting and many other situations.

We will now begin to describe the framework which will be the basis of this book. A differential equation is an equation of the form

$$\frac{d^n x}{dt^n} = F\left(t, x, \frac{dx}{dt}, \ldots, \frac{d^{n-1}x}{dt^{n-1}}\right) \tag{1.6}$$

and (modulo some technical assumptions described in Section 1.2) these equations have solutions given some set of initial conditions at $t = t_0$

$$x(t_0) = c_1, \ \frac{dx}{dt}(t_0) = c_2, \ldots, \frac{d^{n-1}x}{dt^{n-1}}(t_0) = c_n. \tag{1.7}$$

Throughout this book we shall choose to consider differential equations in the form

$$\dot{y} = f(y,t), \quad y \in \mathbf{R}^n, \ f : \mathbf{R}^n \times \mathbf{R} \to \mathbf{R}^n, \tag{1.8}$$

where the dot denotes differentiation with respect to time. We shall not be overly concerned about optimal smoothness conditions on $f$ for results to hold, but will assume that $f$ is sufficiently smooth for the Taylor expansions and other techniques used to be valid. Note that any equation of the form (1.6) can be rewritten as (1.8) by setting

$$\frac{d^k x}{dt^k} = y_{k+1} \tag{1.9}$$

for $0 \leq k \leq n-1$, in which case

$$\dot{y}_k = y_{k+1}, \quad 1 \leq k \leq n-1 \tag{1.10a}$$

$$\dot{y}_n = F(t, y_1, \ldots, y_n) \tag{1.10b}$$

i.e. if $y = (y_1, y_2, \ldots, y_n)$ and

$$f(y,t) = (y_2, \ldots, y_n, F(t, y_1, \ldots, y_n))$$

then $\dot{y} = f(y,t)$. The number $n$ is called the order of the differential equation, and an $n^{th}$ order differential equation needs $n$ initial conditions to specify a solution. These can be thought of as the $n$ constants which arise in the $n$ integrations required to solve the equation.

A particularly simple example, which can be solved in general, is the linear differential equation

$$\dot{y} = Ay \tag{1.11}$$

where $A$ is an $n \times n$ matrix with constant coefficients. If the initial condition at $t = 0$ is $y_0$ then this equation has solutions

$$y = e^{tA} y_0. \tag{1.12}$$

So provided we understand the exponential of a matrix we can solve this differential equation exactly. Unfortunately, these are about the only equations which can be solved exactly, and bitter experience has taught scientists that the world is not linear. Hence, to understand more complicated (nonlinear) models we must learn how to treat nonlinear equations which we are unable to solve. Before doing this we should think a little harder about what it means to solve a differential equation.

## 1.1 Solving differential equations

Let's start with a simple example and see how some of the standard techniques for solving differential equations work. Consider the equation

$$\ddot{x} + x = 0 \tag{1.13}$$

with initial conditions $x(0) = a$ and $\dot{x}(0) = b$. This equation should be familiar; it is the equation for simple harmonic motion, (1.2), with solution

$$x(t) = a\cos t + b\sin t. \tag{1.14}$$

This solution is meaningless unless the properties of the functions sine and cosine of $t$ are well known, which, of course, they are. So, how did we solve this equation? We shall sketch three different methods, at least one of which is, I hope, familiar to you.

*Method 1*

Note that if we set $y = \dot{x}$ then (1.13) becomes

$$\begin{pmatrix} \dot{x} \\ \dot{y} \end{pmatrix} = \begin{pmatrix} 0 & 1 \\ -1 & 0 \end{pmatrix} \begin{pmatrix} x \\ y \end{pmatrix}$$

with initial conditions $x(0) = a$ and $y(0) = b$. In matrix notation with $w = (x, y)^T$ and the matrix on the righthand side of this equation denoted by $A$, this becomes $\dot{w} = Aw$, with $w(0) = (a, b)^T$. As pointed out in the preamble to this chapter, in equation (1.12), this has solutions $\exp(tA)w(0)$ and so we need to calculate the matrix $\exp(tA)$, which we do by means of the series definition

$$\exp(tA) = \sum_{k=0}^{\infty} \frac{t^n A^n}{n!}.$$

It is a simple exercise to show that

$$A^{2n} = \begin{pmatrix} (-1)^n & 0 \\ 0 & (-1)^n \end{pmatrix} \text{ and } A^{2n+1} = \begin{pmatrix} 0 & (-1)^n \\ (-1)^{n+1} & 0 \end{pmatrix}$$

and so

$$\exp(tA) = \begin{pmatrix} \sum_n t^{2n}(-1)^n/(2n)! & \sum_n t^{2n+1}(-1)^n/(2n+1)! \\ \sum_n t^{2n+1}(-1)^{n+1}/(2n+1)! & \sum_n t^{2n}(-1)^n/(2n)! \end{pmatrix},$$

which we recognise as being series solutions for sine and cosine of $t$ to give

$$\exp(tA) = \begin{pmatrix} \cos t & \sin t \\ -\sin t & \cos t \end{pmatrix}.$$

Hence $x(t) = a\cos t + b\sin t$ and $y(t) = -a\sin t + b\cos t$.

*Method 2*

Try a trial solution of the form $x = e^{ct}$ and solve for $c$. Then note that since (1.13) is linear, if $x_1$ and $x_2$ are independent solutions then the general solution is a sum of $x_1$ and $x_2$. Substituting $x = e^{ct}$ into the differential equation gives $(c^2 + 1)e^{ct} = 0$ so $c = \pm i$. The solution is therefore $x(t) = c_1 e^{it} + c_2 e^{-it}$ where the complex coefficients $c_1$ and $c_2$ are determined from the initial conditions.

*Method 3*

Note that the differential equation can be written as

$$\left(\frac{d}{dt}+i\right)\left(\frac{d}{dt}-i\right)x=0.$$

Set $v(t)=(\frac{d}{dt}-i)x$ so

$$\left(\frac{d}{dt}+i\right)v=0 \text{ or } e^{-it}\frac{d}{dt}(ve^{it})=0.$$

Hence $ve^{it}=c_1$, or $v=c_1e^{-it}$. Now replace $v$ by the definition of $v$ in terms of $x$ and solve another linear first order differential equation to obtain $x$ as a sum of $e^{\pm it}$ as in method 2.

In all three methods we have assumed and used properties of the exponential function and of sine and cosine in order to solve integrals or guess solutions. But what happens if the differential equation is more complicated and, in particular, if it is nonlinear? As an example consider

$$\ddot{x}+x-x^3=0. \tag{1.15}$$

It does not take much effort to see that none of the methods described above can be applied to this equation; in all cases we are either unable to start or end up with integrals that we cannot solve. Nonetheless these equations do have solutions (snoidal functions, which are defined in terms of elliptic integrals). So, if you knew about elliptic functions you could solve the differential equation (i.e. write down the solution in terms of these functions). To what extent is this useful? Old fashioned books of mathematical functions will often have elliptic functions in tabulated form, so in principle it would be possible to find the solution at a given time approximately using these. There are also formulae for these functions which are valid in particular regimes. However, because we are not familiar with these functions the closed form solution is not, on its own, very helpful.

As a further example consider

$$\ddot{x}+x+x^3+x^7=0. \tag{1.16}$$

Once again, these have solutions given initial values of $x$ and $\dot{x}$ although to the best of my knowledge they are not tabulated anywhere. So, instead of writing this book I could define the solutions of these equations to be the functions $Gl(t)$ ($Gl$ for Glendinning, perhaps) and write a book exploring the properties of these functions and giving tabulated approximations. I fear that such a book would bring me neither fame nor fortune. The problem is that the functions would not have sufficiently wide application to be interesting and, as we shall see, many properties of solutions can be deduced without resorting to the tiresome exercise of solving the differential equation either numerically or in certain limits to obtain approximations to exact solutions.

Let us pause to take stock for a minute. I hope that these examples illustrate the point that closed form solutions are not always possible to find, and that even when they can be found they may not be particularly useful. This suggests that we need an alternative way of looking at the solutions of differential equations. To develop this possibility we need to be clear about what we consider to be the truly important feature of solutions. Perhaps the most important feature of the linear differential equation $\ddot{x}+x=0$ is that all solutions are periodic; that is, they repeat themselves after each period of $2\pi$ (we can see this from the $2\pi$ periodicity of the functions sine and cosine). There is one special case. If the initial condition is $(a,b)=(0,0)$ then $x=0$ for all time. Hence the qualitatively useful information which we deduce from the exact solutions is that if $x$ and $\dot{x}$ are initially both zero then they remain zero for all time (this is called a stationary point) whilst otherwise the solutions are periodic. Since we don't know enough about elliptic functions or Glendinning functions we cannot say what we might consider to be important for the other two examples. It may seem reasonable to look at the asymptotic behaviour, i.e. what happens as $t\to\infty$ and see whether we can understand that motion in some way. For example, if a solution to some equation gives $x(t)=3/(2+x_0e^{-t})$ then, as $t\to\infty$, $x$ tends to the constant value $3/2$, and the way in which it approaches this constant value is, for many purposes, less important than the fact that for large enough values of $t$ the solution is arbitrarily close to $3/2$. Thus we might transfer attention from the exact solution to having a general picture of the type of behaviour observed after some time has elapsed.

Another way of looking at the equations described above is to multiply through by $\dot{x}$ and note that (by the chain rule)

$$\frac{d}{dt}(\tfrac{1}{2}\dot{x}^2)=\dot{x}\ddot{x} \quad \text{and} \quad \frac{d}{dt}(\tfrac{1}{n}x^n)=\dot{x}x^{n-1}.$$

Hence for the linear equation of simple harmonic motion we find

$$\dot{x}(\ddot{x}+x)=0=\frac{d}{dt}(\tfrac{1}{2}\dot{x}^2+\tfrac{1}{2}x^2). \tag{1.17}$$

Therefore we can integrate once to obtain

$$\tfrac{1}{2}\dot{x}^2+\tfrac{1}{2}x^2=C \tag{1.18}$$

for some positive constant $C$ which depends upon the initial conditions. This implies that solutions plotted in the $(x, \dot{x})$ plane lie on concentric circles centred at the origin as shown in Figure 1.3a, which is called a phase portrait or phase space diagram of the system. These concentric circles represent the periodic solutions, and the arrows on the curves indicate the direction of time. The arrows of time can easily be deduced by noting that the equation can be written as $\dot{x} = y$, $\dot{y} = -x$ and so $\dot{x} > 0$ in $y > 0$ and $\dot{x} < 0$ in $y < 0$. Hence the $x$ coordinate increases on solutions when $y > 0$, and $x$ decreases when $y < 0$. In this description then, we have lost the precise parametrization by time, but we retain the important geometric information that solutions lie on closed curves (periodic orbits) unless $C = 0$, which gives the single point at the origin. Note that we could go on (Method 4) to solve for the solution explicitly by solving

$$\frac{dx}{dt} = \sqrt{2C - x^2}, \; C \geq 0,$$

that is,

$$\int_{x_0}^{x(t)} \frac{dz}{\sqrt{2C - z^2}} = \int_0^t d\tau, \tag{1.19}$$

which can be solved without a great deal of sophistication. However, this final step is, at least to some extent, unnecessary, since we have already been able to deduce the important features of the solutions from the structure of solution curves in the $(x, \dot{x})$ plane.

The function $\frac{1}{2}\dot{x}^2 + \frac{1}{2}x^2$ is called a first integral of the problem, and it is (again, unfortunately) rare to be able to obtain first integrals with such ease. However, both the nonlinear examples of this section can be approached in this way. Multiplying $\ddot{x} + x - x^3 = 0$ by $\dot{x}$ gives

$$\dot{x}(\ddot{x} + x - x^3) = 0 = \frac{d}{dt}\left(\tfrac{1}{2}\dot{x}^2 + \tfrac{1}{2}x^2 - \tfrac{1}{4}x^4\right)$$

and so solutions lie on the curves

$$\tfrac{1}{2}\dot{x}^2 + \tfrac{1}{2}x^2 - \tfrac{1}{4}x^4 = C \tag{1.20}$$

in the $(x, \dot{x})$ plane. These are sketched in Figure 1.3b, from which we see immediately that there is a bounded family of periodic solutions about the origin, and a family of unbounded solutions. We shall describe the limiting solutions which separate these two families of solutions later on. If we wanted to give the complete solution of these equations we would

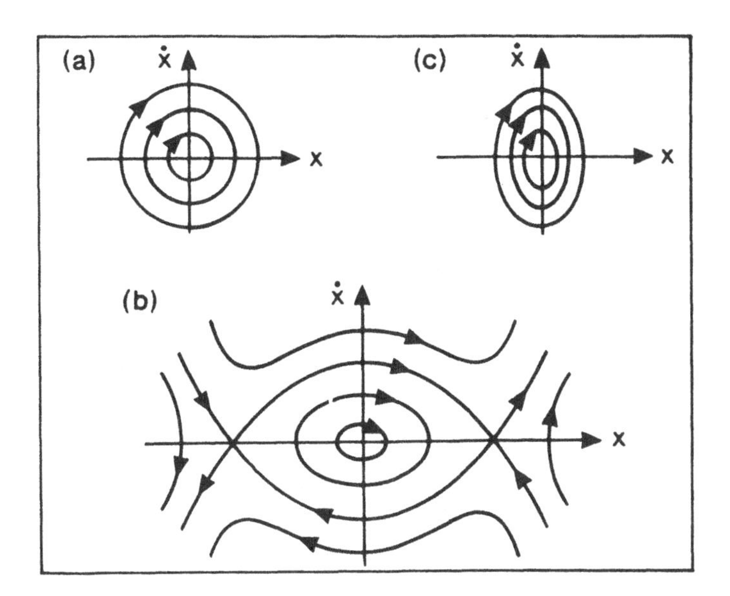


Fig. 1.3 Phase portraits for (a) $\ddot{x}+x=0$; (b) $\ddot{x}+x-x^3=0$; and (c) $\ddot{x}+x+x^3+x^7=0$.

need to be able to solve the integral

$$\int_{x_0}^{x} \frac{dz}{\sqrt{4C-2z^2+z^4}},$$

which is an elliptic integral and has solution in terms of special functions (the cnoidal and snoidal functions referred to earlier).

Similarly, the differential equation $\ddot{x}+x+x^3+x^7=0$ has the first integral

$$\tfrac{1}{2}\dot{x}^2+\tfrac{1}{2}x^2+\tfrac{1}{4}x^4+\tfrac{1}{8}x^8=C. \tag{1.21}$$

The solutions therefore lie on concentric closed curves in the $(x,\dot{x})$ plane, looking like a squeezed version of the simple harmonic oscillator (Fig. 1.3c). We can therefore deduce immediately that all solutions of the differential equation are periodic except for the stationary point at the origin. The analysis of the Glendinning integral

$$\int_{x_0}^{x} \frac{dz}{\sqrt{8C-4z^2-2z^4-z^8}}$$

is clearly as redundant as it is horrible. These three examples are special, in that they all have particularly simple first integrals, but the general message is clear: rather than worry about obtaining exact solutions a great deal of information can be deduced from the geometry of solution curves in some appropriate space. We shall generalise this idea to higher dimensions and more general second order systems in Section 3 of this chapter, but first we should make sure that solutions to differential equations really do exist!

## 1.2 Existence and uniqueness theorems

Most existence and uniqueness theorems for solutions of differential equations are obtained by simple applications of the contraction mapping theorem. We shall not give proofs here (see Hartman, 1964 for details) but we shall give statements of the relevant results and some examples of what can go wrong.

(1.1) THEOREM (LOCAL EXISTENCE AND UNIQUENESS)

*Suppose* $\dot{x} = f(x,t)$ *and* $f : \mathbf{R}^n \times \mathbf{R} \to \mathbf{R}^n$ *is continuously differentiable. Then there exists maximal* $t_1 > 0$ *and* $t_2 > 0$ *such that a solution* $x(t)$ *with* $x(t_0) = x_0$ *exists and is unique for all* $t \in (t_0 - t_1, t_0 + t_2)$.

As an example which shows the non-uniqueness of solutions if the continuously differentiable property is relaxed consider $x \in \mathbf{R}$ with $\dot{x} = f(x)$, $x(0) = 0$, where

$$f(x) = \begin{cases} \sqrt{x} & \text{if } x > 0 \\ 0 & \text{if } x \leq 0 \end{cases}.$$

It is a straightforward exercise to verify that

$$x_\tau(t) = \begin{cases} 0 & \text{if } t < \tau \\ \frac{1}{4}(t-\tau)^2 & \text{if } t \geq \tau \end{cases}$$

is a solution for all $\tau > 0$.

An example which shows that $t_2$ is not necessarily infinity, however smooth the function $f$ may be, is

$$\dot{x} = x^2$$

which has solutions with $x(0) = x_0$ given by

$$x(t) = \frac{1}{x_0^{-1} - t}.$$

This blows up in finite time if $x_0 > 0$ and in finite negative time if $x_0 < 0$. However, the solutions that do exist are unique for the given initial conditions.

(1.2) THEOREM (CONTINUITY OF SOLUTIONS)

*Suppose that $f$ is $C^r$ ($r$ times continuously differentiable) and $r \geq 1$, in some neighbourhood of $(x_0, t_0)$. Then there exists $\epsilon > 0$ and $\delta > 0$ such that if $|x' - x_0| < \epsilon$ there is a unique solution $x(t)$ defined on $[t_0 - \delta, t_0 + \delta]$ with $x(t_0) = x'$. Solutions depend continuously on $x'$ and on $t$.*

This theorem might appear to contradict the possibility of chaotic motion on the grounds of loss of continuity. The point is that although solutions are continuous in space and in time, this does not imply that solutions are uniformly continuous in both variables together: double limits can be tricky.

## 1.3 Phase space and flows

In Section 1.1 we saw that solutions to differential equations could be represented as curves in some appropriate space. This is an extremely important idea and we want to formalise it with greater precision in this section. Suppose that

$$\dot{x} = f(x), \quad x \in \mathbf{R}^n \tag{1.22}$$

so time does not appear explicitly on the right hand side of the equation. Such equations are called autonomous differential equations, whilst if time does appear explicitly they will be called non-autonomous. The solution to the differential equation with $x(0) = x_0$ can be thought of as a continuous curve in $\mathbf{R}^n$ parametrized by time. $\mathbf{R}^n$ is called the phase space of the differential equation and the continuous curve is called an integral curve, orbit or trajectory through $x_0$. Since time only appears as a parametrization of the curve it is less important than the curve itself and so we simply indicate the direction of time by putting an arrow on the curve in the direction of positive time. If $x = (x_1, \ldots, x_n)$ then the differential equation can be written component by component as

$$\dot{x}_i = f_i(x_1, \ldots, x_n), \ i = 1, \ldots, n. \tag{1.23}$$

The integral curves can then be found by dividing through this set of equations by any single one of them. So, assume that $f_1(x_1, \ldots, x_n) \neq 0$.

Then we find

$$\frac{dx_k}{dx_1} = \frac{f_k(x_1, \ldots, x_n)}{f_1(x_1, \ldots, x_n)} \tag{1.24}$$

for $k = 2, 3, \ldots, n$. In this way we have reduced the dimension of the problem by one at the expense of losing the explicit time dependence. Of course, if $f_1(x_1, \ldots, x_n) = 0$ at some value of $x$ we have to change the 'independent' variable to one of the other components and work with a different set of equations. However, provided $f_i(x_1, \ldots, x_n) \neq 0$ for some $i$ we can do this trick. In the case that $f_i(x_1, \ldots, x_n) = 0$ for all $i$ we see that $\dot{x}_i = 0$ and the motion is trivial anyway.

*Example 1.1*

Consider simple harmonic motion again:

$$\dot{x} = y, \ \dot{y} = -x.$$

There are a number of equivalent ways of approaching this.

(i) Consider the function $E(x, y) = x^2 + y^2$. Then

$$\frac{dE}{dt} = \dot{x}\frac{\partial E}{\partial x} + \dot{y}\frac{\partial E}{\partial y} = 2xy - 2yx = 0$$

so $E(x, y)$ is constant along solutions giving integral curves that are concentric circles.

(ii) On integral curves,

$$\frac{dy}{dx} = \frac{-x}{y}$$

provided $y \neq 0$ and

$$\frac{dx}{dy} = \frac{-y}{x}$$

provided $x \neq 0$. In both cases we obtain $ydy + xdx = 0$, i.e.

$$x^2 + y^2 = constant$$

as the equation for the integral curves except at the point $(x, y) = (0, 0)$, which is stationary.

Before giving the final approach we give the formal definition of an integral curve.

(1.3) DEFINITION

*The curve* $(x_1(t), \ldots, x_n(t))$ *in* $\mathbf{R}^n$ *is an integral curve of equation* (1.22) *iff*

$$(\dot{x}_1(t), \ldots, \dot{x}_n(t)) = f(x_1(t), \ldots, x_n(t))$$

*for all* $t$. *Thus the tangent to the integral curve at* $(x_1(t_0), \ldots, x_n(t_0))$ *is* $f(x_1(t_0), \ldots, x_n(t_0))$.

It is easy to see that if $(x_1(t), \ldots, x_n(t))$ is an integral curve, then it is a solution of the differential equation $\dot{x} = f(x)$. Let us return to the example.

*Example 1.1 continued*

(iii) Integral curves are everywhere tangential to $f(x, y) = (y, -x)^T$, i.e. they are perpendicular to the radius vector $(x, y)^T$. Hence they are circles centred at the origin.

This discussion should now give the intuitive picture of solutions of differential equations as parametrized curves in phase space. We can see this as a *flow* on phase space in the following way.

(1.4) DEFINITION

*Suppose that* $\dot{x} = f(x)$. *The solutions of this differential equation define a flow,* $\varphi(x, t)$, *such that* $\varphi(x, t)$ *is the solution of the differential equation at time* t *with initial value (at* t $= 0$*)* $x$. *Hence*

$$\frac{d}{dt}\varphi(x, t) = f(\varphi(x, t)) \tag{1.25}$$

*for all* t *such that the solution through* $x$ *exists and* $\varphi(x, 0) = x$. *In terms of the previous notation, the solution* $x(t)$ *with* $x(0) = x_0$ *is* $\varphi(x_0, t)$.

The flow can be thought of as the motion of a fluid in phase space, (1.25) giving the equation of motion of a particle at $x$ suspended in the fluid. By Theorem 1.2 the flow is a continuous function of $x$ and $t$. Some properties of the flow $\varphi : \mathbf{R}^n \times \mathbf{R} \to \mathbf{R}^n$ are simple consequences of this definition.

(1.5) LEMMA (PROPERTIES OF THE FLOW)

i) $\varphi(x, 0) = x$;
ii) $\varphi(x, t+s) = \varphi(\varphi(x,t), s) = \varphi(\varphi(x,s), t) = \varphi(x, s+t)$.

In terms of the flow we can now define some simple objects in phase space.

(1.6) DEFINITION

*A point $x$ is a stationary point of the flow iff $\varphi(x,t) = x$ for all $t$.*

Thus, at a stationary point $f(x) = 0$ and so they are relatively easy to find.

*Example 1.2*

In the introduction the idea of hyperbolicity was introduced as a sort of robustness property of solutions. The examples we have met so far all have nested sequences of closed curves and some function $E(x, y)$ which is conserved (rather like an energy function). To show that this structure is not hyperbolic consider

$$\ddot{x} + 2\epsilon\dot{x} + x = 0 \tag{1.26}$$

or

$$\dot{x} = y, \qquad \dot{y} = -2\epsilon y - x \tag{1.27}$$

for some small $\epsilon > 0$. Since (1.26) is linear it is easy to write down the solution:

$$x(t) = e^{-\epsilon t}(A\cos(1-\epsilon^2)^{\frac{1}{2}}t + B\sin(1-\epsilon^2)^{\frac{1}{2}}t) \tag{1.28}$$

and so we see that all solutions tend to zero as $t \to \infty$; the infinite collection of closed curves which exist for $\epsilon = 0$ is destroyed by an arbitrarily small perturbation of the defining differential equations. This can be seen in another way: let $E(x, y) = \frac{1}{2}(x^2 + y^2)$, the first integral of (1.26) when $\epsilon = 0$. Then, differentiating as in Example 1.1(i),

$$\frac{dE}{dt} = \dot{x}\frac{\partial E}{\partial x} + \dot{y}\frac{\partial E}{\partial y} = \dot{x}x + \dot{y}y.$$

Substituting for $\dot{x}$ and $\dot{y}$ from (1.27) we find

$$\frac{dE}{dt} = -2\epsilon y^2 \leq 0$$

and so the function $E$ decreases along trajectories. Therefore it seems reasonable to expect that all trajectories tend to minima of $E$ and since the only minimum is $(x, y) = (0, 0)$ trajectories tend to this point. This idea will be justified in Chapter 2, but note once again that we have been able to deduce features of the flow *without* solving any differential equations.

*Example 1.3*

Consider the differential equation

$$\dot{x} = x(1 - x). \tag{1.29}$$

This has two stationary points, one at $x = 0$ and the other at $x = 1$. The flow can be deduced without solving the differential equation by looking at the graph of $\dot{x}$ against $x$ (Fig. 1.4). This shows that $\dot{x} > 0$ (and hence $x$ increases with time) if $x \in (0, 1)$, and $\dot{x} < 0$ if $x < 0$ or $x > 1$. Hence solutions with initial conditions in $x > 0$ tend to the stationary point $x = 1$ whilst solutions with initial conditions in $x < 0$ diverge to $-\infty$. Note once again that this information has been obtained without solving the differential equation, although in this case it is easy to verify these statements by integrating the equations explicitly.

(1.7) DEFINITION

*A point $x$ is periodic of (minimal) period $T$ iff $\varphi(x, t + T) = \varphi(x, t)$ for all $t$, and $\varphi(x, t + s) \neq \varphi(x, t)$ for all $0 < s < T$. The curve $\Gamma = \{y | y = \varphi(x, t), 0 \leq t < T\}$ is called a periodic orbit of the differential equation and is a closed curve in phase space.*

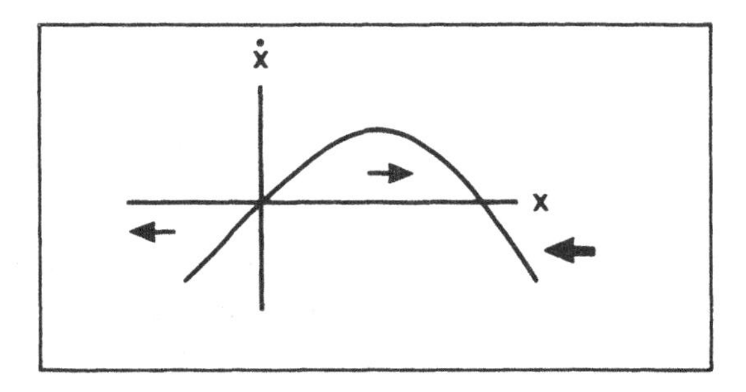


Fig. 1.4 A plot of $x$ against $\dot{x}$ showing the direction of motion. This is *not* a phase portrait, simply the graph of the function $x(1 - x)$.

Periodic orbits are significantly more difficult to find and some sections of this book are devoted to proving that periodic orbits do exist for certain differential equations.

*Example 1.4*

Consider the differential equation

$$\dot{r} = r(1-r), \quad \dot{\theta} = 1 \tag{1.30}$$

where $(r, \theta)$ are polar coordinates, $0 \leq r < \infty$ and $0 \leq \theta < 2\pi$. The only stationary point is the origin, $r = 0$, and when $r = 1$ there is a periodic orbit with period $2\pi$.

Stationary points and periodic orbits are examples of invariant sets. These are, as the name suggests, sets which do not change with time.

(1.8) DEFINITION

*A set $M$ is invariant iff for all $x \in M$, $\varphi(x,t) \in M$ for all $t$. A set is forward (resp. backward) invariant if for all $x \in M$, $\varphi(x,t) \in M$ for all $t > 0$ (resp. $t < 0$).*

One can also ask what other types of invariant sets exist (we have already defined periodic orbits and stationary points), and what the corresponding geometric objects in phase space look like. The two examples described above have very simple geometric structures, but this is by no means the case in general as we shall see in Chapter 12, where some so-called strange attractors are described, but there is one further definition which can be given here.

(1.9) DEFINITION

*A solution is quasi-periodic with $n$ independent frequencies if it can be written as $\varphi(x,t) = g(\omega_1 t, \ldots, \omega_n t)$ where $g$ is periodic of period 1 in each of its arguments and the set of frequencies $(\omega_i)$ are rationally independent. These solutions define an $n$-torus in phase space.*

## 1.4 Limit sets and trajectories

In this section we shall fix some of the terminology used throughout this book and derive some properties of invariant sets associated with integral curves (or trajectories). Suppose we are given the differential equation $\dot{x} = f(x)$ with the flow $\varphi(x,t)$. We shall assume that the flow is defined for all $x \in \mathbf{R}^n$ and $t \in \mathbf{R}$; if this is not the case then the definitions carry over after suitable restrictions are made on the domains of $x$ and $t$.

(1.10) DEFINITION

*The trajectory through $x$ is the set*

$$\gamma(x) = \bigcup_{t \in \mathbf{R}} \varphi(x,t)$$

*and the positive semi-trajectory, $\gamma^+(x)$, and negative semi-trajectory, $\gamma^-(x)$ are defined as*

$$\gamma^+(x) = \bigcup_{t \geq 0} \varphi(x,t) \text{ and } \gamma^-(x) = \bigcup_{t \leq 0} \varphi(x,t).$$

The trajectory through $x$ is precisely the integral curve through $x$, and from now on we shall use the term trajectory instead of integral curve. Note that there is a slight difference in that we defined the integral curve (Definition 1.1) in terms of the parametrization by time, whereas the trajectory is just a set of points. So the trajectory through $x$ and the integral curve through $x$ are the same when seen as sets in phase space. In Section 1.1 we suggested that it might be sensible to concentrate upon the long term behaviour of solutions. In order to define these sets we need a couple of consequences of the property of invariance.

(1.11) LEMMA

i) *$M$ is invariant iff $\gamma(x) \subset M$ for all $x \in M$;*
ii) *$M$ is invariant iff $\mathbf{R}^n \backslash M$ is invariant;*
iii) *Let $(M_i)$ be a countable collection of invariant subsets of $\mathbf{R}^n$. Then $\cup_i M_i$ and $\cap_i M_i$ are also invariant.*

We leave the proof of these results as an exercise. We can now define two invariant sets from the trajectory of $x$, the $\omega$-limit set, $\Lambda(x)$, which is the set of points which $x$ tends to (i.e. the limit points of $\gamma^+(x)$) and the $\alpha$-limit set, $A(x)$, which is the set of points that the trajectory through $x$ tends to in backward time.

(1.12) DEFINITION

*The $\omega$-limit set of $x$, $\Lambda(x)$, and the $\alpha$-limit set of $x$, $A(x)$, are the sets*

$$\Lambda(x) = \{y \in \mathbf{R}^n | \ \exists \ (t_n) \textit{ with } t_n \to \infty \text{ and } \varphi(x, t_n) \to y \textit{ as } n \to \infty\}$$

*and*

$$A(x) = \{y \in \mathbf{R}^n | \ \exists \ (s_n) \textit{ with } s_n \to -\infty \textit{ and } \varphi(x, t_n) \to y \text{ as } n \to \infty\}$$

*Example 1.5*

Consider the differential equation on $\mathbf{R}^2$ in polar coordinates

$$\dot{r} = r(1 - r^2), \quad \dot{\theta} = 1. \tag{1.31}$$

Since $\dot{r} > 0$ if $r \in (0, 0)$ and $\dot{r} < 0$ if $r > 0$ trajectories which start in $r > 0$ tend to $r = 1$ and since $\dot{\theta} \neq 0$ this solution is a periodic orbit. The origin ($r = 0$) is a stationary point and so the $\alpha$-limit set and the $\omega$-limit set of the origin is the origin itself. If $r \neq 0$

$$\Lambda(r, \theta) = \{(r, \theta) | r = 1\}$$

and

$$A(r, \theta) = \begin{cases} \{(r, \theta) | r = 0\} & \text{if } r < 1 \\ \text{undefined} & \text{if } r > 1 \end{cases}.$$

The phase portrait of this system is sketched in Figure 1.5.

Finally, we want to establish two properties of the $\omega$-limit set of $x$. These properties are important, but the proofs are somewhat different in style to most of the mathematics in this book and can be omitted.

(1.13) LEMMA

$$\Lambda(x) = \bigcap_{y \in \gamma(x)} cl(\gamma^+(y))$$

*where cl(X) denotes the closure of the set $X$.*

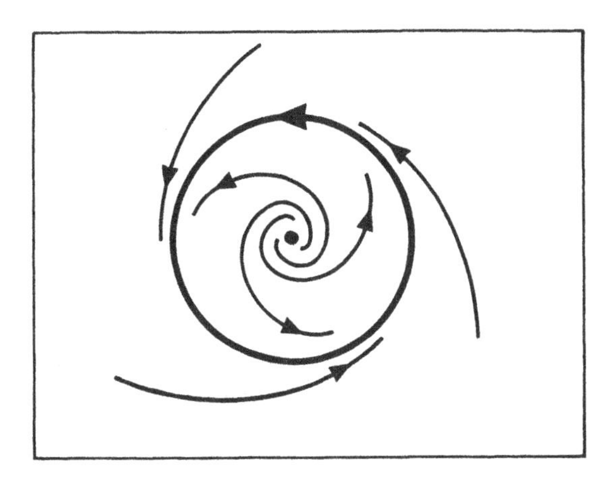

Fig. 1.5 Phase portrait for Example 1.5.

*Proof*: We begin by claiming that

$$cl(\gamma^+(x)) = \gamma^+(x) \cup \Lambda(x).$$

This can be proved in two steps.
(a) It is obvious that $\gamma^+(x) \subset cl(\gamma^+(x))$. Furthermore, from the definition of $\Lambda(x)$ and the closure of a set, $\Lambda(x) \subseteq cl(\gamma^+(x))$. Hence $\gamma^+(x) \cup \Lambda(x) \subseteq cl(\gamma^+(x))$.
(b) To complete the proof of the claim we need to show that $cl(\gamma^+(x)) \subseteq \gamma^+(x) \cup \Lambda(x)$. Let $x^* \in cl(\gamma^+(x))$. Then by the definition of the closure of a set there exist $t_1, t_2, \ldots$ with $t_i \geq 0$ such that $\varphi(x, t_i) \to x^*$ as $i \to \infty$. If $t_i \to \infty$ as $i \to \infty$ then $x^* \in \Lambda(x)$ and we are done. If not then the sequence $(t_i)$ is bounded above and so it has a convergent subsequence $(t_i')$ with $t_i' \to T$, say, as $i \to \infty$. But since $(t_i')$ is a subsequence of $(t_i)$ we have that $\varphi(x, t_i') \to x^*$ as $i \to \infty$, and hence, by the continuity of $\varphi(x, t)$ in $t$,

$$\varphi(x, t_i') \to \varphi(x, T) = x^* \text{ as } i \to \infty.$$

Since, by assumption, $T < \infty$ this implies that $x^* \in \gamma^+(x)$. This completes the proof of the claim.

We now use this claim to write

$$\bigcap_{y \in \gamma(x)} cl(\gamma^+(y)) = \bigcap_{y \in \gamma(x)} [\gamma^+(y) \cup \Lambda(y)].$$

But $\Lambda(y) = \Lambda(x)$ for all $y \in \gamma(x)$ and so

$$\bigcap_{y\in\gamma(x)} cl(\gamma^+(y)) = \left(\bigcap_{y\in\gamma(x)} \gamma^+(y)\right) \cup \Lambda(x).$$

Now, the intersection in square brackets is empty unless $\gamma(x)$ is a closed curve, in which case $\gamma(x) = \Lambda(x)$. In either case we obtain the desired result.

The last result shows that $\Lambda(x)$ is invariant, and that if $\gamma^+(x)$ is bounded then $\Lambda(x)$ is non-empty and compact (i.e. closed and bounded). This will be important in subsequent chapters.

(1.14) THEOREM

*The set $\Lambda(x)$ is invariant, and if $\gamma^+(x)$ is bounded then $\Lambda(x)$ is non-empty and compact.*

*Proof*: We begin with the invariance property. Choose $y \in \Lambda(x)$. Then we need to show that $\varphi(y,t) \in \Lambda(x)$ for all $t$. By the definition of $\Lambda(x)$ there exists a sequence $(t_i)$, with $t_i \to \infty$ as $i \to \infty$ such that $\varphi(x,t_i) \to y$ as $i \to \infty$. Now fix $t$ and choose $t_i$ sufficiently large so that $t + t_i > 0$, then

$$\varphi(x, t + t_i) = \varphi(\varphi(x, t_i), t) \to \varphi(y, t) \text{ as } i \to \infty$$

and so $\varphi(y,t) \in \Lambda(x)$. Hence $\Lambda(x)$ is invariant.

Now suppose that $\gamma^+(x)$ is bounded, so $cl(\gamma^+(x))$ is compact (closed and bounded). Now choose an increasing sequence $(t_i)$ with $t_i \to \infty$ as $i \to \infty$ and let $x_i = \varphi(x, t_i)$. Then

$$\bigcap_{y\in\gamma(x)} cl(\gamma^+(y)) = \bigcap_{i\geq 0} cl(\gamma^+(x_i))$$

and $cl(\gamma^+(x_i)) \subset cl(\gamma^+(x_j))$, for all $i > j$. Hence by Lemma 1.11 $\Lambda(x)$ is non-empty and compact since it is the countable intersection of a decreasing sequence of non-empty compact sets.

## Exercises 1

1. By solving the differential equation

$$\dot{r} = r(1 - r^2), \quad \dot{\theta} = 1$$

explicitly find the $\alpha$- and $\omega$-limit sets of all points in the plane.

2. For autonomous differential equations show that if $\gamma(x) \cap \gamma(y) \neq \emptyset$ then $\gamma(x) = \gamma(y)$. (This has the important consequence that trajectories cannot cross.)

3. Sketch the phase portrait of the differential equation

$$\dot{x} = -x, \quad \dot{y} = y.$$

Show that both the $x$-axis and the $y$-axis are invariant and that the union of these axes is the union of five disjoint trajectories.

4. Classify all possible flows on the line given by

$$\dot{x} = a_0 + a_1 x + a_2 x^2 + x^3.$$

[Hint: do not solve any equations, but consider how many stationary points may exist and the behaviour of the system by plotting a graph of the right hand side of the equation.]

5. Show that the solutions of

$$\dot{x} = \begin{cases} 0 & \text{if } x \leq 0 \\ x^{1/n} & \text{if } x > 0 \end{cases}$$

with $x(0) = 0$ are not unique for $n = 2, 3, 4, \ldots$.

6. Find a first integral for the equation

$$\ddot{x} + \sin x = 0$$

and hence sketch the form of trajectories in phase space.

7. Sketch the phase space of

$$\ddot{x} = ax + bx^3$$

in the $(x, \dot{x})$ plane for $(a, b) = (1, 1), (-1, 1), (1, -1)$ and $(-1, -1)$. In each case determine the stationary points and mark these clearly on your sketch.

8. Burger's equation for wave formation is

$$\frac{\partial U}{\partial t} = \frac{\partial^2 U}{\partial x^2} + U\frac{\partial U}{\partial x}.$$

Travelling waves are solutions $U(x,t) = u(x-ct)$. By setting $\xi = x - ct$ show that $u(\xi)$ satisfies

$$u_{\xi\xi} + (c+u)u_{\xi} = 0$$

where the subscripts denote differentiation. Hence find a first integral and sketch the phase portrait of this system. Interpret the effect of different values of $c$.

9. Sketch the phase portrait of $\ddot{x} + x = ax^2$ for $a = 1$ and $a = -1$.

10. Find the stationary points of $\dot{x} = y$, $\dot{y} = (y+1)(1-x^2)$. By considering $\frac{dy}{dx}$ on trajectories find the equation of the trajectories.

11. The ideal pendulum equation is

$$\ddot{x} + \sin x = 0.$$

Sketch the phase portraits of

i) $\ddot{x} + x = 0$;
ii) $\ddot{x} + x - \frac{1}{3!}x^3 = 0$;
iii) $\ddot{x} + x - \frac{1}{3!}x^3 + \frac{1}{5!}x^5 = 0$;
iv) $\ddot{x} + \sin x = 0$;

describe, without further justification, the differences in the phase portraits you would expect to find for

$$\ddot{x} + \sum_{n=0}^{N} \frac{(-1)^n x^{2n+1}}{(2n+1)!} = 0$$

in the case of $N$ odd and $N$ even.

# 2

# Stability

The stability of solutions of differential equations can be a very difficult property to pin down. Rigorous mathematical definitions are often too prescriptive and it is not always clear which properties of solutions or equations are most important in the context of any particular problem. In practice, different definitions are used (or defined) according to the problem being considered. The effect of this confusion is that there are more than 57 varieties of stability to choose from! We will concentrate on three of the most commonly used definitions. Before going into any detail consider a simple example which demonstrates the problems involved.

*Example 2.1*

The linear differential equation

$$\dot{x} = -y, \quad \dot{y} = x \tag{2.1}$$

can be solved exactly:

$$x = r_0 \cos(t + \varphi), \quad y = r_0 \sin(t + \varphi). \tag{2.2}$$

Closed form solutions are not always especially useful (although this one is so simple that it would be enough for our purposes) and the geometric information that we really want is that solutions lie on concentric circles. This has already been illustrated in the previous chapter, but let us go through the calculation in a slightly more general way. Set $x(t) = r(t)\cos\theta(t)$ and $y(t) = r(t)\sin\theta(t)$, the usual transformation for polar coordinates. Then $x^2 + y^2 = r^2$ and so, differentiating through with respect to time,

$$2x\dot{x} + 2y\dot{y} = 2r\dot{r}$$

or

$$\dot{r} = \frac{1}{r}(x\dot{x} + y\dot{y}). \tag{2.3}$$

Also, differentiating the definitions of $x(t)$ and $y(t)$ we get

$$\dot{x} = \dot{r}\cos\theta - r\dot{\theta}\sin\theta$$

$$\dot{y} = \dot{r}\sin\theta + r\dot{\theta}\cos\theta.$$

Multiplying the first equation by $-\sin\theta$, the second by $\cos\theta$ and adding gives

$$-\dot{x}\sin\theta + \dot{y}\cos\theta = r\dot{\theta}.$$

Dividing through by $r$ and using $\sin\theta = y/r$, $\cos\theta = x/r$ we obtain

$$\dot{\theta} = \frac{x\dot{y} - y\dot{x}}{r^2}. \tag{2.4}$$

Equations (2.3) and (2.4) for $\dot{r}$ and $\dot{\theta}$ are important, since they allow us to work in either Cartesian or polar coordinates, choosing whichever is easiest for the calculations. Substituting $\dot{x} = -y$ and $\dot{y} = x$ we find that the differential equation in polar coordinates is

$$\dot{r} = 0, \quad \dot{\theta} = 1 \tag{2.5}$$

with solution

$$r(t) = r_0, \quad \theta(t) = t + \varphi, \tag{2.6}$$

where $r(0) = r_0$ and $\theta(0) = \varphi$. This shows immediately that solutions are circles on which the motion has constant angular velocity, 1. The origin, $r = 0$, is a stationary point. To talk about its stability we need to know the context of the problem: what sort of stability are we after?

Suppose that the equation represents the motion of a particle which we want to position at the origin. If it is not in the correct position, then it will not get any better over time (since it remains the same distance from the origin). Hence if we want to say that a stationary point is stable if all nearby initial conditions tend to it, the origin is not stable. However, if we know that we can put the particle reasonably close to the origin and that a small error will not affect the outcome of the experiment too much, then the system is stable, since if the particle starts near the origin then it stays near the origin. On the other hand, if we only know that the differential equations are an approximation to the true equations of motion, we may be more interested in discovering whether the solutions of slightly different equations give the same basic solution structure (concentric circles in this case). Thus we could investigate the effect of adding small nonlinear terms to the equations.

*Example 2.2*

Consider the small perturbation of Example 2.1 defined by

$$\dot{r} = \epsilon r^2, \quad \dot{\theta} = 1, \tag{2.7}$$

where $\epsilon$ is a small positive constant. In this case, $\dot{r} > 0$ for all $r > 0$ and so small errors are amplified in time. Indeed, the $r$ equation can be integrated explicitly, and if $r(0) \neq 0$ solutions tend to infinity in *finite* time. Hence in terms of this idea of stability the system is certainly unstable, not because solutions diverge, but because the solutions of the original equations are different in character from those of the perturbed equations.

In these two examples we have identified three types of stability. For stationary points we have two ideas: it could be stable either because nearby solutions tend to it, or because nearby solutions remain nearby for all (positive) time. The third idea was for perturbations of the defining equations; systems which are close behave in a similar manner. This third idea, called *structural stability*, is left to Chapter 4; in this chapter we concentrate upon ideas of stability which do not involve changing the defining differential equation.

## 2.1 Definitions of stability

The two types of stability described above for stationary points have names: a point is Liapounov stable if points which start nearby stay nearby, and it is quasi-asymptotically stable if nearby points tend to it. If a point is both Liapounov stable and quasi-asymptotically stable then it is said to be asymptotically stable. Although we have motivated the definitions in terms of stationary points, there is no reason to restrict the definitions in this way. The proper mathematical definitions are given below. Consider the differential equation

$$\dot{x} = f(x,t), \quad x \in \mathbf{R}^n. \tag{2.8}$$

(2.1) DEFINITION

*A point $x$ is Liapounov stable ('start near stay near') iff for all $\epsilon > 0$ there exists $\delta > 0$ such that if $|x - y| < \delta$ then*

$$|\varphi(x,t) - \varphi(y,t)| < \epsilon$$

*for all* $t \geq 0$.

This is illustrated (in the case where $x$ is a stationary point) in Figure 2.1a.

(2.2) DEFINITION

*A point $x$ is quasi-asymptotically stable ('tends to eventually') iff there exists $\delta > 0$ such that if $|x-y| < \delta$ then $|\varphi(x,t)-\varphi(y,t)| \to 0$ as $t \to \infty$.*

Note that this definition only states what happens in the limit as $t$ tends to infinity. The solution can go all over the place before tending to the point.

(2.3) DEFINITION

*A point $x$ is asymptotically stable ('tends to directly') iff it is both Liapounov stable and quasi-asymptotically stable.*

Quasi-asymptotic stability and asymptotic stability are illustrated in Figure 2.1b, c. These definitions are particularly useful when $x$ is a stationary point, so $\varphi(x,t) = x$. The problem with more general solutions (periodic orbits for example) is that the simple definitions are uniform in $t$, so phase information can make a difference to one's naive impression of what should be meant by stability.

*Example 2.3*

Consider the system

$$\dot{r} = 0, \quad \dot{\theta} = 1 + r \tag{2.9}$$

with solutions

$$r(t) = r_0, \quad \theta(t) = (1 + r_0)t + \theta_0. \tag{2.10}$$

Arguments similar to those for Example 2.1 show that the origin, $r = 0$, is Liapounov stable but not quasi-asymptotically stable and that solutions are concentric circles about the origin. In some sense, one might like to say that these circles are also stable in that if one starts near a given circle (with radius $r_0$) the solution will stay near that circle. Unfortunately, there is a phase lag on each circle that means that no points on the circle are stable in either a Liapounov or a quasi-asymptotic

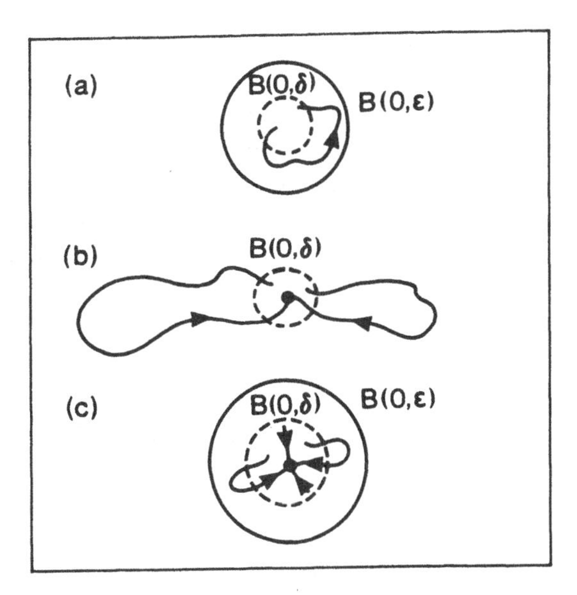


Fig. 2.1 (a) Liapounov stability; (b) quasi-asymptotic stability; (c) asymptotic stability.

sense. To see this consider two nearby initial conditions, $(r_0, 0)$ and $(r_0 + \delta, 0)$. After a time $t$ the difference in angle (or phase) between the two solutions is $\Delta\theta$ where

$$\Delta\theta = (1 + r_0)t - (1 + r_0 + \delta)t = -\delta t. \tag{2.11}$$

So when $t = (2n + 1)\pi/\delta$, the two solutions are diametrically opposite and the distance between the solutions is $2r_0 + \delta$. Hence there is always a time at which the two solutions are further away from each other than any given $\epsilon$ sufficiently small and the solution is neither Liapounov nor quasi-asymptotically stable.

(2.4) EXERCISE

*Show that all points are Liapounov stable for Example 2.1, but none are quasi-asymptotically stable.*

(2.5) EXERCISE

*Consider Example 2.3. Show that in new coordinates $(R, \Theta)$ defined by*

$$R = r, \quad \Theta = \theta - rt \tag{2.12}$$

*all points are Liapounov stable.*

As remarked earlier these definitions are particularly useful when applied to stationary points. If a stationary point is asymptotically stable then there must exist a neighbourhood of the point such that all points in this neighbourhood tend to the stationary point. The largest neighbourhood for which this is true is called the domain of (asymptotic) stability (or basin of attraction) of this point.

(2.6) DEFINITION

*Let $x$ be an asymptotically stable stationary point of the equation $\dot{x} = f(x)$, so*

$$\text{for all } \epsilon > 0 \text{ there exists } \delta > 0 \text{ such that}$$
$$|y - x| < \delta \Rightarrow |\varphi(y,t) - x| < \epsilon \quad \forall t \geq 0$$

*and*

$$\exists\, \delta > 0 \text{ such that } |y - x| < \delta \Rightarrow |\varphi(y,t) - x| \to 0 \text{ as } t \to \infty,$$

*then*

$$D_x = \left\{ y \in \mathbf{R}^n \,|\, \lim_{t\to\infty} |\varphi(y,t) - x| = 0 \right\}$$

*is called the domain of asymptotic stability of $x$. If $D_x = \mathbf{R}^n$ then $x$ is globally asymptotically stable.*

Usually it is very hard to determine the domain of stability of a stationary point, and we will refer to any subset of the domain of stability as a domain of stability. The next few sections are devoted to techniques for determining whether a stationary point is Liapounov stable or quasi-asymptotically stable, but we end this section with an example which shows that a stationary point can be quasi-asymptotically stable but not Liapounov stable. Example 2.1 shows that a stationary point can be Liapounov stable but not quasi-asymptotically stable.

*Example 2.4*

Consider the equation

$$\dot{x} = x - y - x(x^2 + y^2) + \frac{xy}{\sqrt{x^2 + y^2}} \tag{2.13a}$$

$$\dot{y} = x + y - y(x^2 + y^2) - \frac{x^2}{\sqrt{x^2 + y^2}} \tag{2.13b}$$

or, using (2.3) and (2.4) to rewrite (2.13) in polar coordinates

$$\dot{r} = r(1 - r^2), \quad \dot{\theta} = 2\sin^2(\tfrac{1}{2}\theta). \tag{2.14}$$

To analyse the behaviour of this system first note that $\dot{r} = 0$ when $r = 0$ or $r = 1$ and $\dot{\theta} = 0$ when $\theta = 0$. Also, $\dot{\theta} > 0$ for all $\theta \neq 0$. Thus $\theta = 0$ is an invariant half-line and trajectories move around to approach this half-line from below. The $\dot{r}$ equation is familiar from the previous chapter: in the $r$ direction trajectories tend to $r = 1$ (unless $r = 0$ initially). Hence the system has two stationary points, the origin ($r = 0$) and the point $(r, \theta) = (1, 0)$, and two invariant curves, $\theta = 0$ and $r = 1$, and almost all trajectories eventually tend to the stationary point $(1, 0)$ as shown in Figure 2.2. Now consider a small neighbourhood of $(1, 0)$. Points in $\theta \leq 0$ will tend to $(1, 0)$ without leaving this neighbourhood, but points in $\theta > 0$ will make a circuit around the invariant curve $r = 1$ before tending to $(1, 0)$ from $\theta < 0$. In particular, a trajectory starting close to $(1, 0)$ on $r = 1$ with $\theta > 0$ will pass through the diametrically opposite point of the circle, $(1, \pi)$, before tending to $(1, 0)$ from $\theta < 0$. Hence, although $(1, 0)$ is quasi-asymptotically stable it is not Liapounov stable.

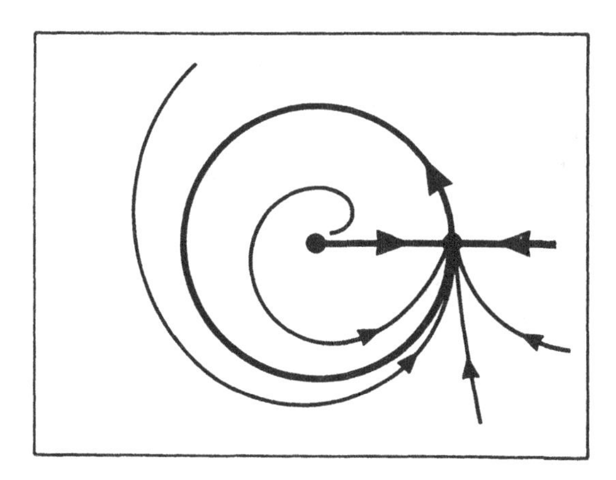

Fig. 2.2 The phase portrait for Example 2.4.

## 2.2 Liapounov functions

Liapounov functions can be thought of as modified energy functions. They are used to prove that a stationary point is Liapounov stable and (with extra conditions) that a stationary point is asymptotically stable. The theory of Liapounov functions is elegant and simple, but proving that a Liapounov function exists or explicitly finding one for a given problem is another question!

(2.7) DEFINITION

*Suppose that the origin, $x = 0$, is a stationary point for the differential equation $\dot{x} = f(x)$, $x \in \mathbf{R}^n$. Let $G$ be an open neighbourhood of 0 and $V : cl(G) \to \mathbf{R}$ a continuously differentiable function. Then we can define the derivative of $V$ along trajectories by differentiating $V$ with respect to time using the chain rule, so*

$$\dot{V} = \frac{dV}{dt} = \dot{x}.\nabla V = f.\nabla V = \sum f_i(x)\frac{\partial V}{\partial x_i} \tag{2.15}$$

*where the subscripts denote the components of $f$ and $x$. Then $V$ is a Liapounov function on $G$ iff $V$ is continuously differentiable on $cl(G)$ and*

i) *$V(0) = 0$ and $V(x) > 0$ for all $x \in cl(G) \setminus \{0\}$;*
ii) *$\dot{V} \leq 0$ for all $x \in G$.*

The reason that Liapounov functions are so nice (when they can be found) is sketched in Figure 2.3. The idea is that if $V$ is a Liapounov function then $V$ decreases along trajectories, and hence (since $V$ is strictly positive except at 0) trajectories tend to zero, which is a minimum value of $V$. Of course, life is never quite that simple (although $V$ is decreasing it is not necessarily strictly decreasing), so let us see precisely what the existence of a Liapounov function implies. The first step is to show that at least some trajectories stay in $G$ for all time if a Liapounov function exists on $G$.

(2.8) LEMMA (BOUNDING LEMMA)

*Suppose that $G$ is some open bounded domain in $\mathbf{R}^n$ with boundary $\partial G$, and that $V : cl(G) \to \mathbf{R}$ is a Liapounov function. If there exists $x_0 \in G$*

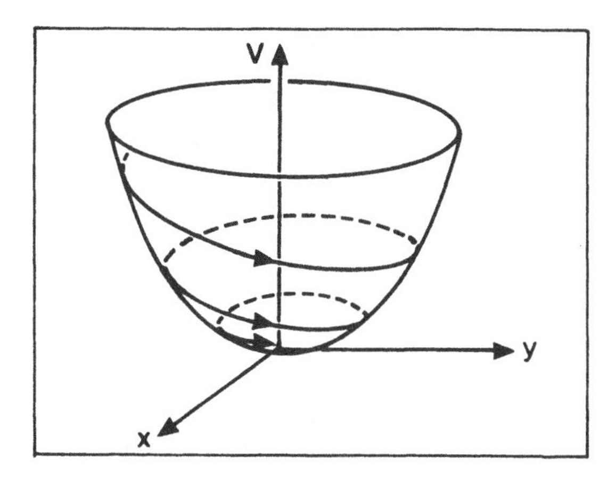


Fig. 2.3 $V(x)$ decreases along trajectories.

*such that* $V(x) > V(x_0)$ *for all* $x \in \partial G$ *then*

$$S(x_0) = \{x \in cl(G) |\ V(x) \leq V(x_0)\}$$

*is a bounded set in* $G$ *and* $\varphi(x_0, t) \in S(x_0)$ *for all* $t \geq 0$.

*Proof*: Since $V$ is continuous in $G$ and $V(x) > V(x_0)$ on $\partial G$ it is obvious that $S(x_0)$ is in $G$ and $x_0 \in S(x_0)$, so $S(x_0)$ is non-empty. By the existence results of the previous chapter, (1.2), we see that there are three possible fates for a trajectory through $x_0$. Either

A) there exists $t' \geq 0$ such that $\lim_{t \to t'} \varphi(x_0, t)$ is infinite; or
B) there exists $t' \geq 0$ such that $\varphi(x_0, t') \in \partial G$; or
C) $\varphi(x_0, t) \in G$ for all $t \geq 0$.

Now, $V$ is a continuous function on $cl(G)$ and $\varphi(x_0, t)$ is a continuous function of $t$ on $[0, t')$. This, together with the non-increasing property of $V$, implies that

$$V(\varphi(x_0, t')) = \lim_{t \to t'} V(\varphi(x_0, t)) \leq V(x_0).$$

Hence (B) gives a contradiction as $V(x) > V(x_0)$ on $\partial G$ and (A) gives a contradiction as $G$ is bounded and $\varphi(x_0, t)$ is continuous for $t < t'$ (so if $\varphi(x_0, t)$ tends to infinity ther must exist $t''$ such that $\varphi(x_0, t'') \in \partial G$). This leaves (C) as the only possibility and $V(\varphi(x_0, t)) \leq V(x_0)$ for all $t \geq 0$ so the trajectory through $x_0$ stays in $S(x_0)$.

This proof is unnecessarily long. Really, all that the proof requires is that we note that if $S(x_0)$ is bounded and in $G$ then since $V(x)$ decreases on trajectories, the trajectory through $x_0$ stays in $S(x_0)$ (otherwise $V$ would have to increase). This is obvious, but from time to time it is worth spelling things out in detail.

To prove that a stationary point is Liapounov stable given the existence of a Liapounov function we need to show that given any $\epsilon > 0$ there is some region $B(0, \delta)$, where $B(y, r)$ denotes the open ball of radius $r$ centred at $y$, such that if a trajectory starts in $B(0, \delta)$ it stays in $B(0, \epsilon)$. In terms of the Bounding Lemma this means that we must find $B(0, \delta)$ such that for all $x_0$ in $B(0, \delta)$, $S(x_0)$ is contained in $B(0, \epsilon)$. This will give us Liapounov's First Stability Theorem.

(2.9) THEOREM (LIAPOUNOV'S FIRST STABILITY THEOREM)

*Suppose that a Liapounov function can be defined on a neighbourhood of the origin, $x = 0$, which is a stationary point of the differential equation $\dot{x} = f(x)$. Then the origin is Liapounov stable.*

*Proof*: Choose $\epsilon > 0$ sufficiently small so that $B(0, \epsilon) \subset G$. Then we need to find $\delta > 0$ such that if $x \in B(0, \delta)$ then $\varphi(x, t) \in B(0, \epsilon)$ for all $t \geq 0$.

Let $\mu = min\{V(x)|x \in \partial B(0, \epsilon)\}$. Since $V$ is a Liapounov function on $G$, and $\epsilon$ has been chosen small enough, $B(0, \epsilon)$ is contained in $G$ and $\mu > 0$. Also, since $\partial B(0, \epsilon)$ is compact and $V$ is continuous there exists $y \in \partial B(0, \epsilon)$ such that $V(y) = \mu$ (see Fig. 2.4). Now let $C = B(0, \epsilon) \cap \{x|V(x) < \mu\}$. Since $\mu > 0$ and $V(0) = 0$, the origin is contained in $C$. Now choose $\delta > 0$ such that $B(0, \delta) \subset C$ and apply the Bounding Lemma with $x_0 \in B(0, \delta)$. Then $S(x_0)$ must lie inside $C$, which lies inside $B(0, \epsilon)$ and the result is proved.

*Example 2.5*

Consider the nonlinear oscillator

$$\ddot{x} + c\dot{x} + ax + bx^3 = 0 \tag{2.16}$$

where $a$, $b$ and $c$ are positive constants. This can be rewritten as a differential equation in two variables by setting $y = \dot{x}$, so $\dot{y} = \ddot{x}$ giving

$$\dot{x} = y \tag{2.17a}$$

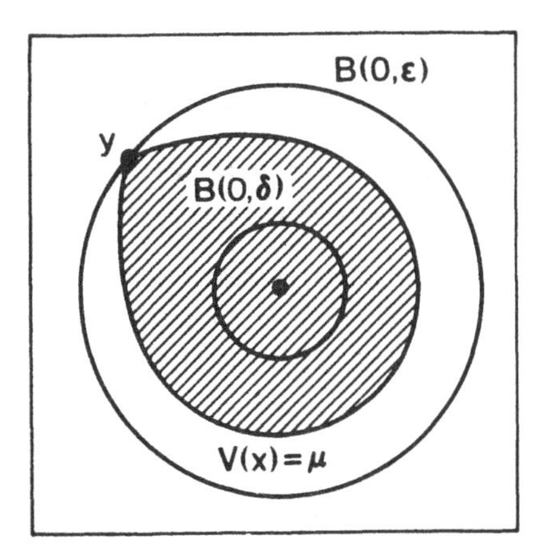


Fig. 2.4 C is the shaded region.

$$\dot{y} = -cy - ax - bx^2. \tag{2.17b}$$

Since $a$ and $b$ are positive constants, the only stationary point of this system is $(x, y) = (0, 0)$ so we now want to find a Liapounov function for the origin. This is very much a matter of luck or judgement, hard work or a stroke of genius. Let us try a function of the form

$$V(x, y) = \alpha x^2 + \beta x^4 + \gamma y^2 \tag{2.18}$$

for positive constants $\alpha$, $\beta$ and $\gamma$. It is obvious that $V(x, y) > 0$ for $(x, y) \neq (0, 0)$ and $V(0, 0) = 0$. We shall now try to choose the constants $\alpha$, $\beta$ and $\gamma$ so that $\dot{V}(x, y)$ is non-positive on some (as yet to be determined) neighbourhood of $(0, 0)$. Differentiating $V$ with respect to time gives

$$\dot{V}(x, y) = (2\alpha x + 4\beta x^3)\dot{x} + 2\gamma y\dot{y} \tag{2.19}$$

and replacing $\dot{x}$ and $\dot{y}$ by (2.17) we find

$$\dot{V}(x, y) = (2\alpha x + 4\beta x^3)y + 2\gamma y(-cy - ax - bx^3)$$

or

$$\dot{V}(x, y) = (2\alpha - 2\gamma a)xy + (4\beta - 2\gamma b)x^3 y - 2\gamma c y^2. \tag{2.20}$$

Since $\gamma$ and $c$ are both positive the last term is certainly non-positive, but the first two terms are more difficult. However, we are free to choose $\alpha$, $\beta$ and $\gamma$ provided they are positive, so setting

$$\gamma = \frac{\alpha}{a} \quad \text{and} \quad \beta = \frac{\gamma b}{2} = \frac{\alpha b}{2a} \tag{2.21}$$

the problem terms disappear leaving

$$\dot{V}(x,y) = -2\gamma c y^2 \leq 0. \tag{2.22}$$

Hence $V(x,y)$ is a Liapounov function on any open bounded subset of $\mathbf{R}^2$ which contains $(0,0)$, and $(0,0)$ is Liapounov stable.

It is rare in examples to find Liapounov functions on the whole space; usually it is only possible to show that a function is a Liapounov function on some small and often unspecified neighbourhood of the stationary point. This is illustrated in the following example.

*Example 2.6*

Consider the differential equation

$$\dot{x} = -x + x^2 - 2xy \tag{2.23a}$$

$$\dot{y} = -2y - 5xy + y^2. \tag{2.23b}$$

The origin $(0,0)$ is clearly a stationary point, but is it Liapounov stable? A trial Liapounov function $V(x,y) = \frac{1}{2}(x^2+y^2)$ (a standard guess) gives

$$\dot{V}(x,y) = x\dot{x} + y\dot{y} = x(-x + x^2 - 2xy) + y(-2y - 5xy + y^2)$$

or

$$\dot{V}(x,y) = -x^2(1 - x + 2y) - y^2(2 + 5x - y). \tag{2.24}$$

Hence $V(x,y)$ is a Liapounov function provided both $(1 - x + 2y)$ and $(2 + 5x - y)$ are positive on some neighbourhood of $(x,y) = (0,0)$. This is obviously the case and we can set $G$ to be any open bounded subset containing $(0,0)$ which lies in the wedge-shaped region

$$\{(x,y) | x - 2y < 1,\ 5x - y > -2\}$$

as shown in Figure 2.5. Thus $(0,0)$ is Liapounov stable.

The next question which comes to mind is whether a Liapounov function can help in proving that a stationary point is quasi-asymptotically stable (and hence asymptotically stable, since it must be Liapounov stable if a Liapounov function exists). Looking at the previous results it should be fairly obvious that the main obstacle to proving quasi-asymptotic stability is the fact that $\dot{V}$ can be zero. We might hope, therefore, that if $\dot{V} < 0$ except at $x = 0$ then there are no other places to which trajectories might tend: they must all tend to the stationary point at the origin where $V(0) = 0$. We shall return to this point in a

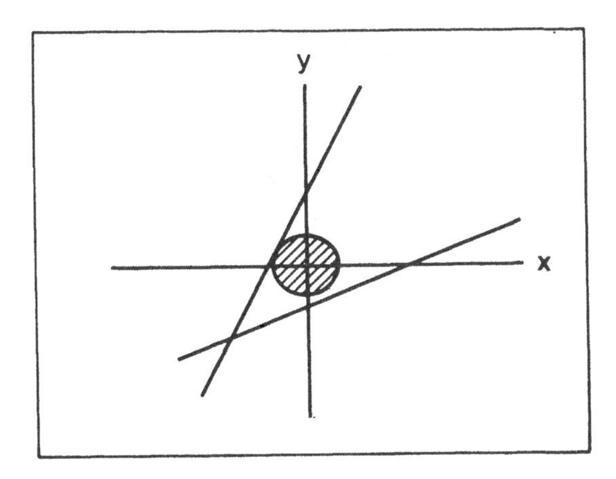


Fig. 2.5 $V$ is a Liapounov function in the wedge-shaped region. The shaded region is that largest set $V(x) = k$ entirely in this region and is therefore a domain of stability.

short while, but having realized that the set of points at which $\dot{V}$ vanishes is potentially interesting we shall see what can be said in general about such a set. It turns out that the zeros of $\dot{V}$ provide a way of identifying the largest invariant set to which trajectories tend *without* calculating trajectories. Since we are often only concerned with asymptotic behaviour (i.e. what trajectories are attracted to) it is a huge advantage to be able to identify these sets without solving the differential equations explicitly. Specifically, we are interested in the set

$$E = \left\{ x \in G | \dot{V}(x) = 0 \right\}. \tag{2.25}$$

(2.10) THEOREM (LA SALLE'S INVARIANCE PRINCIPLE)

*Suppose that $x = 0$ is a stationary point of $\dot{x} = f(x)$ and $V$ is a Liapounov function on some neighbourhood $G$ of $x = 0$. If $x_0 \in G$ has its forward trajectory, $\gamma^+(x_0)$, bounded with limit points in $G$ and $M$ is the largest invariant subset of $E$, then*

$$\varphi(x_0, t) \to M \quad \text{as} \quad t \to \infty.$$

*Proof*: By assumption on $\gamma^+(x_0)$, the $\omega$-limit set of $x_0$, $\Lambda(x_0)$, is non-empty and is contained in $G$ for each $x_0$ in $G$ satisfying the conditions of the theorem (cf. Theorem 1.14). Furthermore, since $V$ is a Liapounov function $V(\varphi(x_0, t))$ is a non-increasing function of $t$ which is bounded below by zero. Hence there exists $c$ (which depends upon $x_0$) such that

$$\lim_{t\to\infty} V(\varphi(x_0,t)) = c.$$

Now take $y \in \Lambda(x_0)$. Then there exists an increasing sequence $(t_k)$ with $t_k \to \infty$ as $k \to \infty$ such that $\varphi(x_0,t_k)$ tends to $y$ as $k$ tends to $\infty$ (this is the definition of a limit point). So by the continuity of $V$, $V(y) = c$ for all $y \in \Lambda(x_0)$.

Now, $\Lambda(x_0)$ is invariant, and so if $y \in \Lambda(x_0)$ then $\varphi(y,t) \in \Lambda(x_0)$ for all $t$ and hence $V(\varphi(y,t)) = c$ for all $t \geq 0$ and all $y \in \Lambda(x_0)$. Hence

$$\dot{V}(y) = 0 \quad \text{for all } y \in \Lambda(x_0)$$

and so

$$\Lambda(x_0) \subset M \subset E.$$

But $\varphi(x_0,t) \to \Lambda(x_0)$ as $t \to \infty$, so $\varphi(z,t) \to M$ as $t \to \infty$ for all $z \in G$ such that $\gamma^+(z) \subset G$.

In the statement of La Salle's Invariance Principle we have referred to the set of points in $G$ for which the forward trajectory lies in $G$. To find out what this set looks like (and hence what subset of $G$ tends to $M$) we simply look back to the Bounding Lemma (Lemma 2.8): this shows that the set of points with forward trajectories in $G$ includes sets of the form

$$V_k = \{x \mid V(x) < k\}$$

which lie entirely inside $G$. It is easy to see that for Example 2.6, $E = \{(x,y) | y = 0\}$ and that the only invariant set in $E$ is the point $(0,0)$, so $(0,0)$ is (globally) asymptotically stable. La Salle's Invariance Principle also allows us to prove a further result about asymptotic stability as suggested earlier. The only thing which gets in the way of proving asymptotic stability is the possibility that $\dot{V}$ may be zero at places other than the stationary point. The solution is simple: remove this possibility.

(2.11) THEOREM (LIAPOUNOV'S SECOND STABILITY THEOREM)

*Suppose $x = 0$ is a stationary point for $\dot{x} = f(x)$ and let $V$ be a Liapounov function on a neighbourhood $G$ of $x = 0$. If $\dot{V}(x) < 0$ for all $x \in G \setminus \{0\}$, then $x = 0$ is asymptotically stable.*

*Proof*: By Liapounov's First Stability Theorem (2.9) the origin is Liapounov stable, i.e. for all $\epsilon > 0$ there exists $\delta > 0$ such that if $x \in B(0, \delta)$ then $\varphi(x, t) \in B(0, \epsilon)$ for all $t \geq 0$.

By La Salle's Invariance Principle with $G = B(0, \epsilon)$, $\varphi(x_0, t) \to M$ as $t \to \infty$, where $M$ is the largest invariant subset of

$$E = \left\{x \in B(0, \epsilon) | \dot{V}(x) = 0\right\},$$

for all $x_0$ with $\gamma^+(x_0) \subset B(0, \epsilon)$ and hence, in particular, for all $x_0 \in B(0, \delta)$.

But $\dot{V}(x) = 0$ for $x \in B(0, \epsilon)$ iff $x = 0$ (by assumption) so $E = M = \{0\}$. Thus for all $x \in B(0, \delta)$, $\varphi(x, t) \to \{0\}$ as $t \to \infty$, and so $x = 0$ is quasi-asymptotically stable. We already know that $x = 0$ is Liapounov stable, so $x = 0$ is asymptotically stable.

*Example 2.7*

Consider the nonlinear differential equation

$$\ddot{z} - a\dot{z}(z^2 - 1) + z = 0. \tag{2.26}$$

We could convert this to a pair of first order equations as in Example 2.5 by setting $y = \dot{z}$, but this is an example where it is easier to work in Liénard coordinates. This choice of coordinates is applicable to any equation of the form

$$\ddot{x} + f(x)\dot{x} + g(x) = 0 \tag{2.27}$$

(here $f(x) = -a(x^2 - 1)$ and $g(x) = x$). The trick with Liénard coordinates is to define

$$F(x) = \int^x f(\xi) d\xi \tag{2.28}$$

and note that $\frac{dF}{dt} = \dot{x}\frac{dF}{dx} = f(x)\dot{x}$. Hence if we define $y = \dot{x} + F(x)$, then $\dot{y} = \ddot{x} + f(x)\dot{x} = -g(x)$. The differential equation can therefore be written as

$$\dot{x} = y - F(x) \tag{2.29a}$$

$$\dot{y} = -g(x). \tag{2.29b}$$

For the particular example considered here this becomes

$$\dot{x} = y + a(\tfrac{1}{3}x^3 - x) \tag{2.30a}$$

$$\dot{y} = -x. \tag{2.30b}$$

Now use the trial Liapounov function $V(x,y) = \frac{1}{2}(x^2 + y^2)$ to get

$$\dot{V}(x,y) = x\dot{x} + y\dot{y} = ax^2(\tfrac{1}{3}x^2 - 1) \tag{2.31}$$

so $\dot{V} \leq 0$ for $x^2 < 3$. The largest domain $V_k = \{x|V(x) < k\}$ which lies entirely in this region is given by $x^2 + y^2 < 3$, i.e. $V_{\frac{3}{2}}$, and $\dot{V} = 0$ on $x = 0$. On $x = 0$, $\dot{x} = y$ and so trajectories which intersect the line $x = 0$ remain on this line if $y = 0$. Hence the largest invariant subset on $x = 0$ is the stationary point $(x, y) = (0, 0)$ and so (by La Salle's Invariance Principle, Theorem 2.10) this stationary point is asymptotically stable and attracts all points in $V_{\frac{3}{2}}$.

(2.12) EXERCISE

*Find a domain of stability of* $(0,0)$ *for Example 2.6; see Figure 2.5.*

Finally, we give a brief outline of the type of approach this section suggests. Suppose that we are given a system $\dot{x} = f(x)$ and told to show (or try to show) that a stationary point $x = a$ is asymptotically stable. First, change coordinates so that this stationary point is at the origin, i.e. set $y = x - a$, so $\dot{y} = g(y)$ with $g(0) = 0$. Then find an open neighbourhood $G$ of $y = 0$ and a continuously differentiable function $V$ such that

i) $V(y) \geq 0$ on $G$ and $V(y) = 0$ if and only if $y = 0$; and
ii) $\dot{V} \leq 0$ on $G$.

Next, find $k$ such that $V_k \subset G$ and so $\varphi(x,t) \to M$ for all $x \in V_k$, where $M$ is the largest invariant subset of the set of points in $G$ for which $\dot{V}$ vanishes. Finally, adjust $V$ and $G$ so that $M = \{0\}$ so $V_k$ is a domain of asymptotic stability for the origin and try different choices of $V$ to maximize the size of $V_k$.

This sequence of operations is sometimes relatively straightforward but, in general, the construction of Liapounov functions is an art. Mastering this art requires a mixture of experience, patience and good luck.

## 2.3 Strong linear stability

Having said that the construction of Liapounov function is often a tricky business, there is one situation in which it is really quite simple to find them. Suppose that $\dot{x} = f(x)$ and $x = a$ is a stationary point, so

$f(a) = 0$. Then we can expand $f$ locally as a Taylor series about $x = a$ to get, in component form with $\xi = x - a$,

$$\dot{\xi}_i = f_i(a) + \frac{\partial f_i}{\partial x_j}(a)\xi_j + o(|\xi|). \tag{2.32}$$

Since $f(a) = 0$ by assumption this can be written as

$$\dot{\xi} = A\xi + o(|\xi|) \tag{2.33}$$

where $A$ is the matrix of partial derivatives of $f$ (the Jacobian matrix) evaluated at $x = a$,

$$A_{ij} = \frac{\partial f_i}{\partial x_j}(a). \tag{2.34}$$

The linear system $\dot{\xi} = A\xi$ is called the linearization of $\dot{x} = f(x)$ at $x = a$. The next two chapters deal with the extent to which the linearized system can give information about the original system, but we can preempt some of these results by explicitly constructing a Liapounov function for $x = a$ when $A$ has $n$ distinct eigenvalues, all of which have strictly negative real parts. To this extent, then, any stationary point whose linearization has this spectral property is asymptotically stable.

*Example 2.8*

Consider the differential equation

$$\dot{x} = x(y - 1) \tag{2.35a}$$

$$\dot{y} = 3x - 2y + x^2 - 2y^2. \tag{2.35b}$$

Looking for stationary points with $\dot{x} = 0$ and $\dot{y} = 0$ we see from (2.35a) that either $x = 0$ or $y = 1$. Substituting $x = 0$ into (2.35b) with $\dot{y} = 0$ gives $y = 0$ or $y = -1$ and similarly, $y = 1$ leads to $x = -4$ or $x = 1$. Hence there are four stationary points: $(x, y) = (0, 0)$, $(0, -1)$, $(-4, 1)$ and $(1, 1)$. The Jacobian matrix is

$$\begin{pmatrix} y - 1 & x \\ 3 + 2x & -2 - 4y \end{pmatrix} \tag{2.36}$$

and so we need to find the eigenvalues of this matrix at the four stationary points. At $(0, 0)$ the matrix becomes

$$\begin{pmatrix} -1 & 0 \\ 3 & -2 \end{pmatrix} \tag{2.37}$$

with eigenvalues $-1$ and $-2$, both of which are strictly negative. Hence we expect the origin to be asymptotically stable. At the other three stationary points one eigenvalue is positive and the other is negative so we cannot expect these points to be asymptotically stable. A more precise description of the flow near these points will be given in Chapters 4 and 5. The last exercise in this chapter describes how to show that such points cannot be Liapounov stable.

(2.13) THEOREM

*Suppose $\dot{x} = f(x)$ has linearization $\dot{x} = Ax$ at $x = 0$. If $A$ has $n$ distinct eigenvalues, each of which has strictly negative real part, then $x = 0$ is asymptotically stable.*

In fact, the theorem remains true even if the eigenvalues of $A$ are not distinct; it is sufficient that $A$ has eigenvalues with strictly negative real parts. However, the proof of the theorem depends upon a little linear algebra which is considerably more straightforward when the eigenvalues of $A$ are distinct (cf. Section 4.5).

### *A little linear algebra*

We need a few results from linear algebra about the orthogonality of eigenvectors in order to prove this result. Since we are only concerned with real matrices, eigenvalues and eigenvectors are either real or they come in complex conjugate pairs. Now, given two (possibly complex) vectors, $x$ and $y$, we define an inner product $< y, x >= y^{*T}x$, where the $*$ denotes complex conjugation and the $T$ denotes the transpose. Then, given a (real, $n \times n$) matrix $A$ we define the adjoint of $A$, $B$, to be the $n \times n$ matrix such that

$$< y, Ax >=< By, x > \quad \text{i.e.} \quad y^{*T}Ax = y^{*T}B^{*T}x. \tag{2.38}$$

Hence $A = B^{*T}$ or equivalently $B = A^{*T}$. Since $A$ is real this gives $B = A^T$. Now, the eigenvalues of $A$ and $A^T$ are the same, so assume that they are all distinct, $(\lambda_i)$, $i = 1, ..., n$. Then there are unique eigenvectors $(e_i)$ and $(f_i)$ such that

$$Ae_i = \lambda_i e_i \quad \text{and} \quad Bf_i = A^T f_i = \lambda_i^* f_i. \tag{2.39}$$

The eigenvectors of $B = A^{*T}$ are usually referred to as the adjoint eigenvectors of $A$. Hence

$$< f_j, Ae_i >= \lambda_i < f_j, e_i >$$

and similarly $< Bf_j, e_i >=< \lambda_j^* f_j, e_i >= \lambda_j < f_j, e_i >$. But by definition $< y, Ax >=< By, x >$ and so

$$\lambda_i < f_j, e_i >= \lambda_j < f_j, e_i > .$$

If $i \neq j$ then $\lambda_i \neq \lambda_j$ so this implies that

$$< f_j, e_i >= 0, \quad \text{for } i \neq j.$$

This is the really important result we need, since after normalizing the eigenvectors it implies that $(e_i)$ and $(f_i)$ can be chosen so that

$$< f_j, e_i >= \delta_{ij}. \tag{2.40}$$

Now, suppose that we write any vector $x$ in the basis of eigenvectors of $A$, so

$$x = \sum_{i=1}^{n} x_i e_i.$$

Then taking the inner product of $x$ with $f_j$ gives

$$< f_j, x >= \sum_{i=1}^{n} x_i < f_j, e_i >= x_j$$

and so

$$x = \sum_{i=1}^{n} < f_i, x > e_i. \tag{2.41}$$

This is all we need to prove the theorem.

*Proof of Theorem 2.13:* Let $A$ be the Jacobian matrix of $f$ at $x = 0$ and let $\lambda_i$, $e_i$ and $f_i$, $i = 1, ..., n$ be the eigenvalues, eigenvectors and adjoint eigenvectors as defined above. Then we can write

$$x = \sum_{i=1}^{n} x_i e_i = \sum_{i=1}^{n} < f_i, x > e_i$$

using (2.41). Hence

$$\frac{dx}{dt} = \sum_{i=1}^{n} \frac{dx_i}{dt} e_i = \sum_{i=1}^{n} \frac{d}{dt} < f_i, x > e_i.$$

But

$$\frac{dx}{dt} = Ax + o(|x|)$$

and

$$Ax = \sum_{i=1}^{n} x_i(Ae_i) = \sum_{i=1}^{n} \lambda_i < f_i, x > e_i$$

so

$$\frac{d}{dt} < f_i, x >= \lambda_i < f_i, x > + \ o(|x|). \tag{2.42}$$

This puts us in a position to define a Liapounov function: let $(v_i)$, $i = 1, ..., n$ be a set of strictly positive real numbers and let

$$V(x) = \sum_{i=1}^{n} v_i < f_i, x >^* < f_i, x > . \tag{2.43}$$

Then $V$ is clearly differentiable (as it is a quadratic function) and positive definite. Furthermore, differentiating (2.43) with respect to time we find that $\dot{V}(x)$ equals

$$\sum_{i=1}^{n} v_i \left[ \left( \frac{d}{dt} < f_i, x >^* \right) < f_i, x > + < f_i, x >^* \left( \frac{d}{dt} < f_i, x > \right) \right] \tag{2.44}$$

which, using (2.42), can be rewritten as

$$\dot{V}(x) = \sum_{i=1}^{n} v_i(\lambda_i^* + \lambda_i) < f_i, x >^* < f_i, x > + \ o(|x|^2). \tag{2.45}$$

Now since $(\lambda_i^* + \lambda_i) < 0$, $i = 1, ..., n$, there is a small neighbourhood, $G$, of $x = 0$ in which the sum dominates the $o(|x|^2)$ terms and hence $\dot{V}$ is strictly less than zero on $G \backslash \{0\}$. Hence, by Theorem 2.11, $x = 0$ is asymptotically stable.

Having constructed a Liapounov function in this way it is possible to vary the coefficients $(v_i)$ in order to maximize the domain of stability which can be deduced from the Liapounov function.

*Example 2.9*

Consider Example 2.7 again:

$$\dot{x} = y + a(\tfrac{1}{3}x^3 - x) \tag{2.46a}$$

$$\dot{y} = -x \tag{2.46b}$$

where $a > 0$. The Jacobian matrix evaluated at $x = 0$ is

$$A = \begin{pmatrix} -a & 1 \\ -1 & 0 \end{pmatrix} \tag{2.47}$$

with characteristic equation $s^2 + as + 1 = 0$. Hence the eigenvalues of $A$ are $s_\pm = (-a \pm \sqrt{a^2 - 4})/2$, both of which have strictly negative real part for all $a > 0$, and which are distinct provided $a^2 \neq 4$. Hence $x = 0$ is asymptotically stable (by Theorem 2.11) for all $a > 0$, $a \neq 2$. In fact, the stationary point is also asymptotically stable in the case of repeated roots, $a = 2$. This is proved in Section 4.5.

*Example 2.10*

Consider the stationary point $(x, y) = (0, 0)$ for the differential equation

$$\dot{x} = 2x - 5y + x^2 - 4xy \tag{2.48a}$$

$$\dot{y} = 2x - 4y + 2x^2 - 3xy + 8y^2. \tag{2.48b}$$

The Jacobian matrix evaluated at $(x, y) = (0, 0)$ is

$$A = \begin{pmatrix} 2 & -5 \\ 2 & -4 \end{pmatrix} \tag{2.49}$$

with characteristic equation $s^2 + 2s + 2 = 0$. Solving this quadratic we find that the eigenvalues of $A$ are $-1 \pm i$ and so the origin is asymptotically stable. However, we would like to construct a Liapounov function for this system: to do this we need the eigenvectors and adjoint eigenvectors of $A$. Set $\lambda_1 = -1 + i$ and $\lambda_2 = -1 - i$ (so that the labelling of the eigenvectors will be clear). Then it is easy to show that the (unnormalized) pair

$$e_1 = \begin{pmatrix} 5 \\ 3 - i \end{pmatrix} \tag{2.50}$$

and

$$e_2 = e_1^* = \begin{pmatrix} 5 \\ 3 + i \end{pmatrix} \tag{2.51}$$

are eigenvectors of $A$ with eigenvalues $-1 + i$ and $-1 - i$ respectively. Now, remembering that the labelling of the adjoint eigenvectors $f_1$ and $f_2$ are such that $A^T f_i = \lambda_i^* f_i$, we obtain

$$f_1 = \begin{pmatrix} 2 \\ -3 - i \end{pmatrix} \tag{2.52}$$

and $f_2 = f_1^*$. The Liapounov function given by (2.43) is independent of the $(e_i)$ so the calculation of $e_1$ and $e_2$ is unnecessary in the construction of the function; however, it is instructive to check that $< f_1, e_2 >$ and $< f_2, e_1 >$ both vanish.

From the construction of the proof of Theorem 2.11,

$$V(w) = v_1 < f_1, w >^* < f_1, w > + v_2 < f_2, w >^* < f_2, w >$$

where $v_1$ and $v_2$ are strictly positive constants and $w$ is the position vector in $\mathbf{R}^2$, i.e. $w = (x, y)^T$. From the definitions of $f_1$ and $f_2$ we have $f_1^* = f_2$ and hence $< f_1, w >^* = < f_2, w >$, so in fact the two terms making up $V(x, y)$ are the same and setting $v = v_1 + v_2$ we have that

$$V(x, y) = v < f_1, w >^* < f_1, w > = v| < f_1, w > |^2, \tag{2.53}$$

so the positive constant, $v$ is simply a scale factor of no importance in this case. Now $< f_1, w > = 2x + (-3 + i)y$ and so $| < f_1, w > |^2 = 4x^2 - 12xy + 10y^2$ giving the Liapounov function

$$V(x, y) = 2x^2 - 6xy + 5y^2 \tag{2.54}$$

(where the constant $v$ has been chosen to equal $1/2$).

(2.14) EXERCISE

*Check that the Liapounov function constructed in the previous example is a Liapounov function.*

## 2.4 Orbital stability

Example 2.2 shows that the definitions of stability introduced so far are not necessarily the most useful if one is interested in an attracting set such as a periodic orbit: two nearby points may move apart due to phase lagging, where the 'angular velocity' varies with the distance from the periodic orbit. Thus the periodic orbit, thought of as a set, may attract nearby points although individual trajectories do not stay close to each other. For this reason it is necessary to define stability in terms of the set of points on the periodic orbit (a closed curve in phase space).

Suppose $x_0 \in \mathbf{R}^n$ is on a periodic orbit of period $T$ for the system $\dot{x} = f(x)$, so $\varphi(x_0, T) = x_0$. Let

$$\Gamma = \{x \in \mathbf{R}^n | x = \varphi(x_0, t) \text{ for some } 0 \le t < T\}.$$

Then we can define a neighbourhood $N(\Gamma, \epsilon)$ of $\Gamma$ in the obvious way: $N(\Gamma, \epsilon) = \{x \in \mathbf{R}^n |$ there exists $y \in \Gamma$ s.t. $|x - y| < \epsilon\}$ (see Fig. 2.6).

(2.15) DEFINITION

*A periodic orbit, $\Gamma$, is Liapounov orbitally stable iff for all $\epsilon > 0$ there exists $\delta > 0$ such that if $x \in N(\Gamma, \delta)$ then $\varphi(x, t) \in N(\Gamma, \epsilon)$ for all $t \geq 0$.*

It is also possible to define orbital quasi-asymptotic stability and orbital asymptotic stability in similar ways. We shall return to questions about the stability of periodic orbits in later chapters.

## 2.5 Bounding functions

The first lemma of this chapter (Lemma (2.8), the Bounding Lemma) gives a simple argument which shows that the existence of a Liapounov function on some open domain $G$ implies that at least some trajectories remain in $G$ for all $t \geq 0$. In many examples there is no stable stationary point and yet all trajectories tend to some bounded region of phase space. What happens in this region can be very complicated (e.g. chaotic) or relatively simple (e.g. periodic) but it is clearly important to identify such regions if possible. It is possible to modify the arguments used in the previous sections to prove this type of result: the function that takes the place of the Liapounov function is called a bounding function.

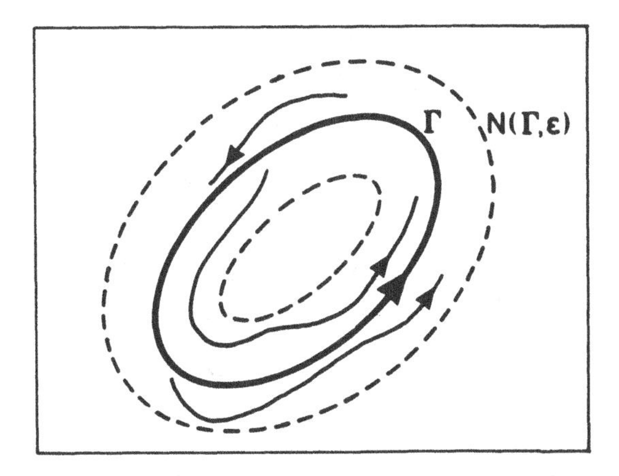


Fig. 2.6 Orbital stability.

Our first result is based on a simple geometric fact: given a function $g : \mathbf{R}^n \to \mathbf{R}$, the gradient of $g$ is normal to surfaces of constant $g(x)$.

(2.16) THEOREM

*Suppose* $\dot{x} = f(x)$, $x \in \mathbf{R}^n$, *and there is a continuously differentiable function* $g : \mathbf{R}^n \to \mathbf{R}$ *such that the set* $D = \{x \in \mathbf{R}^n | g(x) < 0\}$ *is a simply connected bounded domain with smooth boundary* $\partial D$. *If*

$$(\nabla g).f < 0 \quad \text{on } \partial D$$

*then for all* $x \in D$, $\varphi(x, t) \in D$ *for all* $t \geq 0$.

We leave the proof of this theorem as an exercise: the idea is that in order to leave $D$ the trajectory must cross $\partial D$. In order to do this its velocity must have some component in the same direction as (or, at worst, tangential to) $\nabla g$. But by assumption $f$ points in the opposite direction to $\nabla g$ on $\partial D$ so no trajectories can cross $\partial D$ outwards (see Fig. 2.7).

The final result in this section is stronger and can be used to prove that all trajectories are eventually in some bounded region of phase space (but only when this statement is true, of course!).

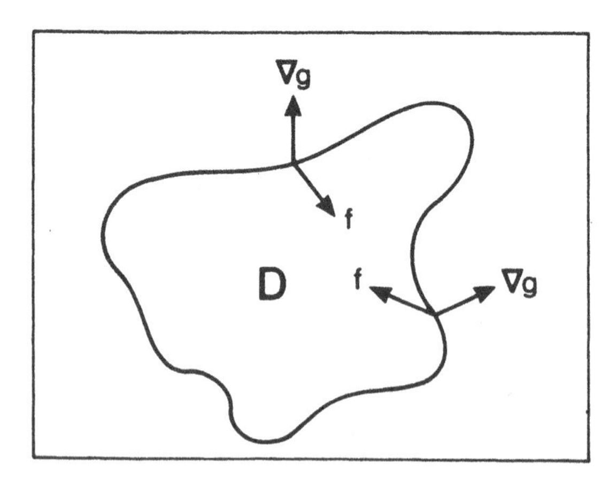


Fig. 2.7 The geometry of bounding functions.

(2.17) THEOREM

*Let $D \subset \mathbf{R}^n$ be a simply connected, compact domain and $V : \mathbf{R}^n \to \mathbf{R}$ a continuously differentiable function. Suppose that for each $k > 0$, $V_k = \{x \in \mathbf{R}^n | V(x) < k\}$ is a simply connected, bounded domain with $V_k \subset V_{k'}$ if $k < k'$. If there exists $\kappa > 0$ such that $D \subset V_\kappa$ and $\delta > 0$ such that $\dot{V}(x) \leq -\delta < 0$ for all $x \in \mathbf{R}^n \backslash D$ then for all $x$ there exists $t(x) \geq 0$ such that $\varphi(x,t) \in V_\kappa$ for all $t > t(x)$.*

*Proof*: If $x \in V_\kappa$ then $\varphi(x,t) \in V_\kappa$ for all $t \geq 0$ by Theorem 2.16 (set $g(x) = V(x) - \kappa$ and note that $f.\nabla g = \dot{V} \leq -\delta < 0$).

Suppose that $y \in \mathbf{R}^n \backslash V_\kappa$. Since $V$ is continuous, there exists $m < \kappa$ such that $D \subset V_m \subset V_\kappa$ and so $\dot{V}(x) \leq -\delta$ for all $x \in \mathbf{R}^n \backslash V_m$. Hence

$$V(\varphi(y,t)) - V(y) = \int_0^t \dot{V}\, dt \leq -\delta t$$

for all $t$ such that $\varphi(y,t) \in \mathbf{R}^n \backslash V_m$. Since $V(y) > m$ this implies that there exists $t(y) \leq (V(y) - m)/\delta$ such that $\varphi(y, t(y)) \in \partial V_m$ (and is therefore also in $V_\kappa$). Hence $\varphi(y,t) \in V_\kappa$ for all $t \geq t(y)$.

*Example 2.11*

The Lorenz equations,

$$\dot{x} = \sigma(-x + y) \tag{2.55a}$$

$$\dot{y} = rx - y - xz \tag{2.55b}$$

$$\dot{z} = -bz + xy \tag{2.55c}$$

will be discussed in more detail in Chapters 11 and 12. The constants $\sigma$, $r$ and $b$ are all positive. We aim to show that all solutions of this equation tend to some bounded ellipsoid in $\mathbf{R}^3$. Consider the functions $V(x,y,z)$ given by

$$2V(x,y,z) = rx^2 + \sigma y^2 + \sigma(z - 2r)^2. \tag{2.56}$$

Surfaces $V(x,y,z) = k$ are bounded ellipsoids for all finite $k > 0$ and $V_k \subset V_{k'}$ if $0 < k < k'$. Differentiating we find that

$$\begin{aligned}\dot{V}(x,y,z) &= r\sigma x(-x+y) + \sigma y(rx - y - xz) + \sigma(-bz + xy)(z - 2r)\\ &= -\sigma(rx^2 + y^2 + bz^2 - 2brz)\\ &= -\sigma(rx^2 + y^2 + b(z-r)^2 - br^2).\end{aligned} \tag{2.57}$$

Now let $D$ be the bounded ellipsoid

$$D = \left\{(x, y, z) | rx^2 + y^2 + b(z - r)^2 < br^2\right\}.$$

Then for sufficiently large $k$, $D \subset V_k$ and $\dot{V}(x, y, z) < 0$ for $(x, y, z) \in \mathbf{R}^3 \backslash D$. Choosing $\kappa$ large enough we can arrange for $\dot{V}(x, y, z) < -1$ for $(x, y, z) \in \mathbf{R}^3 \backslash V_\kappa$ and $D \subset V_\kappa$. Hence all trajectories eventually enter and never leave the bounded ellipsoid $V_\kappa$ by Theorem 2.17. With a little more work it is possible to estimate the size of $\kappa$ with more care, and choose a smaller value of $\delta$ ($\delta = 1$ in this example) to get a smaller region which all trajectories enter.

## 2.6 Non-autonomous equations

Suppose $\dot{x} = f(x, t)$, $x \in \mathbf{R}^n$, and $f(0, t) = 0$ for all $t \geq 0$, so $x = 0$ is a stationary point of the flow. The definitions of Liapounov stability, quasi-asymptotic stability and asymptotic stability apply equally well in this non-autonomous case so we should be able to define Liapounov functions as before. The problem is that if $V$ depends on both $x$ and $t$ then the region $V(x, t) < \mu$ may become unbounded as time increases. We prevent this possibility by underpinning $V$ with a positive definite function $U(x)$.

(2.18) DEFINITION

*Suppose $\dot{x} = f(x, t)$, $x \in \mathbf{R}^n$, and $f(0, t) = 0$ for all $t \geq 0$. Then $V(x, t)$ is a Liapounov function on some neighbourhood $G$ of $x = 0$ iff*

i) *$V$ is continuously differentiable in both $x$ and $t$ for $(x, t) \in G \times \mathbf{R}$;*
ii) *$V(0, t) = 0$ and $V(x, t) \geq U(x) > 0$ for $x \neq 0$;*
iii) *$\dot{V}(x, t) = \frac{\partial V}{\partial t} + f.\nabla V \leq 0$ for $x \in G$ and $t \geq 0$.*

(2.19) THEOREM

*Suppose $\dot{x} = f(x, t)$ and $f(0, t) = 0$ for all $t \in \mathbf{R}$. If a Liapounov function $V(x, t)$ exists on some neighbourhood $G$ of $x = 0$ then $x = 0$ is Liapounov stable.*

*Proof*: Given $\epsilon > 0$ and $t_0$ we need to find $\delta > 0$ such that if $x \in B(0,\delta)$ when $t = t_0$ then $\varphi(x,t) \in B(0,\epsilon)$ for all $t \geq t_0$.

Choose $\epsilon$ sufficiently small so that $B(0,\epsilon) \subset G$ and let $\mu = min\{U(x)|x \in \partial B(0,\epsilon)\}$. Since $\partial B(0,\epsilon)$ is compact there exists $y \in \partial B(0,\epsilon)$ with $U(y) = \mu$ and $\mu > 0$ as $y \neq 0$. Now, $V$ is continuous in $x$ so for $\mu > 0$ we can find $\delta > 0$ such that if $x \in B(0,\delta)$ then $V(x,t_0) < \mu$. Furthermore, $V$ is non-increasing along trajectories, so for $x \in B(0,\delta)$

$$V(\varphi(x,t),t) < \mu \text{ and hence } U(\varphi(x,t)) < \mu$$

for all $t \geq t_0$. But this implies that $\varphi(x,t) \in B(0,\epsilon)$ for all $t \geq t_0$ and the result is proved.

Similarly, Liapounov's second stability theorem needs to be strengthened since $\dot{V} < 0$ is no longer enough to prove asymptotic stability, as the next example shows.

*Example 2.12*

Take the trivial example $\dot{x} = 0$, so the origin is Liapounov stable but not asymptotically stable. Then the function

$$V(x,t) = |x|^2\left(1 + \frac{1}{1+t}\right)$$

is a Liapounov function and $\dot{V} = -|x|^2(1+t)^{-2} < 0$ for all $t > 0$,

(2.20) THEOREM

*Suppose $\dot{x} = f(x,t)$ and $x = 0$ is a stationary point. If a Liapounov function $V(x,t)$ can be defined on a neighbourhood $G$ of $x = 0$ and*

$$-\dot{V}(x,t) \geq W(x) > 0$$

*for $x \neq 0$ then $x = 0$ is asymptotically stable.*

## Exercises 2

1. Show that the origin is asymptotically stable for

$$\dot{x} = xP(x,y), \quad \dot{y} = yQ(x,y)$$

if $P, Q < 0$ in a neighbourhood of the origin.

2. Show that every solution of

$$\dot{x} = -t^2 x, \quad \dot{y} = -ty$$

is asymptotically stable.

3. By using the Liapounov function $V(x, y, z) = x^2 + \sigma y^2 + \sigma z^2$, show that the origin is globally asymptotically stable for the Lorenz equations (Example 2.11) if $0 < r < 1$.

4. By considering $V(x, y) = x^2 + y^2$, show that solutions of

$$\dot{x} = -x(x^2 + y^2 - 2), \quad \dot{y} = -y(x^2 + y^2 - 3x + 1)$$

tend to some bounded region $R$, and determine the smallest region $R$ which can be found using this function.

5. Show that for

$$\dot{x} = x(x^2 + 2y^2 - 1), \; \dot{y} = y(y^2 + 2z^2 - 1), \; \dot{z} = z(z^2 + 2x^2 - 1)$$

the sphere $x^2 + y^2 + z^2 = 1$ is the union of complete trajectories. Deduce that the maximal domain of stability for the origin is $x^2 + y^2 + z^2 < 1$. What happens to trajectories with $x^2 + y^2 + z^2 > 1$ at $t = 0$?

6. Find a Liapounov function for the system

$$\dot{x} = -y - x^3, \quad \dot{y} = x^5$$

and hence show that the origin is asymptotically stable.

7. Determine the values of $k$ for which $x^2 + ky^2$ is a Liapounov function for the system

$$\dot{x} = -x + y - x^2 - y^2 + xy^2, \quad \dot{y} = -y + xy - y^2 - x^2 y.$$

What information about the domain of stability of the origin can be deduced in the special case $k = 1$?

8. (a) By finding the adjoint eigenvectors of the linearized system at the origin, construct a Liapounov function for

$$\dot{x} = -x - 9y + 3x^2 - 24y^2 + 2x^5, \quad \dot{y} = x - y + x^2 - 7xy$$

and hence show that the origin is asymptotically stable.
(b) Use the same method to find Liapounov functions for the points $(0, 0)$ and $(-4, 1)$ of

$$\dot{x} = x(y - 1), \quad \dot{y} = 3x - 2y + x^2 - 2y^2$$

(cf. Example 2.8).

9. Show that the origin is a stationary point of

$$\dot{x} = x(1 - 4x^2 - y^2) - \frac{1}{2}y(1 + x), \quad \dot{y} = y(1 - 4x^2 - y^2) + 2x(1 + x).$$

By considering the function $V(x, y) = (1-4x^2-y^2)^2$ show that the origin is unstable and that all trajectories tend to the ellipse $4x^2 + y^2 = 1$ as $t \to \infty$.

10. Let $f(x, y)$ be a continuous, non-negative function of $x$ and $y$ such that $f(0, 0) = 0$ and $f(x, y) \to \infty$ as $x^2 + y^2 \to \infty$. Prove that the trajectories of

$$\dot{x} = x + y - f(x, y)x, \quad \dot{y} = -x + y - f(x, y)y$$

are eventually bounded. Show that the origin is the only stationary point of this system and, assuming that $f$ can be expanded as a Taylor series about $(0, 0)$, that the origin is completely unstable (i.e. it is asymptotically stable in reverse time). Hence show that all trajectories other than $(0, 0)$ eventually lie in an annular region

$$\{(x, y) |\ R_1^2 \le x^2 + y^2 \le R_2^2\}.$$

11. A stationary point is *unstable* if it is not Liapounov stable. Consider the equation

$$\dot{x} = Ax + g(x)$$

where $A$ is a constant $n \times n$ matrix, $g(0) = 0$ and $\frac{|g(x)|}{|x|} \to 0$ as $|x| \to 0$. Suppose further that the eigenvalues, $(\lambda_i)$, of $A$ are distinct and satisfy

$$Re\lambda_1 \le \ldots \le Re\lambda_k < 0 < Re\lambda_{k+1} \le \cdots \le Re\lambda_n.$$

Let $(f_i)$ be the normalised adjoint eigenvectors of $A$ (cf. (2.40)) and define

$$W(x) = \sum_{i=k+1}^{n} < f_i, x >^* < f_i, x > .$$

Show that in a small neighbourhood of the origin, $|x| < \epsilon$, there exists $\delta > 0$ such that if $W(0) = a > 0$ then $\dot{W} \ge \delta W$. Hence show that $|x|^2$ grows exponentially with time and hence that some trajectories must leave the region $|x| < \epsilon$. This proves that the origin is unstable.

[You may assume that there exists $M > 0$ such that $W(x) \le M|x|^2$.]

# 3

## *Linear differential equations*

Linear differential equations are nice because we can write down solutions in terms of the exponential function, which is both well-behaved and familiar. For example, given the autonomous differential equation

$$\dot{x} = Ax, \quad x \in \mathbf{R}^n \tag{3.1}$$

where $A$ is a constant $n \times n$ matrix, and the initial condition $x(0) = x_0$ the solution is

$$x(t) = e^{tA}x_0. \tag{3.2}$$

Hence if we are able to exponentiate constant matrices the solution is simple. This solution also enables us to solve equations such as

$$\dot{x} = Ax + g(t) \tag{3.3}$$

since if we set $y = e^{-tA}x$ we have that $y(0) = x(0) = x_0$ and, differentiating $y$ with respect to $t$

$$\dot{y} = e^{-tA}\dot{x} - Ae^{-tA}x = e^{-tA}g(t).$$

Hence

$$y - x_0 = \int_0^t e^{-sA}g(s)ds$$

or

$$x(t) = e^{tA}x_0 + e^{tA}\int_0^t e^{-sA}g(s)ds. \tag{3.4}$$

All these results depend upon standard properties of the exponential of a matrix. Equation (3.1) is a particular case of the more general linear differential equation

$$\dot{x} = A(t)x, \quad x \in \mathbf{R}^n. \tag{3.5}$$

A simple property of linear equations is that if $x_1(t)$ and $x_2(t)$ are solutions, then $\lambda x_1(t) + \mu x_2(t)$ is also a solution. This superposition principle no longer holds in general for nonlinear systems but has an important consequence for linear systems. If we can find $n$ independent solutions to a linear differential equation in $\mathbf{R}^n$ then any other solution can be written as a sum of these solutions. In other words, if $x_1(t), \ldots, x_n(t)$ are independent solutions the general solution can be written as

$$x(t) = \Phi(t)c \tag{3.6}$$

where $c \in \mathbf{R}^n$ is constant and $\Phi(t)$ is an $n \times n$ matrix whose columns are the solutions $x_i(t)$, i.e. $\Phi(t) = [x_1(t), \ \ldots \ , x_n(t)]$. $\Phi(t)$ is called a fundamental matrix for the problem, and $e^{tA}$ is simply a natural choice of fundamental matrix for the problem $\dot{x} = Ax$. From the solution in terms of a fundamental matrix, $x(0) = x_0 = \Phi(0)c$, and so $c = \Phi(0)^{-1}x_0$ and $x(t) = \Phi(t)\Phi(0)^{-1}x_0$. Comparing this with (3.2) we see that if $\Phi(t)$ is any fundamental matrix of the autonomous problem, (3.1), then

$$e^{tA} = \Phi(t)\Phi(0)^{-1}. \tag{3.7}$$

The idea of a fundamental matrix will be particularly useful when we come to consider problems with periodic coefficients, $\dot{x} = A(t)x$ with $A(t) = A(t+T)$ for some $T > 0$. In turn, these equations will be useful when we consider the stability of periodic orbits in later chapters. Indeed, one of the main reasons for studying linear systems is that we will be able to use some of the results in later sections on nonlinear systems.

## 3.1 Autonomous linear differential equations

In this section we will verify that the exponential $e^{tA}$ is a fundamental matrix for the differential equation $\dot{x} = Ax$ where $x \in \mathbf{R}^n$ and $A$ is a linear map of $\mathbf{R}^n$ to itself (so $A$ can be thought of as an $n \times n$ matrix with constant coefficients). To begin with we need to define the exponential of a matrix and establish some of its properties.

(3.1) DEFINITION

*Let $A$ be an $n \times n$ matrix with constant coefficients, then the exponential of $A$, $e^A$, is defined by the power series*

$$e^A = \sum_{k=0}^{\infty} \frac{A^k}{k!}.$$

The exponential of a matrix has very similar properties to the exponential of a real number. The following exercises establish some of these properties.

(3.2) EXERCISE

*(For the pure minded.) Given a matrix $A$ (i.e. a linear map of $\mathbf{R}^n$ to itself) define the norm of $A$, $||A||$ by*

$$||A|| = \sup_{v\in\mathbf{R}^n,\ v\neq 0} \frac{|Av|}{|v|}.$$

*Show that*

i) $0 \leq ||A|| < \infty$ *[Hint: note that $A(\lambda v) = \lambda Av$ for all $\lambda \in \mathbf{R}$ and so $|A(\lambda v)|/|\lambda v| = |Av|/|v|$, which implies that we can take $v$ to be on the unit ball in $\mathbf{R}^n$, which is compact.]*

ii)

$$||\lambda A|| = |\lambda|.||A||, \ for\ \lambda \in \mathbf{R}$$

$$||A + B|| \leq ||A|| + ||B||$$

$$||AB|| \leq ||A||.||B||$$

$$||A^k|| \leq ||A||^k.$$

*Let $(A_k)$ be a set of linear maps of $\mathbf{R}^n$. Then a series $\sum_0^\infty A_k$ is absolutely convergent if $\sum_0^\infty ||A_k||$ is absolutely convergent. If $\sum A_k$ is absolutely convergent then the sum exists and addition, multiplication and differentiation can be done term by term. Show that $e^A$ is absolutely convergent and $e^{(s+t)A} = e^{sA}e^{tA}$ for all real numbers $s$ and $t$. If $A$ and $B$ commute (i.e. $AB = BA$) show that $e^{A+B} = e^A e^B$. What is $e^{A+B}$ in terms of $e^A$ and $e^B$ if $A$ and $B$ do not commute? [Hint: define $C = AB - BA$.]*

With these basic properties of the exponential we can prove the existence and uniqueness of solutions to autonomous linear differential equations.

(3.3) THEOREM

*Let $A$ be an $n \times n$ matrix with constant coefficients and $x \in \mathbf{R}^n$. Then the unique solution $x(t)$ of $\dot{x} = Ax$ with $x(0) = x_0$ is $x(t) = e^{tA}x_0$.*

*Proof*: First note that

$$\frac{d}{dt}e^{tA} = \sum_{k=0}^{\infty} \frac{d}{dt}\left(\frac{t^k A^k}{k!}\right) = \sum_{k=1}^{\infty}\left(\frac{t^{k-1}A^k}{(k-1)!}\right) = Ae^{tA}$$

where we have used the result quoted in the second exercise above to differentiate the sum term by term. Hence

$$\frac{d}{dt}x(t) = \frac{d}{dt}e^{tA}x_0 = Ae^{tA}x_0 = Ax(t)$$

and so $e^{tA}x_0$ is a solution of the equation $\dot{x} = Ax$ with $x(0) = x_0$. To prove uniqueness, suppose that $y(t)$ is another solution to the differential equation with the same initial value, $y(0) = x_0$. Set $z(t) = e^{-tA}y(t)$, then

$$\dot{z} = -Ae^{-tA}y(t) + e^{-tA}\dot{y} = -Ae^{-tA}y(t) + Ae^{-tA}y(t) = 0$$

i.e. $z(t)$ is constant. But $z(0) = y(0) = x_0$ and so $z(t) = x_0$. Now, from the definition of $z(t)$, this implies that $y(t) = e^{tA}x_0 = x(t)$, and hence solutions are unique.

This result shows that we can find the solutions of linear differential equations by exponentiating matrices, but it does not tell us *how* to calculate with exponentials of matrices. If we want to be able to write down solutions explicitly we must learn how to take the exponential of a matrix. If $A$ is diagonal, $A = \text{diag}(a_1, \ldots, a_n)$, then it should be obvious that $e^{tA} = \text{diag}(e^{a_1 t}, \ldots, e^{a_n t})$ and solutions in component form are $x_i(t) = e^{a_i t}x_{0i}$, but for more general matrices the solutions are not so obvious. To deal with this problem we need a little more linear algebra and the idea of a Jordan normal form.

## 3.2 Normal forms

Consider a simple change of coordinates $x = Py$ where $P$ is an $n \times n$ invertible matrix ($\det P \neq 0$). Then $\dot{x} = Ax$ implies that

$$\begin{aligned} \dot{y} &= P^{-1}\dot{x} = P^{-1}Ax \\ &= P^{-1}APy = \Lambda y \end{aligned} \tag{3.8}$$

where $\Lambda = P^{-1}AP$ and the initial value $x(0) = x_0$ is transformed to $y(0) = P^{-1}x_0 = y_0$. In these new coordinates the solution is $y(t) = e^{t\Lambda}y_0$, and so transforming back to the $x$ coordinates

$$x(t) = Py(t) = Pe^{t\Lambda}y_0 = Pe^{t\Lambda}P^{-1}x_0. \tag{3.9}$$

Comparing this with the known solution $e^{tA}x_0$ we see that

$$e^{tA} = Pe^{t\Lambda}P^{-1}. \tag{3.10}$$

The strategy in this section is to choose $P$ in such a way that $\Lambda$ takes a particularly simple form which will allow us to calculate $e^{t\Lambda}$ and hence $e^{tA}$ with a minimum of fuss.

We have already seen that it is easy to exponentiate diagonal matrices. So suppose that $A$ has $n$ distinct real eigenvalues, $\lambda_1, \dots, \lambda_n$, with associated eigenvectors $e_i$, so $Ae_i = \lambda_i e_i$, $1 \le i \le n$. Let $P = [e_1, \ \dots \ , e_n]$, the matrix with the eigenvectors of $A$ as columns. Then, since the eigenvectors of distinct real eigenvalues are real and independent, $\det P \neq 0$ and

$$\begin{aligned} AP &= [Ae_1, \ \dots \ , Ae_n] = [\lambda_1 e_1, \ \dots \ , \lambda_n e_n] \\ &= [e_1, \ \dots \ , e_n]\operatorname{diag}(\lambda_1, \dots, \lambda_n) \end{aligned}$$

and so if $\Lambda = \operatorname{diag}(\lambda_1, \dots, \lambda_n)$ we have $AP = P\Lambda$ or $\Lambda = P^{-1}AP$. Hence, with this choice of $P$ we can bring the differential equation into the form

$$\dot{y} = \operatorname{diag}(\lambda_1, \dots, \lambda_n)y, \tag{3.11}$$

which is, of course, particularly easy to solve. Note in particular that if $y_0 = (0, 0, \dots, 0, 1, 0, \dots, 0)^T$, then $y(t) = (0, 0, \dots, 0, e^{\lambda_i t}, 0, \dots, 0)$. Thus each coordinate axis is invariant under the flow: solutions starting on a coordinate axis remain on that coordinate axis for all time. Translating back to the $x$ coordinates the coordinate axis corresponds to the eigenvector $e_i$, so each eigendirection is invariant.

*Example 3.1*

Consider the matrix

$$A = \begin{pmatrix} -2 & 1 \\ 0 & 2 \end{pmatrix}.$$

The eigenvalue equation is $(s+2)(s-2) = 0$ and so the eigenvalues are $\lambda_1 = -2$ and $\lambda_2 = 2$. Solving for the eigenvectors we find

$$e_1 = \begin{pmatrix} 1 \\ 0 \end{pmatrix} \text{ and } e_2 = \begin{pmatrix} 1 \\ 4 \end{pmatrix}$$

and so

$$P = \begin{pmatrix} 1 & 1 \\ 0 & 4 \end{pmatrix}.$$

Now, for a general invertible $2 \times 2$ matrix

$$\begin{pmatrix} a & b \\ c & d \end{pmatrix}^{-1} = \frac{1}{ad-bc}\begin{pmatrix} d & -b \\ -c & a \end{pmatrix}$$

and so

$$P^{-1} = \frac{1}{4}\begin{pmatrix} 4 & -1 \\ 0 & 1 \end{pmatrix}.$$

We leave it as an exercise to verify that $P^{-1}AP = \text{diag}(-2, 2)$ and hence

$$e^{tA} = P\begin{pmatrix} e^{-2t} & 0 \\ 0 & e^{2t} \end{pmatrix}P^{-1} = \begin{pmatrix} e^{-2t} & (e^{2t} - e^{-2t})/4 \\ 0 & e^{2t} \end{pmatrix}.$$

Hence the solution to $\dot{x} = Ax$ with $x(0) = (a, b)^T$ is

$$\begin{pmatrix} ae^{-2t} + b(e^{2t} - e^{-2t})/4 \\ be^{2t} \end{pmatrix}.$$

Now suppose that $A$ is a $2 \times 2$ matrix with a pair of complex conjugate eigenvalues, $\rho \pm i\omega$. Then we claim that there is a real invertible matrix, $P$, such that

$$P^{-1}AP = \Lambda = \begin{pmatrix} \rho & -\omega \\ \omega & \rho \end{pmatrix}. \tag{3.12}$$

This is a convenient form since $\Lambda = D + C$ where

$$D = \begin{pmatrix} \rho & 0 \\ 0 & \rho \end{pmatrix} \text{ and so } e^{tD} = \begin{pmatrix} e^{\rho t} & 0 \\ 0 & e^{\rho t} \end{pmatrix}$$

and $C = \begin{pmatrix} 0 & -\omega \\ \omega & 0 \end{pmatrix}$, so

$$C^{2n} = \begin{pmatrix} \omega^{2n} & 0 \\ 0 & \omega^{2n} \end{pmatrix} \text{ and } C^{2n+1} = \begin{pmatrix} 0 & -\omega^{2n+1} \\ \omega^{2n+1} & 0 \end{pmatrix},$$

which gives, using the series definitions of sine and cosine,

$$e^{tC} = \begin{pmatrix} \cos\omega t & -\sin\omega t \\ \sin\omega t & \cos\omega t \end{pmatrix}. \tag{3.13}$$

Since $D$ and $C$ commute, $e^{t\Lambda} = e^{tD}e^{tC}$, and so

$$e^{t\Lambda} = e^{\rho t}\begin{pmatrix} \cos\omega t & -\sin\omega t \\ \sin\omega t & \cos\omega t \end{pmatrix}. \tag{3.14}$$

We now have to prove this claim, i.e. define a matrix $P$ such that $\Lambda = P^{-1}AP$ given a matrix $A$ with a complex conjugate pair of eigenvalues $\rho \pm i\omega$. As in the previous case we will define $P$ using the eigenvectors of $A$. Suppose that $A$ is a (real) $2 \times 2$ matrix with a pair of complex conjugate eigenvalues $\rho \pm i\omega$. Then there is a (complex) eigenvector $z = (z_1, z_2)^T$ such that $Az = (\rho + i\omega)z$. Consider the real matrix whose columns are made up of the imaginary and real parts of $z$; then, since $A$ is real,

$$\begin{aligned} A[\mathrm{Im}(z),\ \mathrm{Re}(z)] &= [\mathrm{Im}((\rho + i\omega)z),\ \mathrm{Re}((\rho + i\omega)z)] \\ &= [\rho\mathrm{Im}(z) + \omega\mathrm{Re}(z),\ \rho\mathrm{Re}(z) - \omega\mathrm{Im}(z)]. \end{aligned}$$

It is now easy to verify that this equals

$$[\mathrm{Im}(z),\ \mathrm{Re}(z)] \begin{pmatrix} \rho & -\omega \\ \omega & \rho \end{pmatrix} \tag{3.15}$$

and hence $\Lambda = P^{-1}AP$ where $P = [\mathrm{Im}(z),\ \mathrm{Re}(z)]$.

*Example 3.2*

Consider the matrix

$$A = \begin{pmatrix} 2 & 1 \\ -2 & 0 \end{pmatrix}.$$

The eigenvalue equation is $s^2 - 2s + 2 = 0$ and so the eigenvalues are $1 \pm i$. An eigenvector corresponding to the eigenvalue $1 + i$ is $z = (1, -1 + i)^T$ and so

$$P = \begin{pmatrix} 0 & 1 \\ 1 & -1 \end{pmatrix}, \quad P^{-1} = \begin{pmatrix} 1 & 1 \\ 1 & 0 \end{pmatrix}.$$

Once again we leave it as an exercise to verify that $P^{-1}AP = \Lambda$ where

$$\Lambda = \begin{pmatrix} 1 & -1 \\ 1 & 1 \end{pmatrix}, \quad e^{t\Lambda} = e^t \begin{pmatrix} \cos t & -\sin t \\ \sin t & \cos t \end{pmatrix}.$$

Hence $e^{tA} = Pe^{t\Lambda}P^{-1}$, i.e.

$$\begin{aligned} e^{tA} &= e^t \begin{pmatrix} 0 & 1 \\ 1 & -1 \end{pmatrix} \begin{pmatrix} \cos t & -\sin t \\ \sin t & \cos t \end{pmatrix} \begin{pmatrix} 1 & 1 \\ 1 & 0 \end{pmatrix} \\ &= e^t \begin{pmatrix} \cos t + \sin t & \sin t \\ -2\sin t & \cos t - \sin t \end{pmatrix}. \end{aligned}$$

Thus the solution of $\dot{x} = Ax$ with $x(0) = (a, b)^T$ is

$$\begin{pmatrix} ae^t(\cos t + \sin t) + be^t \sin t \\ -2ae^t \sin t + be^t(\cos t - \sin t) \end{pmatrix}.$$

This argument generalizes to higher dimension in the obvious way (described below), which enables us to use a change of coordinate to bring any differential equation $\dot{x} = Ax$, where $A$ has distinct eigenvalues into normal form $\dot{y} = \Lambda y$ where $\Lambda = \mathrm{diag}(B_1, \ldots, B_m)$. The blocks $(B_i)$ are given by $B_i = \lambda_i$ if the $i^{th}$ eigenvalue is real and the $2 \times 2$ matrix

$$B_i = \begin{pmatrix} \rho & -\omega \\ \omega & \rho \end{pmatrix}$$

if the corresponding eigenvalues are a complex conjugate pair, $\rho \pm i\omega$.

(3.4) THEOREM

*Let $A$ be a real $n \times n$ matrix with $k$ distinct real eigenvalues $\lambda_1, \ldots, \lambda_k$ and $m = \frac{1}{2}(n-k)$ pairs of distinct complex conjugate eigenvalues $\rho_1 \pm i\omega_1, \ldots, \rho_m \pm i\omega_m$. Then there exists an invertible matrix $P$ such that*

$$P^{-1}AP = \Lambda = \mathrm{diag}(\lambda_1, \ldots, \lambda_k, B_1, \ldots, B_m)$$

*where the $B_i$, $1 \le i \le m$, are $2 \times 2$ blocks, $B_i = \begin{pmatrix} \rho_i & -\omega_i \\ \omega_i & \rho_i \end{pmatrix}$. Furthermore, $e^{tA} = Pe^{t\Lambda}P^{-1}$ and*

$$e^{t\Lambda} = \mathrm{diag}(e^{\lambda_1 t}, \ldots, e^{\lambda_k t}, e^{tB_1}, \ldots, e^{tB_m})$$

*where*

$$e^{tB_i} = e^{\rho_i t} \begin{pmatrix} \cos \omega_i t & -\sin \omega_i t \\ \sin \omega_i t & \cos \omega_i t \end{pmatrix}.$$

*Proof*: Let $(e_i)$, $1 \le i \le k$, be the real eigenvectors associated with the real eigenvalues $\lambda_i$ and $(z_j)$, $1 \le j \le m$, be the (complex) eigenvectors associated with the eigenvalues $\rho_j + i\omega_j$. Set $P$ to be the matrix whose first $k$ columns are the eigenvectors $e_1, \ldots, e_k$ and the remaining $n-k$ columns are the imaginary and real parts of the eigenvectors $z_j$, i.e.

$$P = [e_1, \ \ldots \ , e_k, \ \mathrm{Im}(z_1), \ \mathrm{Re}(z_1), \ \ldots \ , \mathrm{Im}(z_m), \ \mathrm{Re}(z_m)].$$

Since the eigenvalues are distinct the eigenvectors are independent and so $\det P \neq 0$ and by the arguments rehearsed above

$$AP = P\Lambda.$$

The rest of the theorem is just a restatement of results which have already been proved above.

If we are prepared to work with complex eigenvectors and matrices it is possible to diagonalize any matrix with distinct eigenvalues. To illustrate this suppose that $A$ is a matrix which satisfies the conditions of Theorem 3.4 above, and let $\rho_j \pm i\omega_j = \gamma_j$, $1 \leq j \leq m$. Then each complex eigenvalue $\gamma_j$ has a complex eigenvector $\epsilon_j$, and $\epsilon_j^*$ is an eigenvector of the complex conjugate eigenvalue $\gamma_j^*$. In this case the matrix of eigenvectors (with $e_i$ real and $\epsilon_j$ complex)

$$P = [e_1, \ \dots \ e_k, \ \epsilon_1 \ , \epsilon_1^*, \ \dots \ , \epsilon_m, \ \epsilon_m^*]$$

diagonalizes $A$. So if $\Lambda = \text{diag}(\lambda_1, \dots, \lambda_k, \gamma_1, \gamma_1^*, \dots, \gamma_m, \gamma_m^*)$ then

$$AP = P\Lambda. \tag{3.16}$$

In this case the differential equation $\dot{x} = \Lambda x$ must be interpreted with $x = (x_1, \dots, x_k, z_1, z_1^*, \dots, z_m, z_m^*)$ where the $x_i$ are real variables and the $z_i$ are complex variables. The differential equation for $z_j$ is then $\dot{z}_j = \gamma_j z_j$. We can regain the real form of this equation by setting $z_j = X + iY$ and $\gamma_j = \rho_j \pm i\omega_j$ in which case (equating real and imaginary parts)

$$\dot{X} = \rho_j X - \omega_j Y, \ \ \dot{Y} = \omega_j X + \rho_j Y. \tag{3.17}$$

In some sections of this book it will be more convenient to work in complex notation.

Theorem 3.4 deals with the cases when $A$ has distinct roots, so we should now consider the possibility of multiple roots. Suppose that $A$ is a real $n \times n$ matrix with eigenvalues $\lambda_1, \dots, \lambda_p$, $p \leq n$. Then the characteristic polynomial of $A$ is

$$\prod_{k=1}^{p} (s - \lambda_k)^{n_k} \tag{3.18}$$

where $n_k \geq 1$ and $\sum_k n_k = n$. If $A$ has distinct eigenvalues then $p = n$ and $n_k = 1$, $1 \leq k \leq n$, but if $p < n$ then at least one of the $n_k$ must be greater than 1 and the characteristic polynomial has repeated roots. The number $n_k$ is called the multiplicity of the eigenvalue $\lambda_k$ and the generalized eigenspace of $\lambda_k$ is

$$E_k = \{x \in \mathbf{R}^n | (A - \lambda_k I)^{n_k} x = 0\}. \tag{3.19}$$

The dimension of $E_k$ is $n_k$ and so we can choose a set of basis vectors $(e_k^1, \dots, e_k^{n_k})$ of $E_k$ which we will refer to as generalized eigenvectors of $A$. To go through all the possible cases of repeated roots would take several chapters of linear algebra (see, for example, Hirsch and Smale

(1976) or Arnold (1973)); here we shall stick to the possibilities which arise for differential equations in $\mathbf{R}^2$ and $\mathbf{R}^3$.

Consider first the case where $A$ is a $2 \times 2$ matrix with a repeated real eigenvalue $\lambda$, so the characteristic polynomial of $A$ is $(s-\lambda)^2 = 0$. Since $A$ satisfies its own characteristic equation this implies that

$$(A - \lambda I)^2 x = 0 \tag{3.20}$$

for all $x \in \mathbf{R}^2$. Now, either $(A - \lambda I)x = 0$ for all $x$ or there exists $e_2$ such that $(A - \lambda I)e_2 \neq 0$ and $e_2 \neq 0$. In the first case $A = \mathrm{diag}(\lambda, \lambda)$ for *any* choice of basis vectors $e_1$ and $e_2$, since $Ax = \lambda x$ for all $x \in \mathbf{R}^2$. In the second case define

$$e_1 = (A - \lambda I)e_2. \tag{3.21}$$

Then $(A - \lambda I)^2 e_2 = 0 = (A - \lambda I)e_1$ and so $Ae_1 = \lambda e_1$, whilst $Ae_2 = e_1 + \lambda e_2$ from the definition of $e_1$, (3.21). Hence

$$A[e_1, e_2] = [\lambda e_1, \lambda e_2 + e_1] = [e_1, e_2] \begin{pmatrix} \lambda & 1 \\ 0 & \lambda \end{pmatrix}. \tag{3.22}$$

In other words, if $\lambda$ is a double eigenvalue of $A$ then there is a change of coordinates which brings $A$ into one of the two cases

$$\begin{pmatrix} \lambda & 0 \\ 0 & \lambda \end{pmatrix} \quad \text{or} \quad \begin{pmatrix} \lambda & 1 \\ 0 & \lambda \end{pmatrix}. \tag{3.23}$$

In both cases it is easy to solve the differential equation $\dot{x} = Ax$ in this choice of coordinate system and then translate back to find the solutions in the original coordinate system.

(3.5) EXERCISE

*Show that the solution to*

$$\dot{x} = \lambda x, \quad \dot{y} = \lambda y$$

*with* $(x(0), y(0)) = (x_0, y_0)$ *is*

$$x(t) = x_0 e^{\lambda t}, \quad y(t) = y_0 e^{\lambda t}$$

*and that the solution to*

$$\dot{x} = \lambda x + y, \quad \dot{y} = \lambda y$$

*subject to the same initial condition is*

$$x(t) = e^{\lambda t}(x_0 + y_0 t), \quad y(t) = y_0 e^{\lambda t}.$$

*Show from first principles that if* $\Lambda = \begin{pmatrix} \lambda & 1 \\ 0 & \lambda \end{pmatrix}$ *then*

$$e^{t\Lambda} = \begin{pmatrix} e^{\lambda t} & te^{\lambda t} \\ 0 & e^{\lambda t} \end{pmatrix}.$$

*Example 3.3*

Consider the matrix

$$A = \begin{pmatrix} 5 & -3 \\ 3 & -1 \end{pmatrix}$$

with characteristic polynomial $s^2 - 4s + 4 = (s-2)^2 = 0$. So $\lambda = 2$ is a double eigenvalue. It is easy to see that $e_2 = (1,-1)^T$ does not satisfy the equation $(A - 2I)e_2 = 0$ and that $e_1 = (A - 2I)e_2 = (6,6)^T$. Hence set

$$P = \begin{pmatrix} 6 & 1 \\ 6 & -1 \end{pmatrix} \quad \text{with} \quad P^{-1} = -\frac{1}{12}\begin{pmatrix} -1 & -1 \\ -6 & 6 \end{pmatrix}$$

so

$$P^{-1}AP = \Lambda = \begin{pmatrix} 2 & 1 \\ 0 & 2 \end{pmatrix} \quad \text{and} \quad e^{t\Lambda} = \begin{pmatrix} e^{2t} & te^{2t} \\ 0 & e^{2t} \end{pmatrix}.$$

To find $e^{tA}$ note that

$$e^{tA} = Pe^{t\Lambda}P^{-1} = \begin{pmatrix} e^{2t}(1+3t) & -3te^{2t} \\ 3te^{2t} & e^{2t}(1-3t) \end{pmatrix}$$

which allows us to write down the solution to the differential equation $\dot{x} = Ax$ with initial condition $x(0) = (x_0, y_0)^T$ as

$$\begin{pmatrix} x_0e^{2t}(1+3t) - 3y_0te^{2t} \\ 3x_0te^{2t} + y_0e^{2t}(1-3t) \end{pmatrix}.$$

The case of three repeated real eigenvalues in $\mathbf{R}^3$ is similar, but there are three cases. First note that the characteristic polynomial is $(s-\lambda)^3 = 0$ and so

$$(A - \lambda I)^3 x = 0 \tag{3.24}$$

for all $x \in \mathbf{R}^n$. The three cases we need to consider are

i) there exists $e_3 \neq 0$ such that $(A - \lambda I)^2 e_3 \neq 0$;
ii) $(A-\lambda I)^2 x = 0$ for all $x$ but there exists $e_2 \neq 0$ such that $(A-\lambda I)e_2 \neq 0$; and
iii) $(A - \lambda I)x = 0$ for all $x$.

These three cases give rise (respectively) to the normal forms

$$\begin{pmatrix} \lambda & 1 & 0 \\ 0 & \lambda & 1 \\ 0 & 0 & \lambda \end{pmatrix}, \quad \begin{pmatrix} \lambda & 1 & 0 \\ 0 & \lambda & 0 \\ 0 & 0 & \lambda \end{pmatrix}, \text{ and } \begin{pmatrix} \lambda & 0 & 0 \\ 0 & \lambda & 0 \\ 0 & 0 & \lambda \end{pmatrix}.$$

The last of these three possibilities is as straightforward as the diagonal case in $\mathbf{R}^2$ and so we will not dwell on it. In case (i) define

$$e_1 = (A - \lambda I)^2 e_3, \text{ and } e_2 = (A - \lambda I)e_3.$$

Clearly $Ae_1 = \lambda e_1$ (as $(A - \lambda I)^3 x = 0$ for all $x$) and $Ae_3 = \lambda e_3 + e_2$ (from the definition of $e_2$). Furthermore, the definition of $e_1$ can be rewritten as $e_1 = (A - \lambda I)e_2$ and so $Ae_2 = \lambda e_2 + e_1$. Putting these three relationships together and forming the matrix $[e_1, e_2, e_3]$ as before gives

$$\begin{aligned} A[e_1, e_2, e_3] &= [\lambda e_1, e_1 + \lambda e_2, e_2 + \lambda e_3] \\ &= [e_1, e_2, e_3] \begin{pmatrix} \lambda & 1 & 0 \\ 0 & \lambda & 1 \\ 0 & 0 & \lambda \end{pmatrix} \end{aligned}$$

and so the matrix $P = [e_1, e_2, e_3]$ gives the transformation for the matrix $A$ to have the required form.

The second case is the most difficult to establish. We have that $(A - \lambda I)^2 x = 0$ and there exists $e_2$ such that $(A - \lambda I)e_2 \neq 0$. As in the two-dimensional case define $e_1 = (A - \lambda I)e_2$, so $Ae_1 = \lambda e_1$ and $Ae_2 = \lambda e_2 + e_1$. We now claim that there is another vector, $e_3$, which is independent of $e_1$ and $e_2$ and which also satisfies $Ae_3 = \lambda e_3$ (we shall not prove this claim, but it is easy to verify in examples). The matrix $P = [e_1, e_2, e_3]$ then produces the desired form.

(3.6) EXERCISE

*Show that if*

$$\Lambda = \begin{pmatrix} \lambda & 1 & 0 \\ 0 & \lambda & 1 \\ 0 & 0 & \lambda \end{pmatrix}$$

*then*

$$e^{t\Lambda} = e^{\lambda t} \begin{pmatrix} 1 & t & \frac{1}{2}t^2 \\ 0 & 1 & t \\ 0 & 0 & 1 \end{pmatrix}.$$

In the case of complex repeated eigenvectors similar manipulations yield normal forms

$$\begin{pmatrix} \rho & -\omega & 1 & 0 \\ \omega & \rho & 0 & 1 \\ 0 & 0 & \rho & -\omega \\ 0 & 0 & \omega & \rho \end{pmatrix},$$

or the standard block on the diagonal with the unit matrix on the off-diagonal. In general (for matrices in $\mathbf{R}^n$) there is always a coordinate transformation to bring into normal forms which are blocks of the kind found above.

## 3.3 Invariant manifolds

The manipulations of the previous section show that given the system $\dot{x} = Ax$ we can do a simple change of coordinates to bring the equation into the normal form $\dot{y} = \Lambda y$. In the simplest case, where $A$ has distinct eigenvalues, the matrix $\Lambda$ is

$$\Lambda = \text{diag}(\lambda_1, \ldots, \lambda_k, B_1, \ldots, B_m) \qquad (3.25)$$

where $(\lambda_i)$ are the real eigenvalues and $B_j$ are the matrices

$$\begin{pmatrix} \rho_j & -\omega_j \\ \omega_j & \rho_j \end{pmatrix}$$

associated with the complex conjugate pairs of eigenvalues $\rho_j \pm i\omega_j$. In component form, with $y = (y_1, \ldots, y_k, w_1, \ldots w_{2m})$ the equation $\dot{y} = \Lambda y$ is therefore

$$\dot{y}_i = \lambda y_i, \quad 1 \leq i \leq k \qquad (3.26)$$

and

$$\begin{pmatrix} \dot{w}_{2j-1} \\ \dot{w}_{2j} \end{pmatrix} = \begin{pmatrix} \rho_j & -\omega_j \\ \omega_j & \rho_j \end{pmatrix} \begin{pmatrix} w_{2j-1} \\ w_{2j} \end{pmatrix}, \quad 1 \leq j \leq m, \qquad (3.27)$$

and since they are uncoupled each of these $k+m$ equations can be solved separately. It follows immediately that the real eigenspaces of $\Lambda$ are invariant, since if $y_0 = (0, \ldots, Y_0, \ldots, 0)$ then $y(t) = (0, \ldots, Y_0 e^{\lambda_i t}, \ldots, 0)$, and similarly the two-dimensional eigenspaces corresponding to the complex conjugate pair of eigenvalues $\rho_j \pm i\omega_j$ are also invariant. Returning to the original equation $\dot{x} = Ax$ this implies that the corresponding eigenspaces of $A$ are invariant. Since all the eigenvalues of $A$ are distinct, these eigenspaces are either one-dimensional or two-dimensional depending on whether the corresponding eigenvalue is real or not. This proves the following theorem.

(3.7) THEOREM

*If the eigenvalues of the $n \times n$ real matrix $A$ are distinct then $\mathbf{R}^n$ decomposes into a direct sum of one-dimensional spaces and two-dimensional spaces. Each of these eigenspaces is invariant under the flow defined by $\dot{x} = Ax$.*

If the eigenvalues of $A$ are not distinct then the normal form of $A$ will contain blocks of matrices with 1s down the off-diagonal as described in the previous section. The corresponding generalized eigenspaces are, of course, also invariant but they may be of dimension greater than two.

*Example 3.4*

Consider the matrix

$$A = \begin{pmatrix} 2 & 0 & 0 & 0 & 0 \\ 0 & -3 & 1 & 0 & 0 \\ 0 & 0 & -3 & 0 & 0 \\ 0 & 0 & 0 & -2 & -1 \\ 0 & 0 & 0 & 1 & -2 \end{pmatrix}.$$

$A$ has 3 blocks: a one-dimensional eigenspace corresponding to the eigenvalue 2, a two-dimensional eigenspace corresponding to the multiple eigenvalue $-3$ and a two-dimensional eigenspace corresponding to the complex conjugate pair of eigenvalues $-2 \pm i$. To solve the differential equation $\dot{x} = Ax$ with $x(0) = (y_1, y_2, y_3, y_4, y_5)$ we can work independently in these three eigenspaces. In the first,

$$\dot{x}_1 = 2x_1, \quad x_1(0) = y_1$$

and so

$$x_1(t) = x_0 e^{2t}.$$

In the degenerate eigenspace

$$\begin{pmatrix} \dot{x}_2 \\ \dot{x}_3 \end{pmatrix} = \begin{pmatrix} -3 & 1 \\ 0 & -3 \end{pmatrix} \begin{pmatrix} x_2 \\ x_3 \end{pmatrix}, \quad \begin{pmatrix} x_2(0) \\ x_3(0) \end{pmatrix} = \begin{pmatrix} y_2 \\ y_3 \end{pmatrix}$$

and so

$$\begin{pmatrix} x_2(t) \\ x_3(t) \end{pmatrix} = \begin{pmatrix} y_2 e^{-3t} + y_3 t e^{-3t} \\ y_3 e^{-3t} \end{pmatrix}.$$

Finally

$$\begin{pmatrix} \dot{x}_4 \\ \dot{x}_5 \end{pmatrix} = \begin{pmatrix} -2 & -1 \\ 1 & -2 \end{pmatrix} \begin{pmatrix} x_4 \\ x_5 \end{pmatrix}, \quad \begin{pmatrix} x_4(0) \\ x_5(0) \end{pmatrix} = \begin{pmatrix} y_4 \\ y_5 \end{pmatrix}$$

with solution

$$\begin{pmatrix} x_4(t) \\ x_5(t) \end{pmatrix} = \begin{pmatrix} e^{-2t}(y_4 \cos t - y_5 \sin t) \\ e^{-2t}(y_4 \sin t + y_5 \cos t) \end{pmatrix}$$

(3.8) EXERCISE

*Find the general form of those points which tend to zero as $t \to \infty$ and the points which tend to zero as $t \to -\infty$.*

Since each generalized eigenspace of $A$ is invariant under the flow $\dot{x} = Ax$, the spaces spanned by several generalized eigenspaces are also invariant. There are three dynamically important sets of eigenspaces. Let $(u_i)$, $1 \leq i \leq n_i$, denote the generalized eigenvectors associated with eigenvalues of $A$ with strictly positive real parts, $(c_j)$, $1 \leq j \leq n_c$, denote the generalized eigenvectors associated with eigenvalues of $A$ with zero real parts and $(s_k)$, $1 \leq k \leq n_s$, denote the generalized eigenvectors associated with eigenvalues of $A$ which have strictly negative real parts. Then every generalized eigenvector is in one of these sets and so $n_u + n_c + n_s = n$. The origin $x = 0$ is a stationary point of the flow, and we define the unstable manifold of $x = 0$, $E^u(0)$, to be the invariant space spanned by the eigenvectors $(u_i)$, the centre manifold, $E^c(0)$, to be the invariant manifold spanned by the eigenvectors $(c_j)$ and the stable manifold, $E^s(0)$, to be the invariant manifold spanned by the eigenvectors $(s_k)$.

(3.9) THEOREM

*Suppose $\dot{x} = Ax$, where $A$ is a constant $n \times n$ matrix. If $n_c = 0$ then*

$$E^u(0) = \{x \in \mathbf{R}^n | e^{tA}x \to 0 \text{ as } t \to -\infty\}$$

$$E^s(0) = \{x \in \mathbf{R}^n | e^{tA}x \to 0 \text{ as } t \to \infty\}.$$

The proof of this theorem in the case that $A$ has distinct eigenvalues is straightforward (see Exercise 3.5), but it needs a little more work, and linear algebra, to prove in the general case (see Section 4.5). Much of the effort in the next chapter will be devoted to proving that the stable and unstable manifolds of a stationary point persist when nonlinear terms are added.

## 3.4 Geometry of phase space

The invariant manifolds corresponding to generalized eigenspaces are simply surfaces in $\mathbf{R}^n$, and the motion on each of these surfaces can be obtained by solving one of a set of simple differential equations. The general solution or flow is found by superposing the solutions within each invariant subspace. Consider, for example, the equation

$$\dot{x} = \rho x - \omega y, \quad \dot{y} = \omega x + \rho y \tag{3.28}$$

with initial condition $(x(0), y(0)) = (x_0, y_0)$ (cf. (3.17) and (3.27)). In polar coordinates this equation becomes

$$\dot{r} = \rho r, \quad \dot{\theta} = \omega \tag{3.29}$$

and so the motion is a uniform rotation about the origin together with exponential growth (if $\rho > 0$) or decay (if $\rho < 0$) of the radial component. Trajectories of this two-dimensional system are therefore logarithmic spirals if $\rho \neq 0$ since

$$\frac{dr}{d\theta} = \frac{\rho}{\omega} r \tag{3.30}$$

and so

$$r(\theta) = r(0)\exp\left\{\frac{\rho\theta}{\omega}\right\}$$

or

$$\theta + \varphi = \frac{\omega}{\rho} \log r \tag{3.31}$$

provided $\rho \neq 0$; $\varphi$ is just a constant. If $\rho = 0$ then trajectories are concentric circles. Solutions on the one-dimensional eigenspaces corresponding to real distinct non-zero eigenvalues are unambiguously towards or away from the origin and so it is easy to superpose the various motions.

*Example 3.5*

Consider the flow $\dot{x} = Ax$ for $x \in \mathbf{R}^3$ where $A$ has a pair of complex conjugate eigenvalues with negative real parts and a positive real eigenvalue.

In normal form this equation becomes $\dot{y} = \Lambda y$ where

$$\Lambda = \begin{pmatrix} \rho & -\omega & 0 \\ \omega & \rho & 0 \\ 0 & 0 & \lambda \end{pmatrix}$$

where $\rho < 0$, $\lambda > 0$ and $\omega \neq 0$. In component form this becomes

$$\begin{pmatrix} \dot{y}_1 \\ \dot{y}_2 \\ \dot{y}_3 \end{pmatrix} = \begin{pmatrix} \rho & -\omega & 0 \\ \omega & \rho & 0 \\ 0 & 0 & \lambda \end{pmatrix} \begin{pmatrix} y_1 \\ y_2 \\ y_3 \end{pmatrix}$$

and we see immediately that the unstable manifold of the origin is the $y_3$-axis and the stable manifold of the origin is the two-dimensional plane defined by $y_3 = 0$. On this plane the motion is given by logarithmic spirals into the origin and on the $y_3$-axis the motion is away from the origin. So, putting these two motions together we obtain the phase space diagram depicted in Figure 3.1a. The general motion of the original system $\dot{x} = Ax$ can be obtained from this by applying a linear transformation to the whole picture, giving a phase portrait like the one sketched in Figure 3.1b. We shall return to a more detailed description of the phase portraits for linear systems in two dimensions in Chapter 5.

## 3.5 Floquet Theory

In this section we want to develop a treatment of non-autonomous linear differential equations with periodic coefficients, i.e.

$$\dot{x} = A(t)x, \quad A(t) = A(t+T). \tag{3.32}$$

The treatment in this section follows the book by Iooss and Joseph (1980), and the interested reader is referred to this book for a more complete and technical discussion of the subject. Let us begin with the easiest case, when $x \in \mathbf{R}$. Then

$$\dot{x} = a(t)x, \quad a(t) = a(t+T) \tag{3.33}$$

and we can integrate immediately to find

$$x(t) = x_0 \exp\left\{\int_0^t a(s)ds\right\}. \tag{3.34}$$

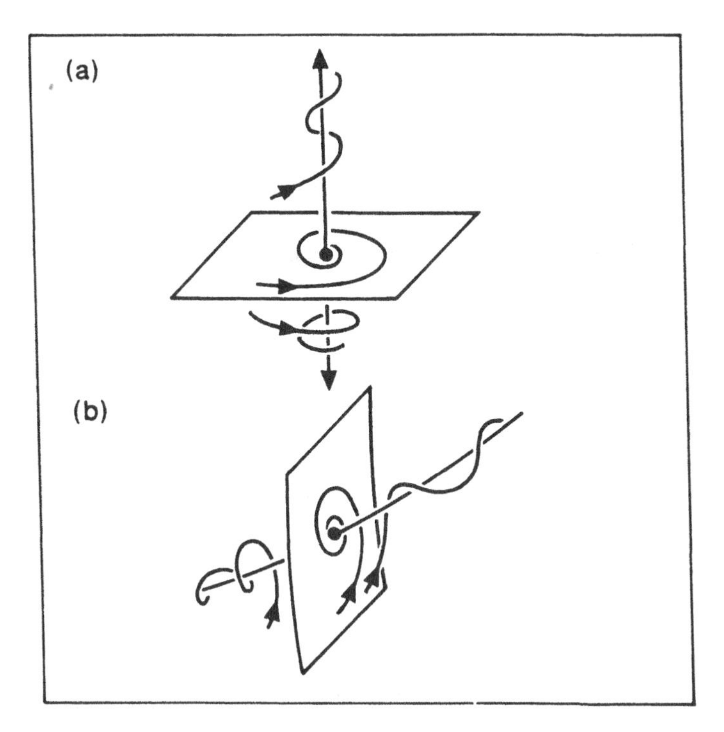


Fig. 3.1 Phase portraits for Example 3.5: (a) in normal form coordinates; (b) in general coordinates.

Hence $\varphi(t) = \exp\left\{\int_0^t a(s)ds\right\}$ is a fundamental matrix for the problem with $x(t) = \varphi(t)x_0$. Now, since

$$\int_0^{t+T} a(s)ds = \int_0^T a(s)ds + \int_T^{t+T} a(s)ds$$

and, using the periodicity of $a(s)$, $a(s) = a(s+T)$,

$$\int_T^{t+T} a(s)ds = \int_0^t a(s)ds,$$

we have

$$\begin{aligned} \varphi(t+T) &= \left(\exp\left\{\int_0^T a(s)ds\right\}\right)\left(\exp\left\{\int_T^{t+T} a(s)ds\right\}\right) \\ &= \varphi(T)\varphi(t) \end{aligned} \tag{3.35}$$

and in particular

$$\varphi(nT) = \varphi(T)^n. \tag{3.36}$$

The number $\varphi(T) = e^{\sigma T}$ is called the Floquet multiplier, whilst $\sigma$ is called a Floquet exponent. Note that $\sigma$ is only determined up to a constant:

$$\sigma = \frac{1}{T}\log\varphi(T) + \frac{2k\pi i}{T} \tag{3.37}$$

for any integer $k$. To determine the stability of the origin (which is a stationary point) define $v(t) = \varphi(t)e^{-\sigma t}$. Then

$$v(t+T) = \varphi(t+T)e^{-\sigma(t+T)} = \varphi(t)e^{-\sigma t}$$

since $\varphi(t+T)e^{-\sigma T} = \varphi(t)$. Hence $v(t) = v(t+T)$, i.e. $v(t)$ is periodic and bounded. Now,

$$x(t) = \varphi(t)x_0 = v(t)e^{\sigma t}x_0 \tag{3.38}$$

so if $\text{Re}\sigma < 0$ solutions tend to zero, whilst if $\text{Re}\sigma > 0$ solutions are unbounded as $t \to \infty$.

*Example 3.6*

Consider the differential equation $\dot{x} = (\delta + \cos t)x$. Then $a(t) = \delta + \cos t$, which is periodic with period $2\pi$. Furthermore

$$\int_0^{2\pi} a(s)ds = \int_0^{2\pi} (\delta + \cos s)ds = 2\pi\delta$$

and so the Floquet exponent is $\sigma = \delta$ and the origin is stable if $\delta < 0$ and unstable if $\delta > 0$.

To generalize the idea of Floquet exponents to $\mathbf{R}^n$ we go through a similar procedure, but the Floquet multipliers become the eigenvalues of the matrix $\Phi(T)$. So, in general, consider the equation

$$\dot{x} = A(t)x, \quad x \in \mathbf{R}^n, \quad A(t) = A(t+T). \tag{3.39}$$

Let $\Phi(t)$ be the fundamental matrix which satisfies $\Phi(0) = I$. So, since $\Phi(t)$ is a fundamental matrix, $\dot{\Phi}(t) = A(t)\Phi(t)$. Hence

$$\dot{\Phi}(t+T) = A(t+T)\Phi(t+T) = A(t)\Phi(t+T) \tag{3.40}$$

and so $\Phi(t+T)$ is a fundamental matrix and (from the discussion of fundamental matrices at the beginning of the chapter)

$$\Phi(t+T) = \Phi(t)C \tag{3.41}$$

where $C$ is a constant matrix. Setting $t = 0$ we see that $C = \Phi(T)$, i.e.

$$\Phi(t+T) = \Phi(t)\Phi(T), \tag{3.42}$$

and, in particular (by induction on $m$)

$$\Phi(mT) = \Phi(T)^m. \tag{3.43}$$

Now let $\lambda_i$ be the eigenvalues of $\Phi(T)$ and $e_i$ the corresponding eigenvectors (we will assume that the eigenvalues are distinct). The $(\lambda_i)$ are called the Floquet multipliers and if $\lambda_i = e^{\sigma_i T}$ then $\sigma_i$ is a Floquet exponent. As in the one-dimensional case, $\sigma_i$ is only defined up to multiples of $\frac{2\pi i}{T}$. Now, let $x_0 = \sum_k a_i e_i$ and remember that $x(t) = \Phi(t)x_0$ for the particular choice of fundamental matrix made above. Hence

$$\begin{aligned} x(t+T) = \Phi(t+T)x_0 &= \Phi(t)\Phi(T)\left\{\sum_k a_i e_i\right\} \\ &= \Phi(t)\left\{\sum_k e^{\sigma_k T} a_i e_i\right\}. \end{aligned} \tag{3.44}$$

Let $x_k(t) = \Phi(t)a_k e_k$, $1 \leq k \leq n$, so $x_k(t+T) = \Phi(t)a_k e^{\sigma_k T} e_k$, and, by analogy with the one-dimensional case, define $v_k(t) = e^{-\sigma_k t} x_k(t)$. Then

$$v_k(t+T) = e^{-\sigma_k(t+T)} x_k(t+T) = e^{-\sigma_k t} x_k(t) = v_k(t)$$

so each of the components $v_k(t)$ is periodic with period $T$. Furthermore, $x(t) = \sum_k x_k(t) = \sum_k e^{\sigma_k t} v_k(t)$, and so we conclude, as in the one-dimensional case, that if all the Floquet exponents have strictly negative real parts then the origin is asymptotically stable, whilst if any of the Floquet exponents has positive real parts then some solutions diverge to infinity.

## Exercises 3

1. Suppose that $\dot{x} = \Lambda x$ where $\Lambda = \text{diag}(\lambda, \lambda, \ldots, \lambda)$. Show that for any linear change of coordinates, $y = Px$, $\det P \neq 0$, the equation remains unchanged.

2. By solving the equation for $\frac{dy}{dx}$ or $\frac{dr}{d\theta}$ on trajectories, as appropriate, sketch the phase portrait of

i) $\dot{x} = x, \quad \dot{y} = 2y;$
ii) $\dot{x} = -3x, \quad \dot{y} = -2y;$
iii) $\dot{x} = -x + y, \quad \dot{y} = -x - y;$
iv) $\dot{x} = 3y, \quad \dot{y} = -3x;$
v) $\dot{x} = 2x, \quad \dot{y} = 2y.$

3. Given the differential equation $\dot{x} = Ax$ for $x \in \mathbf{R}^2$, find the eigenvalue equation of $A$ in terms of $\mathrm{Tr}A$ and $\det A$. Hence find the regions in the $(\mathrm{Tr}A, \det A)$ plane in which $A$ has two negative real eigenvalues, two positive real eigenvalues, a pair of complex conjugate eigenvalues with negative real part, and so on. Find the curve on which $A$ has multiple eigenvalues. If

$$A = \begin{pmatrix} 4 & 1 \\ -1 & d \end{pmatrix}$$

find the two values of $d$ for which $A$ has repeated eigenvalues. How does the behaviour of the differential equation differ at these two values?

4. By setting $y = xz$ show that the differential equation

$$\frac{dy}{dx} = \frac{Cx + Dy}{Ax + By}$$

can be rewritten in the form

$$x\frac{dz}{dx} = -\frac{Bz^2 + (A - D)z - C}{A + Bz}.$$

Describe how you would solve this equation and how the solutions are different for different combinations of the constants $A, B, C$ and $D$. Relate these comments to solutions of the differential equation

$$\dot{x} = Ax + By, \quad \dot{y} = Cx + Dy.$$

5. Suppose $\dot{x} = Ax$ and $A$ has distinct eigenvalues. Let $E^s(0)$ be the space spanned by the eigenvectors of $A$ corresponding to those eigenvalues with negative real part. Show that

$$E^s(0) = \{x | \varphi(x, t) \to 0 \text{ as } t \to \infty\}.$$

6. By transforming the matrices into normal form, solve the differential equation $\dot{x} = Ax$ for the following choices of $A$:

$$\begin{pmatrix} 0 & -2 \\ 2 & 3 \end{pmatrix} \quad \begin{pmatrix} -8 & 27 \\ -6 & 20 \end{pmatrix} \quad \begin{pmatrix} 11 & 25 \\ -4 & -9 \end{pmatrix}.$$

7. Consider the differential equation $\dot{x} = Ax$ for $x \in \mathbf{R}^3$ and

$$A = \begin{pmatrix} 17 & 0 & 41 \\ 0 & 3 & 0 \\ -5 & 0 & -13 \end{pmatrix}.$$

Find a linear transformation which will bring $A$ into normal form and hence solve the differential equation.

8. Consider the differential equation $\dot{x} = Ax$ for $x \in \mathbf{R}^7$ and

$$A = \begin{pmatrix} 1 & 0 & 0 & 0 & 0 & 0 & 0 \\ -3 & -1 & 0 & 0 & 0 & 0 & 0 \\ 0 & 0 & 0 & -2 & 0 & 0 & 0 \\ 0 & 0 & 2 & 0 & 0 & 0 & 0 \\ 0 & 0 & 0 & 0 & -3 & -1 & 0 \\ 0 & 0 & 0 & 0 & 1 & -3 & 0 \\ 0 & 0 & 0 & 0 & 0 & 0 & 1 \end{pmatrix}$$

Find, in coordinate form, the following sets:

i) $E^s(0)$;
ii) $E^c(0)$;
iii) $E^u(0)$; and
iv) the invariant manifold associated with the eigenvalue 1.

9. Suppose $\dot{x} = a(t)x$ with $x \in \mathbf{R}$ and $a(t)$ a smooth $T$-periodic function of time. Find the equation satisfied by $y = p(t)x$. For the choice $p(t) = c(t)\varphi(t)$, where $\varphi(t)$ is a fundamental matrix for the problem, show that $c(t)$ can be chosen such that $\dot{y} = ky$ for any constant $k$. If $k = \sigma$, the Floquet exponent, show that $p(t)$ is periodic with period $T$.

10. Consider the differential equation $\dot{x} = A(t)x$ for $x \in \mathbf{R}^2$ and where $A(t)$ is smooth and $T$-periodic. Let $\Phi(t)$ be a fundamental matrix with $\Phi_{ij}(0) = \delta_{ij}$. By writing the equations out in component form or otherwise, show that

$$\frac{d}{dt}\det\Phi = \mathrm{Tr}A.\det\Phi.$$

Hence show that the modulus of the product of the Floquet multipliers of $A$ equals

$$\exp\left(\int_0^T \mathrm{Tr}A(t)dt\right).$$

11. Suppose $u(t)$ is a periodic solution of the autonomous differential equation $\dot{x} = f(x)$ with least period $T > 0$. By considering small perturbations of this solution, $x = u(t) + \epsilon v(t)$, $\epsilon \ll 1$, show that the evolution of perturbations to lowest order is

$$\dot{v} = A(t)v, \quad A(t) = A(t+T)$$

where $A(t) = Df(u(t))$. Show that $\dot{u}(t)$ is a solution of this equation and hence deduce that one Floquet multiplier is always equal to unity in such situations. Use this result together with question 10 to show that the Floquet multipliers of a periodic orbit in the plane are non-negative.

# 4

## *Linearization and hyperbolicity*

We have just spent a whole chapter describing linear differential equations. Why? First, they are one of the few classes of systems which can be treated in complete generality. More importantly though, we can hope that the results obtained for linear systems can be used (at least locally) in the study of nonlinear systems. In this chapter we want to investigate the extent to which this hope is justified. Suppose that the nonlinear differential equation

$$\dot{x} = f(x), \quad x \in \mathbf{R}^n \tag{4.1}$$

has a stationary point, $x_0$. After a simple shift in the coordinate system we can arrange for this stationary point to be at the origin, so without loss of generality we assume that $f(0) = 0$. Assuming $f$ is smooth, we can expand $f$ about the origin as a Taylor series to obtain

$$\dot{x} = Df(0)x + O(|x|^2) \tag{4.2}$$

where $Df(0)$ is the Jacobian matrix of $f$ evaluated at the origin: $Df(0)_{ij} = \frac{\partial f_i}{\partial x_j}(0)$. Thus if we ignore the terms of order $|x|^2$ and higher we have the linear differential equation

$$\dot{x} = Df(0)x. \tag{4.3}$$

It would be very nice if there were a local change of coordinates, $x = h(y)$, which brings the nonlinear equation, (4.2), into the linear equation, (4.3), in some neighbourhood of the origin. This turns out to be a little too much to hope for, but in Section 4.1 we show that this can be done in some cases. Another possibility would be that the local structure in terms of the stable, unstable and centre manifolds is preserved (with some perturbation), and this gives probably the most important result of this chapter: the stable manifold theorem. This states that provided $Df(0)$ has no zero or purely imaginary eigenvalues (so the linear system does not have a centre manifold) then there are stable and unstable manifolds for the nonlinear system which are tangent to the stable and unstable manifolds of the linear system at the stationary point. To this extent, then, the local structure of the linear system is preserved. In the remainder of this chapter we look at the properties of flows which are preserved under perturbation. This is important because it is rare in physical and chemical contexts to be certain that the model being studied is precisely correct, and so it is important to know which properties of solutions remain true for 'nearby' models.

## 4.1 Poincaré's Linearization Theorem

Suppose that $\dot{x} = f(x)$, $f(0) = 0$, and $f$ is analytic on $\mathbf{R}^n$. What conditions need to be satisfied for there to exist a change of coordinates in some neighbourhood of the origin such that the differential equation in these new coordinates is the linear system $\dot{y} = Df(0)y$? The answer to this question depends upon the eigenvalues of the matrix $Df(0)$; there are problems if the eigenvalues are resonant. Arnold (1983) provides a particularly readable account of this problem, and we follow his method here. For more details and a general exposition of other situations in which these resonances arise the reader should consult this book.

(4.1) DEFINITION

*Suppose that the eigenvalues of $Df(0)$ are $(\lambda_1, \lambda_2, \ldots, \lambda_n)$. Then $Df(0)$ is resonant if there exist non-negative integers $(m_1, m_2, \ldots, m_n)$ with $\sum_k m_k \geq 2$ such that*

$$(m, \lambda) = \sum_{k=1}^{n} m_k \lambda_k = \lambda_s$$

*for some $s \in \{1, 2, \ldots, n\}$. The quantity $|m| = \sum_1^n m_k$ is called the order of the resonance.*

The problem associated with resonance is one of the convergence of the power series expansion of the new coordinates in terms of the old coordinates. We shall try to find the new coordinate system. Rather than solving for the new coordinate system in one go, we shall construct a series of coordinate changes which kill terms of order $j = 2, 3, 4, \ldots$ in turn. At each stage the coefficients of the coordinate change will have denominators of the form $\lambda_s - (m, \lambda)$ with $|m| = j$ and so if this expression vanishes we will be unable to make the coordinate change. This is an example of what is known as a small divisor problem. The technique of using successive coordinate changes of order $2, 3, 4, \dots$ to simplify the differential equation is one which will recur throughout this book, most notably in Chapters 8 and 9 (and also at the end of this chapter), so it is worth going through the argument below quite carefully.

It is always easiest to work initially in coordinates for which $Df(0)$ is in Jordan normal form. So assume that this linear coordinate change has already been made; so if $Df(0)$ has distinct eigenvalues (which may be complex) then the coordinate system $(x_1, x_2, \dots, x_n)$ is such that

$$\dot{x}_i = \lambda_i x_i + \text{higher order terms.}$$

Now group together the higher order terms by order, so

$$\dot{x} = Df(0)x + v_r(x) + v_{r+1}(x) + \dots. \tag{4.4}$$

where $v_j$ contains terms only of order $j$ and $r \geq 2$. Thus, if we let $M_r = \{m \in \mathbf{Z}^n | m_i \geq 0, \sum m_k = r\}$, we can write

$$v_r(x) = \sum_{m \in M_r} a_m x^m \tag{4.5}$$

where $m = (m_1, \dots, m_n)$, $x^m = x_1^{m_1} x_2^{m_2} \dots x_n^{m_n}$ (called a monomial of degree $|m|$) and $a_m$ is an $n$ dimensional vector. For example, if $n = 2$ and $r = 2$ then $M_r = \{(0, 2), (1, 1), (2, 0)\}$ and, in coordinates $x = (x_1, x_2)$,

$$v_2(x) = a_{(2,0)} x_1^2 + a_{(1,1)} x_1 x_2 + a_{(0,2)} x_2^2.$$

We shall try to construct a near identity change of coordinates $y = x + \dots$ such that

$$\dot{y} = Df(0)y + V_{r+1}(y) + \dots \tag{4.6}$$

If we can do this, and then repeat the argument for the terms of order $r+1$ and so on we will have a formal power series for $y$ such that, having killed terms of order $r$, $r+1$, $\dots$ successively, $\dot{y} = Df(0)y$. So, we have

$$\dot{x}_i = \lambda_i x_i + \sum_{m \in M_r} a_{mi} x^m + v_{r+1,i}(x) + \dots \tag{4.7}$$

and we will try a coordinate change

$$y_i = x_i + \sum_{m \in M_r} b_{mi} x^m \tag{4.8}$$

with inverse

$$x_i = y_i - \sum_{m \in M_r} b_{mi} y^m + O(|y|^{r+1}) \tag{4.9}$$

and choose the coefficients $b_{mi}$ such that

$$\dot{y}_i = \lambda_i y_i + V_{r+1,i}(x) + \dots . \tag{4.10}$$

To do this we need to differentiate $y_i$ with respect to time, and so we must differentiate $x^m$ with respect to time. Now,

$$\begin{aligned}\frac{d}{dt}x^m &= \sum_{k=1}^{n} \frac{\dot{x}_k}{x_k} m_k x^m \\ &= \sum_{k=1}^{n} m_k \lambda_k x^m + O(|x|^{r+1})\end{aligned} \tag{4.11}$$

so, differentiating the expression for $y$ in terms of $x$ and writing $(m, \lambda) = \sum_1^n m_k \lambda_k$,

$$\dot{y}_i = \dot{x}_i + \sum_{m \in M_r} b_{mi}(m, \lambda) x^m + O(|x|^{r+1}). \tag{4.12}$$

Substituting for $\dot{x}_i$ from (4.7) gives

$$\dot{y}_i = \lambda_i x_i + \sum_{m \in M_r} a_{mi} x^m + \sum_{m \in M_r} b_{mi}(m, \lambda) x^m + O(|x|^{r+1})$$

and finally, substituting $x$ in terms of $y$ using (4.9)

$$\begin{aligned}\dot{y}_i = \lambda_i (y_i - &\sum_{m \in M_r} b_{mi} y^m) + \sum_{m \in M_r} a_{mi} y^m \\ &+ \sum_{m \in M_r} b_{mi}(m, \lambda) y^m + O(|y|^{r+1}).\end{aligned} \tag{4.13}$$

Putting the first and third sums into a single sum we find that

$$\begin{aligned}\dot{y}_i = \lambda_i y_i + &\sum_{m \in M_r} b_{mi} y^m (-\lambda_i + (m, \lambda)) y^m \\ &+ \sum_{m \in M_r} a_{mi} y^m + O(|y|^{r+1})\end{aligned} \tag{4.14}$$

and hence if we choose the coefficients $b_{mi}$ such that

$$b_{mi} = \frac{a_{mi}}{\lambda_i - (m, \lambda)} \tag{4.15}$$

all the terms of order $r$ disappear and we are left with

$$\dot{y}_i = \lambda_i y_i + O(|y|^{r+1}) \tag{4.16}$$

or

$$\dot{y}_i = \lambda_i y_i + V_{r+1,i}(y) + \dots . \tag{4.17}$$

But of course, we can only choose this value of $b_{mi}$ provided

$$\lambda_i - (m, \lambda) \neq 0 \tag{4.18}$$

i.e. if $\lambda$ is not resonant of order $r$. So provided that $\lambda$ is not resonant of order $r$, we can use a near identity change of coordinates with terms of order $r$ such that all the terms of order $r$ are killed. We can now repeat this argument with the terms of order $r+1$ and so on. This proves the following theorem.

(4.2) THEOREM

*Suppose that $\dot{x} = f(x)$, $f(0) = 0$ and $Df(0)$ is not resonant. Then if $Df(0)$ is diagonal there exists a formal near identity change of coordinates $y = x + \ldots$ for which $\dot{y} = Df(0)y$.*

(4.3) EXERCISE

*Consider the case of systems in $\mathbf{R}^2$ and suppose that $Df(0)$ has eigenvalues $\lambda_1$ and $\lambda_2$. Then if $\lambda_1 = 6$ and $\lambda_2 = -2$, the system is resonant of order 5 since $\lambda_1 = 2\lambda_1 + 3\lambda_2$ (or $\lambda_2 = \lambda_1 + 4\lambda_2$). Are the following choices of $\lambda_1$ and $\lambda_2$ resonant, and if so, what is the order of the resonance?*

i) $\lambda_1 = 1$, $\lambda_2 = -1$;
ii) $\lambda_1 = 10$, $\lambda_2 = 5$;
iii) $\lambda_1 = 10$, $\lambda_2 = -5$;
iv) $\lambda_1 = 7$, $\lambda_2 = 9$;
v) $\lambda_1 = 3$, $\lambda_2 = -18$.

In fact, the restriction that $Df(0)$ should be diagonal is not necessary; the argument works perfectly well in the case of degenerate eigenvalues, but the calculations are a little more complicated and we leave this as an exercise for the interested reader.

The change of coordinate is given, implicitly, by a formal power series. This only gives a true change of coordinate if the power series converges in some neighbourhood of the origin. Proving the convergence of this power series is a subtle piece of analysis which we will not go into, but some results are easily stated.

Poincaré was able to prove that the power series of Theorem 4.2 converges if the eigenvalues $(\lambda_i)$ are non-resonant and either $\mathrm{Re}\,\lambda > 0$ for $i = 1, ..., n$ or $\mathrm{Re}\,\lambda_i < 0$ for $i = 1, ..., n$. If some eigenvalues have negative real parts and others positive real parts then the statement is more complicated. This convergence result is due to Siegel.

(4.4) DEFINITION

*The $n$-tuple $\lambda = (\lambda_1, ..., \lambda_n)$ satisfies the Siegel condition if there exists $C > 0$ and $\nu$ such that for all $i = 1, ..., n$*

$$|\lambda_i - (m, \lambda)| \geq \frac{C}{|m|^\nu}$$

*for all $m = (m_1, ..., m_n)$, where $(m_i)$ are non-negative integers with $|m| = \Sigma_1^n m_i \geq 2$.*

Thus the eigenvalue satisifies Siegel's condition (sometimes called a diophantine condition) if $|\lambda_i - (m, \lambda)|$ is sufficiently far from zero. The results of Poincaré and Siegel can be summarized in the following theorem.

(4.5) THEROEM (POINCARÉ'S LINEARIZATION THEOREM)

*If the eigenvalues $(\lambda_i), i = 1, ..., n$, of the linear part of an analytic vector field at a stationary point are non-resonant and either* $\mathrm{Re}\,\lambda_i > 0$, $i = 1, ..., n$, *or* $\mathrm{Re}\,\lambda_i < 0$, $i = 1, ..., n$, *or $(\lambda_i)$ satisfies a Siegel condition, then the power series of Theorem 4.2 converges on some neighbourhood of the stationary point.*

If the conditions of Theorem 4.5 hold, then there is an analytic change of coordinates in a neighbourhood of the stationary point which brings the differential equation into linear form. Note that provided the eigenvalues are non-resonant, it is always possible to make local polynomial changes of coordinate which will kill nonlinear terms to arbitrarily high order.

The next example shows the extent to which a resonant vector field is different from its linearization. We shall see that the difference is really not that great, indeed, there is no difference topologically. This will motivate the next section, in which we concentrate on those topological features of the local flow which are preserved when going from a vector field to its linearization.

*Example 4.1*

Consider

$$\dot{x} = x, \quad \dot{y} = 2y + x^2$$

with linearization at the origin given by

$$\dot{x} = x, \quad \dot{y} = 2y.$$

The eigenvalues of the linearization are $(\lambda_1, \lambda_2) = (1, 2)$ and this is resonant of order two since $2\lambda_1 = \lambda_2$. Solution curves of the linearized equation lie on solutions of

$$\frac{dy}{dx} = \frac{2y}{x},$$

which we can easily solve to obtain a family of parabolae,

$$y = kx^2. \tag{4.19}$$

Similarly, the full problem has solutions which lie on curves defined by

$$\frac{dy}{dx} = \frac{2y}{x} + x,$$

which gives, multiplying through by the integrating factor $x^{-2}$,

$$x^{-2}\frac{dy}{dx} = \frac{d}{dx}(x^{-2}y) + \frac{2y}{x^3} = \frac{2y}{x^3} + x^{-1}.$$

Hence the solution curves are of the form

$$y = x^2(\log|x| + k). \tag{4.20}$$

## 4.2 Hyperbolic stationary points and the stable manifold theorem

If we are only interested in topological properties of the flows the non-resonance conditions of Theorem 4.5 can be replaced by much weaker conditions.

(4.6) DEFINITION

*A stationary point $x$ is hyperbolic iff $Df(x)$ has no zero or purely imaginary eigenvalues.*

There are two important theorems for hyperbolic stationary points, the stable manifold theorem and Hartman's theorem. The first shows that the local structure of hyperbolic stationary points of nonlinear flows, in terms of the existence and transversality of local stable and unstable manifolds, is the same as the linearized flow, and the second asserts that there is a continuous invertible map in some neighbourhood of the stationary point which takes the nonlinear flow to the linear flow preserving the sense of time.

Let $U$ be some neighbourhood of a stationary point, $x$. Then, by analogy with the definition of the invariant manifolds for linear systems we can define the local stable manifold of $x$, $W^s_{loc}(x)$, and the local unstable manifold of $x$, $W^u_{loc}(x)$, by

$$W^s_{loc}(x) = \{y \in U | \varphi(y,t) \to x \ as \ t \to \infty, \ \varphi(y,t) \in U \ for \ all \ t \geq 0\}$$

and

$$W^u_{loc}(x) = \{y \in U | \varphi(y,t) \to x \ as \ t \to -\infty, \ \varphi(y,t) \in U \ for \ all \ t \leq 0\}.$$

The stable manifold theorem states that these manifolds exist and are of the same dimension as the stable and unstable manifolds of the linearized equation $\dot{y} = Df(x)y$ if $x$ is hyperbolic, and that they are tangential to the linearized manifolds at $x$.

(4.7) THEOREM (STABLE MANIFOLD THEOREM)

*Suppose that the origin is a hyperbolic stationary point for* $\dot{x} = f(x)$ *and* $E^s$ *and* $E^u$ *are the stable and unstable manifolds of the linear system* $\dot{x} = Df(0)x$. *Then there exist local stable and unstable manifolds* $W^s_{loc}(0)$ *and* $W^u_{loc}(0)$ *of the same dimension as* $E^s$ *and* $E^u$ *respectively. These manifolds are (respectively) tangential to* $E^s$ *and* $E^u$ *at the origin and as smooth as the original function* $f$.

Note that in Chapter 3 the centre manifold of a stationary point was also defined, but that for a hyperbolic stationary point the centre manifold is empty. We shall return to the problems of finding nonlinear centre manifolds in Chapters 7 and 8, where the basic ideas of bifurcation theory are introduced. The content of this theorem is illustrated in Fig. 4.1. The proof is, unfortunately, long and technical and we leave this to the end of this chapter, since we will be much more concerned with the use of this theorem.

One further point can be made without difficulty: suppose that $x_0$ is a hyperbolic stationary point, then there are three possibilities. Either $W^s_{loc}(x_0) = \emptyset$, or $W^u_{loc}(x_0) = \emptyset$, or both manifolds are non-empty. These three possibilities are given names: $x_0$ is called a source, sink or saddle respectively. From the definition of the linear stable and unstable manifolds and the stable manifold theorem it should be obvious that these definitions can be made in terms of the eigenvalues of the Jacobain matrix at $x_0$ in the following way.

(4.8) DEFINITION

*Suppose that $x_0$ is a hyperbolic stationary point of $\dot{x} = f(x)$ and let $Df(x_0)$ denote the Jacobian matrix of $f$ evaluated at $x_0$. Then $x_0$ is a sink if all the eigenvalues of $Df(x_0)$ have strictly negative real parts and a source if all the eigenvalues of $Df(x_0)$ have strictly positive real parts. Otherwise $x_0$ is a saddle.*

We shall see in Sections 4.3 and 4.5 that for small perturbations of the defining equations, a source remains a source, a sink remains a sink and a saddle remains a saddle. Furthermore, as one would expect, if $x_0$ is a sink then it is asymptotically stable.

If we choose a coordinate system for which the linear part of the differential equation at the origin is in normal form we can always arrange

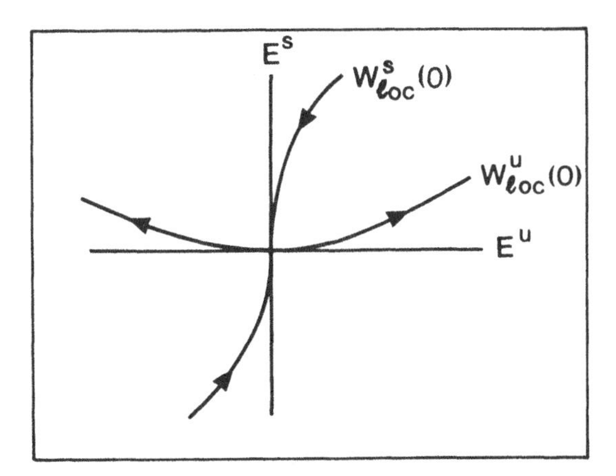


Fig. 4.1 Stable and unstable manifolds.

for the differential equation to be of the form

$$\dot{x} = Ax + g_1(x, y) \quad \dot{y} = -By + g_2(x, y) \tag{4.21}$$

where $x \in \mathbf{R}^{n_u}$, $y \in \mathbf{R}^{n_s}$ (where $n_u$ is the dimension of the local unstable manifold and $n_s$ is the dimension of the local stable manifold, $n_u + n_s = n$) and both the square matrices $A$ and $B$ have eigenvalues with positive real parts. The functions $g_i(x, y)$, $i = 1, 2$, contain the nonlinear parts of the equation, so they vanish, together with their first derivatives at the origin, $(x, y) = (0, 0)$. Hence

$$E^s(0,0) = \{(x, y)|x = 0\} \text{ and } E^u(0,0) = \{(x, y)|y = 0\}.$$

Since the stable and unstable manifolds are smooth and are tangential to these manifolds at the origin they can be described as the graphs of functions, so the stable manifold is given by

$$x_i = S_i(y), \quad i = 1, \ldots, n_u \tag{4.22}$$

where

$$\frac{\partial S_i}{\partial y_j}(0) = 0, \ 1 \le i \le n_u, 1 \le j \le n_s \tag{4.23}$$

since the manifold is tangential to $E^s$ at 0. Similarly we can write the unstable manifold (again locally) as

$$y_j = U_j(x), \quad \frac{\partial U_j}{\partial x_i}(0) = 0, \ 1 \le i \le n_u, 1 \le j \le n_s. \tag{4.24}$$

This observation allows us to approximate the stable and unstable manifolds by expanding the functions $S_i$ and $U_j$ as power series. Consider $U_j$ (the argument is the same for $S_i$). We begin by expanding $U_j$ as a power series in $x$, so

$$U_j(x) = \sum_{r \ge 2} \sum_{m \in M_r} u_{mj} x^m \tag{4.25}$$

where the notation is as in the previous section. If $B$ is diagonal then with eigenvalues $(\lambda_i)$, $i = 1, 2, \ldots, n_s$ then

$$\dot{y}_j = -\lambda_j y_j + g_{2j}(x, y) \tag{4.26}$$

and on the unstable manifold $y = U(x)$ so

$$\dot{y}_j = -\lambda_j U_j(x) + g_{2j}(x, U(x)). \tag{4.27}$$

On the other hand

$$\dot{y}_j = \frac{d}{dt} U_j(x) = \sum_{k=1}^{n_u} \dot{x}_k \frac{\partial}{\partial x_k} U_j(x). \tag{4.28}$$

Comparing the right hand sides of (4.27) and (4.28) we find that

$$-\lambda_j U_j(x) + g_{2j}(x, U(x)) = \sum_{k=1}^{n_u} \dot{x}_k \frac{\partial}{\partial x_k} U_j(x) \tag{4.29}$$

and we can now substitute the series expansion for the functions $U_j$ into these equations and equate coefficients of powers of $x$ in order to get a set of simultaneous equations for the coefficients $u_{mi}$ which can be solved to arbitrary order. An example may make this clearer.

*Example 4.2*

Consider the equations

$$\dot{x} = x, \quad \dot{y} = -y + x^2.$$

This has a unique stationary point at $(x, y) = (0, 0)$ and the equation is already in normal form near the stationary point. The linearized equation is

$$\dot{x} = x, \quad \dot{y} = -y,$$

giving a saddle at the origin with invariant linear subspaces

$$E^s(0,0) = \{(x,y)|x = 0\} \text{ and } E^u(0,0) = \{(x,y)|y = 0\}.$$

By the stable manifold theorem we know that the nonlinear system has a local unstable manifold of the form

$$y = U(x), \quad \frac{\partial U}{\partial x}(0) = 0$$

and so we try a series expansion for $U$,

$$U(x) = \sum_{k \geq 2} u_i x^i.$$

Now,

$$\dot{y} = -y + x^2 = -\sum_{k \geq 2} u_i x^i + x^2$$

on the unstable manifold and also

$$\dot{y} = \dot{x} \frac{\partial U}{\partial x}(x) = \sum_{k \geq 2} k u_k x^k.$$

Equating terms of order $x^2$, $x^3$ and so on gives

$$-u_2 + 1 = 2u_2, \text{ and } -u_k = k u_k, \; k \geq 3.$$

Hence $u_2 = \frac{1}{3}$, $u_k = 0$ for $k \geq 3$ and so

$$W^u_{loc}(0,0) = \{(x,y) | y = \tfrac{1}{3}x^3\}.$$

A similar exercise shows that $W^s_{loc}(0,0) = E^s(0,0)$.

Later in this book (Chapter 12) we will see that a great deal of interesting dynamics is controlled by the behaviour of the stable and unstable manifolds of stationary points; for this we need to extend the local manifolds to obtain global stable and unstable manifolds defined by

$$W^u(0) = \bigcup_{t \leq 0} \varphi(W^u_{loc}(0), t) \text{ and } W^s(0) = \bigcup_{t \geq 0} \varphi(W^s_{loc}(0), t).$$

The second result of this section is associated with a weakening of the requirements of Poincaré's Linearization Theorem. In the previous section we looked for a change of variable such that the equation in the new variable is locally just the linear flow. This turned out to be quite a tough condition to meet, but in Example 4.1 we saw that even when the linearization has resonant eigenvalues the flow was remarkably similar to the linear flow (at least for the hyperbolic stationary point considered). This suggests that an alternative strategy might be to look for a map from the nonlinear flow to the linear flow in a neighbourhood of the stationary point, which takes trajectories of the nonlinear flow to trajectories of the linear flow.

(4.9) THEOREM (HARTMAN'S THEOREM)

*If $x = 0$ is a hyperbolic stationary point of $\dot{x} = f(x)$ then there is a continuous invertible map, $h$, defined on some neighbourhood of $x = 0$ which takes orbits of the nonlinear flow to those of the linear flow $exp(tDf(0))$. This map can be chosen so that the parametrization of orbits by time is preserved.*

Note that the map is only continuous (not necessarily differentiable) and so it does not distinguish between, for example, a logarithmic spiral (cf. (3.31)) and the phase portrait obtained when the Jacobian at the stationary point has real eigenvalues. If we want greater smoothness we find ourselves involved once again in problems of resonance.

## 4.3 Persistence of hyperbolic stationary points

Another important feature of hyperbolic stationary points is the fact that they persist under small perturbations of the defining differential equations. Hence if the origin is a hyperbolic stationary point of $\dot{x} = f(x)$ and $v$ is any smooth vector field on $\mathbf{R}^n$ then for sufficiently small $\epsilon$ the equation

$$\dot{x} = f(x) + \epsilon v(x) \tag{4.30}$$

has a hyperbolic stationary point near the origin of the same type as the hyperbolic point of the unperturbed equation. This robustness, together with the results of the previous section, shows that the dynamics in a neighbourhood of a hyperbolic stationary point is not radically altered by small perturbations. This will be of crucial importance when we come to consider bifurcation theory in Chapters 7 and 8. To see this, suppose that $f(0) = 0$ and look for stationary points of the perturbed system. They satisfy

$$f(x) + \epsilon v(x) = 0. \tag{4.31}$$

Expanding this equation about $x = 0$ (or using the implicit function theorem) gives

$$[Df(0) + \epsilon Dv(0)]x + \epsilon v(0) + O(|x|^2) = 0 \tag{4.32}$$

with solutions

$$x = -\epsilon[Df(0) + \epsilon Dv(0)]^{-1}v(0) + O(\epsilon^2) \tag{4.33}$$

provided $[Df(0) + \epsilon Dv(0)]$ is invertible. Now, if $x = 0$ is a hyperbolic stationary point, the eigenvalues of $Df(0)$ are bounded away from zero and hence the eigenvalues of $[Df(0) + \epsilon Dv(0)]$ are bounded away from zero for sufficiently small $\epsilon$. So $\det[Df(0) + \epsilon Dv(0)] \neq 0$ for sufficiently small $\epsilon$ and hence this matrix is invertible. We now want to show that the stationary point of the perturbed equation is also hyperbolic. By continuity in $\epsilon$, there is a neighbourhood of $\epsilon = 0$ for which the real parts of the eigenvalues of $[Df(x) + \epsilon Dv(x)]$ are all non-zero for sufficiently small $x$. In particular, no eigenvalue can cross the imaginary axis and so the number of eigenvalues on the right of the imaginary axis and on the left of the imaginary axis is the same for all $x$ sufficiently small in this neighbourhood of $\epsilon = 0$. Now simply choose $\epsilon$ small enough so that it is in this neighbourhood of $\epsilon = 0$ and the stationary point of the perturbed equation has sufficiently small $|x|$. Then for all values of $\epsilon$ sufficiently small the stationary point of the perturbed equation is hyperbolic.

For the application of these results to bifurcation theory later it is worth pointing out the different roles of zero and purely imaginary eigenvalues in the non-hyperbolicity conditions. To ensure the existence of a stationary point for small enough perturbations of the original equations it is enough that no eigenvalue of $Df(0)$ is zero. However, to ensure that no eigenvalue passes through the imaginary axis (so stability properties of the perturbed stationary point are the same as those for the original stationary point) we need the unperturbed stationary point to be hyperbolic, i.e. no eigenvalues of the linearized flow are zero or purely imaginary.

## 4.4 Structural stability

In the previous section we established that sufficiently small perturbations of a flow with a hyperbolic stationary point have a hyperbolic stationary point of the same type. One could ask a deeper question: when are all sufficiently small perturbations of a differential equation equivalent to the original or unperturbed equation? Of course, this involves global rather than local perturbations and also depends upon how the notion of equivalence is defined and how wide a class of perturbations is allowed. We shall not go into these questions in any great depth, but sketch some of the ideas and major results which arise. The idea of equivalence for flows is basically the same as for Hartman's theorem, except that the parametrization of solutions by time is not necessarily preserved although the sense of time is preserved.

(4.10) DEFINITION

*Two smooth vector fields $f$ and $g$ are flow equivalent iff there exists a homeomorphism, $h$, (so both $h$ and its inverse exist and are continuous) which takes trajectories under $f$, $\varphi_f(x,t)$ to trajectories of $g$, $\varphi_g(x,t)$, and which preserves the sense of parametrization by time.*

A simple consequence of this definition is that for any $x$ and $t_1$ there exists $t_2$ such that

$$h(\varphi_f(x,t_1)) = \varphi_g(h(x),t_2).$$

The fact that time parametrization is not necessarily preserved reflects the fact that we are more interested in the geometry of solutions than the amount of time along trajectories, so we want periodic orbits of different periods to be (at least potentially) flow equivalent. We now restrict the class of perturbations allowed with the aim of retaining as much generality as possible whilst being able to prove interesting results.

(4.11) DEFINITION

*A vector field $f : \mathbf{R}^n \to \mathbf{R}^n$ is structurally stable if for all twice differentiable vector fields $v : \mathbf{R}^n \to \mathbf{R}^n$ there exists $\epsilon_0 > 0$ such that $f$ is flow equivalent to $f + \epsilon v$ for all $\epsilon \in (0, \epsilon_0)$.*

The idea of structural stability was introduced and refined in the 1960s and 1970s. The hope was that the space of differential equations would split into regions where the flow is structurally stable, bounded by codimension one surfaces which would form the boundary between different structurally stable classes of vector fields. In this way 'typical' behaviour of systems could be classified (the structurally stable ones). The major triumph of this programme was the complete description of structurally stable flows on compact manifolds of dimension two. To state this result we need a definition of hyperbolicity for periodic orbits. This will be described in more detail in Chapter 6, but for the moment, from question 11 of the exercises to Chapter 3, we shall use the following definition.

(4.12) DEFINITION

*Suppose $u(t)$ is a periodic orbit of the system $\dot{x} = f(x)$ with least period $T$. Then $u(t)$ is hyperbolic if all the Floquet multipliers of the $T$-periodic equation*

$$\dot{v} = Df(u(t))v$$

*lie off the unit circle except one, which must equal unity.*

In other words, the multipliers play a similar role to the eigenvalues of the Jacobian matrix of a stationary point, but the condition $\text{Re}(\lambda_i) \neq 0$ for stationary points is replaced by $|\lambda_i| \neq 1$ for the Floquet multipliers of the periodic orbit. A consequence of this definition (which will be elaborated in Chapter 6) is that for $x \in \mathbf{R}^2$ if $\Gamma$ denotes the periodic orbit in phase space then there is a tubular neighbourhood $N$ of $\Gamma$ such that for all $x \in N$, $\varphi(x, t) \to \Gamma$ as $t \to \infty$ or as $t \to -\infty$.

(4.13) THEOREM (PEIXOTO'S THEOREM)

*Let* $f : \mathbf{R}^2 \to \mathbf{R}^2$ *be a twice differentiable vector field and let* $D$ *be a compact, connected subset of* $\mathbf{R}^2$ *bounded by the simple closed curve* $\partial D$ *with outward normal* $n$. *Suppose that* $f.n \neq 0$ *on* $\partial D$. *Then* $f$ *is structurally stable on* $D$ *iff*

i) *all stationary points are hyperbolic;*
ii) *all periodic orbits are hyperbolic;*
iii) *if* $x$ *and* $y$ *are hyperbolic saddles (with* $x = y$ *possibly) then* $W^s(x) \cap W^u(y) = \emptyset$.

*Furthermore, the set of structurally stable vector fields is open and dense in the set of twice continuously differentiable vector fields satisfying the conditions of the theorem.*

The restriction to flows on compact subsets of $\mathbf{R}^2$ is needed to avoid awkward things happening at infinity. $D$ can be replaced by any compact two-manifold, but we have restricted the result to the plane for simplicity. The hope was that a similar theorem would hold in general (this was called Thom's $\omega\Sigma$ conjecture). In the late 1970s work on chaos showed that even for flows in $\mathbf{R}^3$ this theorem does not generalise nicely. We shall see an example of how much more complicated life can be in Chapter 12. Ideas of structural stability, using further refined definitions of equivalence, are still an active topic of research.

## 4.5 Nonlinear sinks

In this section we want to extend the results of Section 2.3 (about the stability of stationary points) to nonlinear systems. We will also use the results about normal forms to treat linear systems with repeated eigenvalues.

The result we want to prove is that any sink is asymptotically stable. In some sense this follows directly from Hartman's Theorem (Theorem 4.9), but since we did not prove the theorem we shall give an independent proof of the stability result.

(4.14) THEOREM

*Suppose* $x_0$ *is a sink of the differential equation* $\dot{x} = f(x)$, *where* $x \in \mathbf{R}^n$ *and* $f$ *is a smooth function. Then* $x_0$ *is asymptotically stable.*

*Proof*: We begin by making a number of obvious coordinate transformations. Without loss of generality we can assume that $x_0$ is the origin and that the linear part of the differential equation is in normal form. Hence we consider

$$\dot{x} = Ax + g(x) \tag{4.34}$$

where $A$ is a $n \times n$ matrix with eigenvalues all having strictly negative real parts and $g$ is a smooth function which vanishes together with its first derivatives at the origin (i.e. $g$ contains the nonlinear terms). Since $A$ is in normal form

$$A = \mathrm{diag}(A_1, \dots, A_k) \tag{4.35}$$

where the $A_i$ are Jordan blocks. We need to make one further change of coordinates. Suppose that $A_q$ is the $r \times r$ block

$$\begin{pmatrix} \lambda & 1 & 0 & \dots & 0 \\ 0 & \lambda & 1 & \dots & 0 \\ \vdots & \vdots & \ddots & \ddots & \vdots \\ 0 & 0 & \dots & \lambda & 1 \\ 0 & 0 & \dots & \dots & \lambda \end{pmatrix} \tag{4.36}$$

involving the variables $x_m$ to $x_{m+r-1}$. Let $y_{m+s} = \epsilon^{-s} x_{m+s}$ for $0 \le s \le r-1$. Then the linear part of the equation is transformed from

$$\dot{x}_m = \lambda x_m + x_{m+1} \quad \text{to} \quad \dot{y}_m = \lambda y_m + \epsilon y_{m+1}$$

and so on. Hence the block is transformed to

$$\begin{pmatrix} \lambda & \epsilon & 0 & \dots & 0 \\ 0 & \lambda & \epsilon & \dots & 0 \\ \vdots & \vdots & \ddots & \ddots & \vdots \\ 0 & 0 & \dots & \lambda & \epsilon \\ 0 & 0 & \dots & \dots & \lambda \end{pmatrix} \tag{4.37}$$

in the new coordinates $(y_m, \dots, y_{m+r-1})$.

Similarly, for a $2r \times 2r$ block of the form

$$\begin{pmatrix} C & I & 0 & \dots & 0 \\ 0 & C & I & \dots & 0 \\ \vdots & \vdots & \ddots & \ddots & \vdots \\ 0 & 0 & \dots & C & I \\ 0 & 0 & \dots & \dots & C \end{pmatrix} \tag{4.38}$$

where

$$C = \begin{pmatrix} \lambda & -\omega \\ \omega & \lambda \end{pmatrix} \quad \text{and} \quad I = \begin{pmatrix} 1 & 0 \\ 0 & 1 \end{pmatrix}$$

the transformation of the corresponding coordinates $(x_p, \dots, x_{p+2r-1})$ to $(y_p, \dots, y_{p+2r-1})$ where

$$\epsilon^s y_{p+2s} = x_{p+2s}, \quad \epsilon^s y_{p+2s+1} = x_{p+2s+1}$$

for $0 \le s \le r-1$ gives the new block

$$\begin{pmatrix} C & \epsilon I & 0 & \dots & 0 \\ 0 & C & \epsilon I & \dots & 0 \\ \vdots & \vdots & \ddots & \ddots & \vdots \\ 0 & 0 & \dots & C & \epsilon I \\ 0 & 0 & \dots & \dots & C \end{pmatrix}. \tag{4.39}$$

With this preparation the differential equation (4.34) becomes

$$\dot{y} = By + h(y) \tag{4.40}$$

where $B$ is the normal form matrix with the $\epsilon$ modifications described above and $h(y)$ is smooth and vanished together with it's first derivatives at the origin.

Now consider the standard first guess for a Liapounov function:

$$V(y) = \sum_{i=1}^{n} y_i^2 \tag{4.41}$$

with

$$\dot{V}(y) = 2 \sum_{i=1}^{n} y_i \dot{y}_i. \tag{4.42}$$

If $y_i$ corresponds to a single real block in the matrix $B$ then

$$y_i \dot{y}_i = \lambda_i y_i^2 + O(|y|^3) \tag{4.43}$$

where the eigenvalue $\lambda_i$ is strictly negative and the cubic order terms come from $h(y)$. Similarly, if $(y_i, y_{i+1})$ corresponds to a complex conjugate pair of eigenvalues in a single block,

$$\begin{pmatrix} \lambda_i & -\omega_i \\ \omega_i & \lambda_i \end{pmatrix},$$

then

$$y_i \dot{y}_i + y_{i+1} \dot{y}_{i+1} = \lambda_i (y_i^2 + y_{i+1}^2) + O(|y|^3). \tag{4.44}$$

Furthermore, if $(y_i, \dots, y_{i+r-1})$ correspond to a block like (4.37) then

$$\sum_{s=0}^{r-1} y_{i+s}\dot{y}_{i+s} = \lambda \sum_{s=0}^{r-1} y_{i+s}^2 + \epsilon(y_i y_{i+1} + \dots + y_{i+r-2}y_{i+r-1}) + O(|y|^3). \quad (4.45)$$

Noting that

$$y_{i+s}y_{i+s+1} = \tfrac{1}{2}\left((y_{i+s} + y_{i+s+1})^2 - y_{i+s}^2 - y_{i+s+1}^2\right)$$

equation (4.45) can be tidied up to give

$$\sum_{s=0}^{r-1} y_{i+s}\dot{y}_{i+s} = (\lambda - \tfrac{1}{2}\epsilon)(y_i^2 + y_{i+r-1}^2) + \sum_{s=1}^{r-2}(\lambda - \epsilon)y_{i+s}^2 + \tfrac{1}{2}\epsilon \sum_{s=0}^{r-2}(y_{i+s} + y_{i+s+1})^2 + O(|y|^3). \quad (4.46)$$

(4.14) EXERCISE

*If $(y_i, \dots, y_{i+2r-1})$ is associated with a block like (4.39) show that*

$$\sum_{s=0}^{2r-1} y_{i+s}\dot{y}_{i+s} = (\lambda - \tfrac{1}{2}\epsilon)(y_i^2 + y_{i+1}^2 + y_{i+2r-2}^2 + y_{i+2r-1}^2) + (\lambda - \epsilon)\sum_{s=2}^{2r-3} y_{i+s}^2 + \tfrac{1}{2}\epsilon \sum_{s=0}^{2r-2}(y_{i+s} + y_{i+s+1})^2 + O(|y|^3). \quad (4.47)$$

Now, all the eigenvalues of $B$ have negative real parts, so choosing $\epsilon < 0$ such that

$$(Re(\lambda_i) - \epsilon) < \mu < 0 \quad (4.48)$$

for all eigenvalues $\lambda_i$ of $B$ we find (using (4.43), (4.44), (4.46) and (4.47)) that

$$\dot{V}(y) \leq 2\mu \sum_{i=1}^{n} y_i^2 + O(|y|^3). \quad (4.49)$$

Since $h$ and its derivatives are smooth, there exists $K > 0$ such that

$$|O(|y|^3)| \leq K|y|^3 \quad (4.50)$$

on some neighbourhood of $y = 0$. So, choosing $|y| \leq K^{-1}|\mu|$ we find that

$$K|y|^3 \leq |\mu||y|^2. \tag{4.51}$$

From (4.49) this implies that

$$\dot{V}(y) \leq \mu \sum_{i=1}^{n} y_i^2. \tag{4.52}$$

This completes the proof: $V$ is a Liapounov function on some neighbourhood of $y = 0$ and $\dot{V} < 0$ for $y \neq 0$. Hence, by Theorem 2.11 (Liapounov's second stability theorem), $y = 0$ is asymptotically stable.

(4.15) EXERCISE

*Show that trajectories in a sufficiently small neighbourhood of $y = 0$ tend to $y = 0$ exponentially fast.*

Theorem 4.14 provides a very simple test for the stability of stationary points: simply evaluate the Jacobian matrix at the stationary point and show that all the eigenvalues are negative. Note, however, that a stationary point at which the Jacobian matrix has an eigenvalue with zero real part can be asymptotically stable (depending on the nonlinear terms), but if there is an eigenvalue with strictly positive real part then the stationary point cannot be asymptotically stable.

## 4.6 The proof of the stable manifold theorem

There are many proofs of the stable manifold theorem, all of them involving rather more pure mathematical machinery than we are assuming for the purposes of this book. The reader who would like to see a complete proof should consult Devaney (1989), Carr (1981), Irwin (1980) or Marsden and Scheurle (1986), each of which gives a different proof. The title of this section is therefore a little misleading: instead of giving a proof we shall suggest a demonstration of a proof based on the same type of arguments as those used in Poincaré's Linearization Theorem in Section 4.1. In particular, we will restrict attention to analytic differential equations.

The idea of this sketch is simple: given the differential equation

$$\dot{x} = Ax + g_1(x, y) \tag{4.53a}$$

$$\dot{y} = -By + g_2(x, y) \tag{4.53b}$$

where the eigenvalues of both $A$ and $B$ are all positive and both matrices are in normal form and the functions $g_1$ and $g_2$ are real analytic and contain only nonlinear terms (cf. (4.21)) we see immediately that for the linearization of (4.53),

$$\dot{x} = Ax, \quad \dot{y} = -By, \tag{4.54}$$

the stable and unstable manifolds of $(x, y) = (0, 0)$ are given by

$$E^s(0,0) = \{(x, y) | \; x = 0\}$$

and

$$E^u(0,0) = \{(x, y) | \; y = 0\}.$$

Hence we expect the stable and unstable manifolds of the nonlinear problem (4.53) to be perturbations of these axes if they exist. Thus, for example, we might look for a stable manifold of (4.53) of the form $x = S(y)$ for some function $S$. Hence, after the change of variable $\xi = x - S(y)$, the axis $\xi = 0$ should be invariant. To prove (sort of) the result we will show that there exists a function $S(y)$, defined as a formal power series, such that after the change of variable

$$\xi = x - S(y) \tag{4.55}$$

the differential equation, (4.53), is transformed into

$$\dot{\xi} = A\xi + \xi F_1(\xi, y) \tag{4.56a}$$

$$\dot{y} = -By + F_2(\xi, y). \tag{4.56b}$$

From (4.56a) it should be immediately obvious that the axis $\xi = 0$ is invariant (since $\xi = 0$ implies that $\dot{\xi} = 0$) and the motion on $\xi = 0$ is given by

$$\dot{y} = -By + F_2(0, y). \tag{4.57}$$

Since $F_2$ contains only nonlinear terms, Theorem 4.14 implies that $y = 0$ is an asymptotically stable solution for (4.57) and the stable manifold theorem follows.

The trick, then, is to prove that there exists a function $S(y)$ such that the change of coordinates (4.55) gives the differential equation (4.56). To do this we begin by noting that for real analytic $g_1(x, y)$ (4.53a) can be written as

$$\dot{x} = Ax + xG_1(x, y) + \sum_{r \geq 2} v_r(y) \tag{4.58}$$

where, as in Section 4.1, $v_r(y)$ is a sum of monomials of order $r$. More explicitly, if $y \in \mathbf{R}^s$, $x \in \mathbf{R}^u$ (i.e. $n_s = s$ and $n_u = u$ in the nomenclature of (4.21)), $m = (m_1, \ldots, m_s)$ and

$$y^m = y_1^{m_1} y_2^{m_2} \ldots y_s^{m_s}$$

then

$$v_r(y) = \sum_{m \in M_r} a_m y^m \tag{4.59}$$

where $a_m \in \mathbf{R}^u$ and, as in Section 4.1,

$$M_r = \left\{ (m_1, \ldots, m_s) | m_i \geq 0, \sum_1^s m_i = r \right\}.$$

We now need to define a coordinate transformation to get rid of the terms $v_r(y)$. As in the proof of Poincaré's Linearization Theorem we do this by induction on $r$. Suppose that we have used a sequence of transformations

$$\xi^{(i+1)} = \xi^{(i)} - S_i(y), \quad y = y \tag{4.60}$$

with $\xi^{(2)} = x$, such that at the $k^{th}$ step the functions $S_2(y)$ to $S_{k-1}(y)$ have been chose such that in the new coordinate $\xi^{(k)}$ equation (4.58) is

$$\begin{aligned} \dot{\xi}^{(k)} &= A\xi^{(k)} + \xi^{(k)} G_k(\xi^{(k)}, y) \\ &\quad + V_k(y) + \sum_{r > k} V_r(y) \end{aligned} \tag{4.61}$$

where the $V_j(y)$ are sums of monomials of order $j$ as the $v_j(y)$ were. Now try the coordinate change

$$\xi^{(k+1)} = \xi^{(k)} - S_k(y) \tag{4.62}$$

with

$$S_k(y) = \sum_{m \in M_k} b_m y^m \tag{4.63}$$

and so, in coordinate form with $1 \leq i \leq u$,

$$\xi_i^{(k)} = \xi_i^{(k+1)} + \sum_{m \in M_k} b_{mi} y^m. \tag{4.64}$$

Now, as in Section 4.1, we will assume that the eigenvalues of $A$ and $B$ are real and distinct (there is no real problem in including the more general cases, but the algebra gets too messy). Then, from (4.11) we have that for $m \in M_k$

$$\frac{d}{dt}y^m = \left(\sum_{p=1}^{s} m_p\beta_p\right) y^m + O(|y|^{k+1}) \tag{4.65}$$

where the $\beta_i$ are the eigenvalues of $-B$, so $\beta_i < 0$. Now recall the convention that

$$\sum_{p=1}^{s} m_p\beta_p = (m, \beta).$$

Differentiating (4.62) in component form and using (4.61) and (4.64) we find that

$$\begin{aligned}\dot{\xi}_i^{(k+1)} &= \alpha_i\xi_i^{(k+1)} + \xi_i^{(k+1)}G_{k+1}(\xi^{(k+1)}, y) \\ &\quad + \sum_{m\in M_k} (a_{mi} + (\alpha_i - (m,\beta))b_{mi})\, y^m \\ &\quad + O(|y|^{k+1}).\end{aligned} \tag{4.66}$$

So, choosing

$$b_{mi} = \frac{a_{mi}}{(m,\beta) - \alpha_i} \tag{4.67}$$

we find that the new equation satisfied by $\xi_i^{(k+1)}$ is

$$\dot{\xi}^{(k+1)} = A\xi^{(k+1)} + \xi^{(k+1)}G_{k+1}(\xi^{(k+1)}, y) + \sum_{r>k} V_r'(y) \tag{4.68a}$$

$$\dot{y} = -By + F_{k+1}(\xi^{(k+1)}, y) \tag{4.68b}$$

where $V_r'(y)$ are modified sums of monomials of order $r$. This completes the proof since we can now use induction on $k$ to get rid of all the terms in $y^m$ of order $k$ in (4.68a) for each $k \geq 2$. Note that since $\beta_i < 0$ and $\alpha_i > 0$ the denominator of (4.67) is never zero and so the problem of small divisors and resonance is avoided. Hence there is a sequence of coordinate changes (4.64) such that we obtain, in the limit, a coordinate change

$$\xi = x - S(y) \tag{4.69}$$

where $S$ is defined as a formal power series,

$$S(y) = \sum_{k\geq 2} s_k(y), \tag{4.70}$$

such that in the coordinates $(\xi, y)$ the equation (4.53) becomes (4.56). The theorem then follows from the remarks at the beginning of this section (assuming, as we do, that the formal power series defined all converge on some neighbourhood of $(\xi, y) = (0,0)$).

## Exercises 4

1. Although the linearization of

$$\dot{x} = x, \quad \dot{y} = 3y + x^2$$

is resonant (of order 3), show that there is a near identity change of coordinates which removes the quadratic term. By solving the equation for trajectories exactly comment on the correspondence between trajectories of the nonlinear equation and its linearization.

2. Find the stable and unstable manifolds, correct to cubic order, for the following systems. Also, find the equation of motion on the unstable manifold correct to quadratic terms in each case.

(i)

$$\dot{x} = 3x + 2y^2 + xy, \quad \dot{y} = -y + 3y^2 + x^2y - 4x^3;$$

(ii)

$$\dot{x} = -2x - 3y - x^2, \quad \dot{y} = x + 2y + xy - 3y^2.$$

3. Suppose $\dot{x} = Ax + f(x)$ for $x \in \mathbf{R}^2$, where $A$ is a $2 \times 2$ matrix and $f$ vanishes with its first derivatives at $x = 0$. Find sufficient conditions on the coefficients of $A$ for the origin to be asymptotically stable.

4. Consider systems as in question 3 but for $x \in \mathbf{R}^3$. If

$$A = \begin{pmatrix} a_{11} & a_{12} & a_{13} \\ a_{21} & a_{22} & a_{23} \\ a_{31} & a_{32} & a_{33} \end{pmatrix}$$

show that the origin is asymptotically stable if

$$a_{11} < 0, \quad \begin{vmatrix} a_{11} & a_{12} \\ a_{21} & a_{22} \end{vmatrix} > 0, \quad \text{and} \quad \det A < 0.$$

5. For what values of $r$, $\sigma$ and $b$ do non-trivial stationary points of the Lorenz equations,

$$\dot{x} = \sigma(-x + y)$$

$$\dot{y} = rx - y - xz$$

$$\dot{z} = -bz + xy,$$

exist? When are they asymptotically stable?

6. Suppose that $\dot{x} = Ax + f(x)$, $x \in \mathbf{R}^n$, where $A$ is an $n \times n$ matrix and $f$ vanishes together with its first derivatives at the origin. Use the Stable Manifold Theorem to show that if one or more of the eigenvalues of $A$ has strictly positive real part then $x = 0$ is not Liapounov stable.

7. Suppose that $x = 0$ is a sink for $\dot{x} = Ax + f(x)$ for $x \in \mathbf{R}^2$ (where $A$ and $f$ are as in question 2) and that $y = 0$ is a sink for $\dot{y} = By + g(y)$, $y \in \mathbf{R}^2$ (with the obvious notation). Show how to construct a map $h$ from a neighbourhood of $x = 0$ to a neighbourhood of $y = 0$ mapping trajectories of one system to trajectories of the other. [Hint: Define $h$ to be the identity on sufficiently small circles containing the points $x = 0$ and $y = 0$ respectively. Then define $h$ inside these circles by following the flow near $x = 0$ back to the circle, across to the circle in the $y$-plane and into this circle following the flow near $y = 0$.]

8. Construct the map $h$ of question 7 explicitly for the linear systems defined by $f(x) = 0$, $g(y) = 0$,

$$A = \begin{pmatrix} -1 & 0 \\ 0 & -2 \end{pmatrix} \quad \text{and} \quad B = \begin{pmatrix} -1 & -1 \\ 1 & -1 \end{pmatrix}.$$

9. Rewrite the differential equation

$$\ddot{x} = x - x^3$$

as a system in the plane by setting $y = \dot{x}$. Show that every criterion for structural stability in the plane is violated.

10. Sketch the locus of stationary points of the one-dimensional differential equation

$$\dot{x} = (x - \mu)(2x^2 - \mu)$$

as a function of $\mu$. Show that the stationary points are hyperbolic except at the intersection of different branches of solutions.

# 5

## *Two-dimensional dynamics*

This chapter is devoted entirely to autonomous differential equations in the plane. Due to various topological properties of the plane it is possible to develop techniques for analyzing these two-dimensional systems which are not applicable in higher dimensions. For example, the fact that a closed curve (i.e. a periodic orbit in phase space) separates the plane into two regions, an inside and an outside, in two dimensions will enable us to deduce properties of periodic orbits of planar systems. By Peixoto's Theorem (Theorem 4.13) we see that the dynamics of autonomous differential equations in the plane are typically relatively simple and indeed, there is no possibility of chaotic behaviour; the fact that trajectories cannot cross imposes severe restrictions on the limit sets of these systems. We begin this chapter with a discussion of the stationary points of two-dimensional equations and then go on to develop methods for proving the existence (and non-existence) of periodic orbits in such systems. More detail of the results in the first two sections can be found in the classic books by Hartman (1964) and Coddington and Levinson (1955), but be warned: these are serious reference books!

### 5.1 Linear systems in $\mathbf{R}^2$

Consider the linear differential equation

$$\begin{pmatrix} \dot{x} \\ \dot{y} \end{pmatrix} = \begin{pmatrix} a & b \\ c & d \end{pmatrix} \begin{pmatrix} x \\ y \end{pmatrix} \tag{5.1}$$

where $a$, $b$, $c$ and $d$ are real constants. Then, using the results of Chapter 3, there is a linear change of coordinates such that in the new coordinates the differential equation is defined by one of the three cases

$$A = \begin{pmatrix} \lambda_1 & 0 \\ 0 & \lambda_2 \end{pmatrix}, \quad B = \begin{pmatrix} \lambda & 1 \\ 0 & \lambda \end{pmatrix} \quad \text{or} \quad C = \begin{pmatrix} \rho & -\omega \\ \omega & \rho \end{pmatrix}. \tag{5.2}$$

Each of these three possibilities will give rise to different phase portraits, and, in the case of $A$, several different phase portraits. Note that the only stationary point is at the origin provided the determinant of the matrix is non-zero, i.e. the matrix has no zero eigenvalues, and so we shall consider those cases first.

**(i) Nodes.** Suppose that $\lambda_1 > \lambda_2 > 0$ in the case of $A$ above. Then the differential equation is

$$\dot{x} = \lambda_1 x, \quad \dot{y} = \lambda_2 y \tag{5.3}$$

or

$$\frac{dy}{dx} = \frac{\lambda_2 y}{\lambda_1 x}, \tag{5.4}$$

which we can integrate immediately to obtain trajectories through $(x_0, y_0)$ giving

$$\log\left(\frac{x}{x_0}\right) = \frac{\lambda_1}{\lambda_2}\log\left(\frac{y}{y_0}\right)$$

or

$$\left(\frac{x}{x_0}\right) = \left(\frac{y}{y_0}\right)^{\lambda_1/\lambda_2}. \tag{5.5}$$

Since $\lambda_1 > \lambda_2 > 0$ this curve describes a generalized parabola tangential to the $y$-axis at the origin as shown in Figure 5.1a unless $x_0 = 0$, in which case $x = 0$ for all time, so the $y$-axis is a union of trajectories (it is invariant). The arrows of time are easy to obtain as both $|x|$ and $|y|$ increase with time ($x(t) = x_0 e^{\lambda_1 t}$). This is called an unstable node. If $-\lambda_1 > -\lambda_2 > 0$ then we can describe the flow in exactly the same way by reversing the arrow of time to give a stable node (Fig. 5.1b). Note that in both cases trajectories tend to the origin (as $t \to -\infty$ for the unstable node and $t \to \infty$ for the stable node) tangential to the $y$-axis, which is the eigenvector associated with the eigenvalue of smallest modulus.

*Example 5.1*

Suppose

$$\begin{pmatrix} \dot{x} \\ \dot{y} \end{pmatrix} = \begin{pmatrix} -1 & 2 \\ 0 & -3 \end{pmatrix} \begin{pmatrix} x \\ y \end{pmatrix}.$$

The eigenvalues of the matrix are $\lambda_1 = -3$ and $\lambda_2 = -1$ and so the origin is a stable node. The eigenvectors corresponding to these eigenvalues are

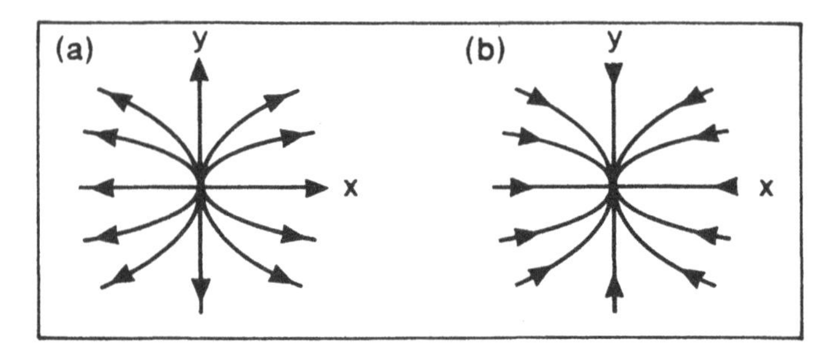


Fig. 5.1 (a) An unstable node; (b) a stable node.

$e_1 = (1, -1)^T$ and $e_2 = (1, 0)^T$ and so trajectories approach the origin tangential to the $e_2$ eigenvector as shown in Figure 5.2.

**(ii) Saddles.** Continuing with case $A$ suppose that $\lambda_2 < 0 < \lambda_1$. Then, integrating the equation as before we obtain a family of generalized hyperbolae

$$\left(\frac{x}{x_0}\right) = \left(\frac{y}{y_0}\right)^{\lambda_1/\lambda_2} \tag{5.6}$$

provided $x_0$ and $y_0$ are not equal to zero (remember that $\lambda_2 < 0$ and so $\lambda_1/\lambda_2 < 0$). Both axes are invariant and trajectories approach the origin along the $y$-axis, which is the $e_2$ eigenvector, and leave the origin along the $x$-axis, which is the $e_1$ eigenvector. This gives the phase portrait shown in Figure 5.3a. Figure 5.3b gives the general phase portrait when

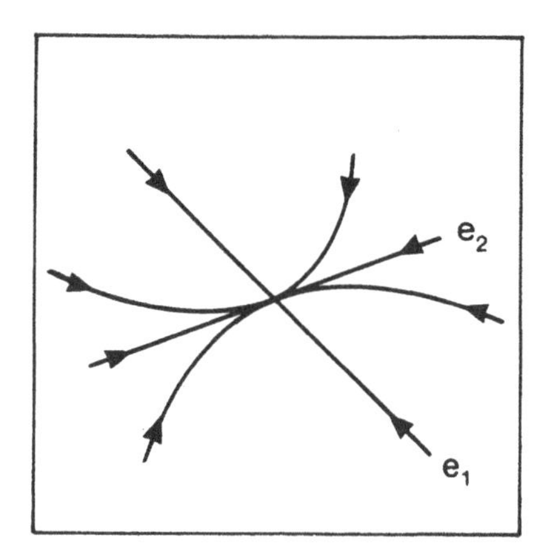


Fig. 5.2

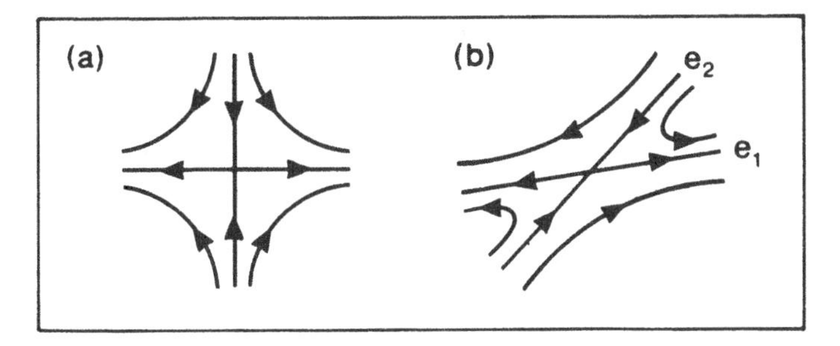


Fig. 5.3 Saddles.

the eigenvectors are not orthogonal. Stationary points with a pair of real eigenvectors of different sign are called saddles.

**(iii) Stars.** The final example of case $A$ is when $\lambda_1 = \lambda_2 = \lambda \neq 0$. Integrating the equation for trajectories again we have $\lambda_1/\lambda_2 = 1$ and so trajectories lie on straight lines through the origin. This gives the two cases in Figure 5.4: an unstable star if $\lambda > 0$ and a stable star if $\lambda < 0$. These phase portraits are the same after any linear change of coordinates since every direction satisfies the eigenvector equation.

**(iv) Improper nodes.** For case $B$ of the normal forms

$$\dot{x} = \lambda x + y, \quad \dot{y} = \lambda y. \tag{5.7}$$

To sketch the phase portrait of this system take $\lambda > 0$ and note that the $x$-axis is invariant with trajectories leaving the origin. To get a more complete description of the phase portrait we use the method of *isoclines.* The idea is to determine, at least roughly, the direction of the flow at points in phase space and to use this information to build up the picture. As we have already pointed out $\dot{y} = 0$ when $y = 0$ and

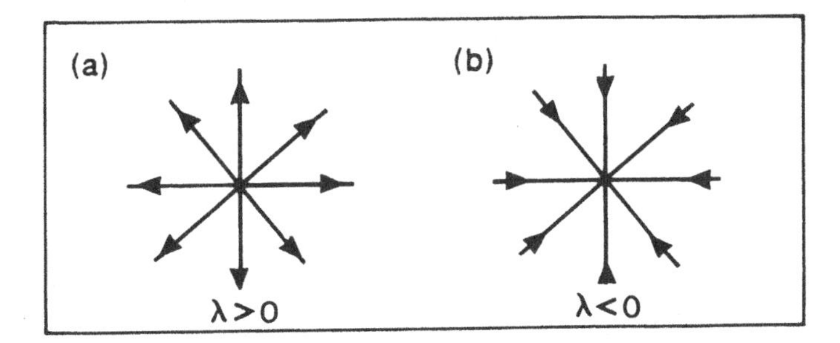


Fig. 5.4 Stars.

so the $x$-axis is invariant. Furthermore, $\dot{y} < 0$ if $y < 0$ and $\dot{y} > 0$ if $y > 0$. Now, $\dot{x} = 0$ when $y = -\lambda x$ and so on this line trajectories move vertically upwards if $y > 0$ and downwards if $y < 0$. To the right of this line $\lambda x + y > 0$ and so $\dot{x} > 0$ and to the left of this line $\dot{x} < 0$. This means that the plane can be divided into four regions (see Fig. 5.5): if $y > 0$ and $\lambda x + y > 0$ then both $\dot{x}$ and $\dot{y}$ are positive, and if $y < 0$ and $\lambda x + y < 0$ then both $\dot{x}$ and $\dot{y}$ are negative. If $y > 0$ and $\lambda x + y < 0$ then $\dot{y} > 0$ and $\dot{x} < 0$ and (finally) if $y < 0$ and $\lambda x + y > 0$ then $\dot{y} < 0$ and $\dot{x} > 0$. Putting this information together as shown in Figure 5.5a we see that trajectories must take the form shown in Figure 5.5b. This is called an unstable improper node. Note that unlike the previous example simply reversing the direction of time does not give the case $\lambda < 0$. To see this set $\tau = -t$ so $\frac{d}{d\tau} = -\frac{d}{dt}$. In this new variable the equation becomes

$$\frac{dx}{d\tau} = -\lambda x - y, \quad \frac{dy}{d\tau} = -\lambda y \tag{5.8}$$

and so we get a $-y$ in the $x$ equation instead of $+y$. Hence it is not enough just to reverse the arrows on Figure 5.5b. But if we define $Y = -y$ then

$$\frac{dx}{d\tau} = -\lambda x + Y, \quad \frac{dY}{d\tau} = -\lambda Y \tag{5.9}$$

and we get the desired equation. To recover the phase portrait of this from the original phase portrait we therefore need to reflect in the $x$-axis (which reverses the sign of $y$) and then change the direction of the arrows ($t \to -t$). This gives Figure 5.5c, an improper node. This technique of reflection and reversal of time can be very useful when considering equations.

**(v) Foci.** The last remaining case is the matrix $C$, which we have already discussed in Chapter 3. Assume $\rho > 0$ and $\omega > 0$. In polar coordinates the equation becomes

$$\dot{r} = \rho r, \quad \dot{\theta} = \omega$$

and so $\frac{dr}{d\theta} = \frac{\rho}{\omega} r$ and

$$\theta = \psi + \frac{\omega}{\rho} \log r \tag{5.10}$$

where $\psi$ is a constant. Hence trajectories are logarithmic spirals into the origin if $\rho < 0$ (a stable focus) and out of the origin if $\rho > 0$ (an unstable focus). The spirals are anti-clockwise since $\omega > 0$; if $\omega < 0$

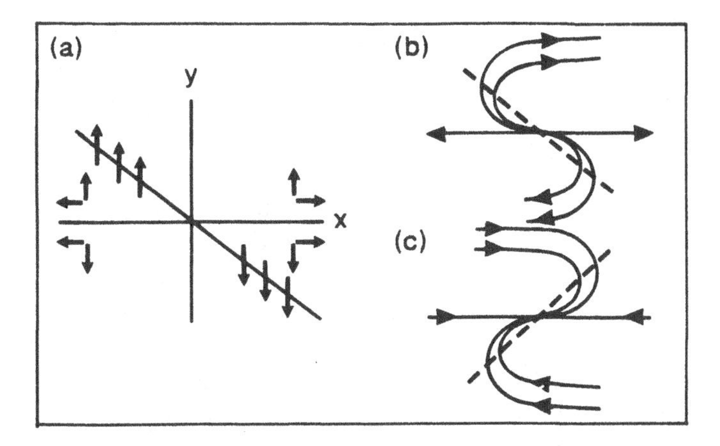


Fig. 5.5 Improper nodes: (a) isoclines ($\lambda > 0$); (b) $\lambda > 0$; (c) $\lambda < 0$.

they are spiral about the origin clockwise as can be seen directly from the $\dot{\theta}$ equation. An example is shown in Figure 5.6a.

**(vi) Centres.** Suppose now that $\rho = 0$ in case $C$. Then $\dot{r} = 0$ and trajectories lie on concentric circles about the origin with the arrow of time pointing anti-clockwise if $\omega > 0$ and clockwise if $\omega < 0$ as shown in Figure 5.6b.

In all the cases above, the origin is the unique stationary point of the linear differential equation. To complete the classification of the behaviour of these linear systems we need to consider the degenerate cases when this is no longer true. Consider first the matrix $A$ again. If both $\lambda_1$ and $\lambda_2$ equal zero then the whole space is fixed. If one of these eigenvalues is zero, $\lambda_1 = 0$, $\lambda_2 \neq 0$ say, then $\dot{x} = 0$ and so $x$ is

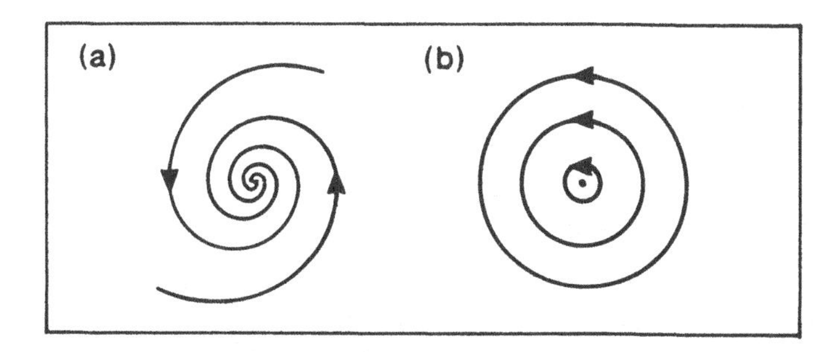


Fig. 5.6 (a) A stable focus; (b) a centre.

constant on trajectories, so trajectories lie on vertical lines. Stationary points are given by $y = 0$, so the entire $x$-axis is stationary and if $\lambda_2 < 0$ a point $(x_0, y_0)$ tends to $(x_0, 0)$ as $t \to \infty$ along the straight line $x = x_0$ (see Fig. 5.7a). The final degenerate possibility is for the matrix $B$ with $\lambda = 0$ so $\dot{x} = y$ and $\dot{y} = 0$. Stationary points are given by $y = 0$, i.e. the $x$-axis is a union of stationary points and since $\dot{y} = 0$ trajectories lie on horizontal lines $y = y_0$. Furthermore, $\dot{x} > 0$ if $y > 0$ and $\dot{x} < 0$ if $y < 0$ giving the phase portrait shown in Figure 5.7b.

The general procedure for describing a two-dimensional linear differential equation $\dot{x} = Lx$ is therefore to find the eigenvalues and eigenvectors of $L$, then determine the nature of the stationary point, which is simple unless there is a repeated or zero eigenvalue, and finally draw the phase portrait in the basis of eigenvectors. In general if

$$L = \begin{pmatrix} a & b \\ c & d \end{pmatrix} \tag{5.11}$$

the eigenvalue equation is $s^2 - (a + d)s + ad - bc = 0$, or if Tr$L$ denotes the trace of $L$, Tr$L = a + d$, and det$L$ is the determinant of $L$ then

$$s^2 - \mathrm{Tr}Ls + \det L = 0$$

which has eigenvalues

$$s = \frac{-\mathrm{Tr}L \pm \sqrt{(\mathrm{Tr}L)^2 - 4\det L}}{2}. \tag{5.12}$$

Hence we see immediately that

if det$L < 0$ then the eigenvalues of $L$ are real and have opposite signs and hence the origin is a saddle;

if $(\mathrm{Tr}L)^2 > 4\det L > 0$ then the eigenvalues of $L$ are real and have the same sign as $-$Tr$L$ and so

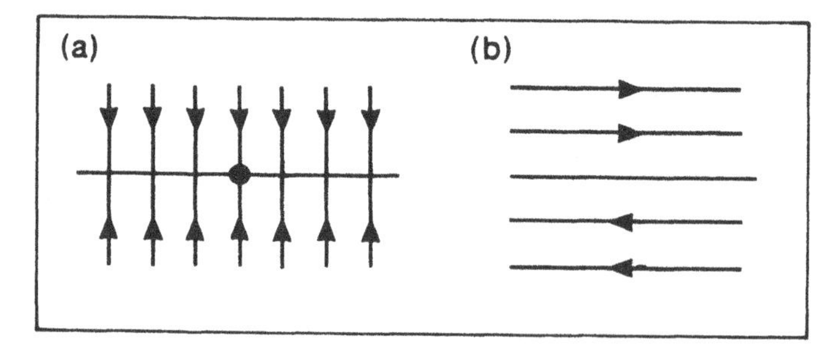


Fig. 5.7 Degenerate cases: (a) $\lambda_1 = 0$, $\lambda_2 < 0$; (b) $\lambda = 0$.

if $\mathrm{Tr}L > 0$ the origin is a stable node;

if $\mathrm{Tr}L < 0$ the origin is an unstable node;

if $4\det L > (\mathrm{Tr}L)^2 \geq 0$ then $L$ has a complex conjugate pair of eigenvalues with real part $-\mathrm{Tr}L$ and so

if $\mathrm{Tr}L > 0$, the origin is a stable focus;

if $\mathrm{Tr}L = 0$, the origin is a centre;

if $\mathrm{Tr}L < 0$, the origin is an unstable focus;

if $(\mathrm{Tr}L)^2 - 4\det L = 0$ then there is a double eigenvalue and

if $\mathrm{Tr}L \neq 0$ then the origin is an improper node, stable if $\mathrm{Tr}L > 0$ and unstable if $\mathrm{Tr}L < 0$, unless $b = c = 0$ when they are stars;

if $\mathrm{Tr}L = 0$ then we have one of the degenerate situations with non-isolated stationary points described above.

These various possibilities are conveniently represented in Figure 5.8.

## 5.2 The effect of nonlinear terms

If the study of linear differential equations in the plane is to be useful when considering nonlinear differential equations we need to use the results of the previous chapter to find out when the linearization about a stationary point can give a valid description of the flow near the stationary point. To some extent we have already done this in Chapter 4; if the origin is hyperbolic (a node, improper node, a star or a focus) then the

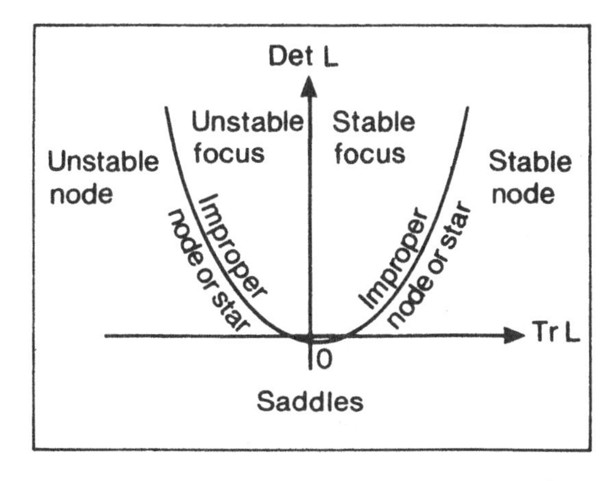


Fig. 5.8

nonlinear flow is equivalent (up to homeomorphism) to its linearization in some neighbourhood of the origin. Put another way, if the linearization of the nonlinear flow at the origin is hyperbolic then the local stable and unstable manifolds of the origin for the nonlinear system have the same dimension as for the linearized system. The potentially misleading term in the general theory is *up to homeomorphism*, since as we have already pointed out nodes and foci are equivalent in this description. For the two-dimensional flows considered here we can try to do better, and determine the extent to which the local structure of the nonlinear system mirrors the corresponding linear system.

Consider the equation

$$\dot{z} = Lz + F(z) \tag{5.13}$$

for $z \in \mathbf{R}^2$, $L$ a constant $2 \times 2$ matrix and $F(z)$ a function which represents the nonlinear terms in the equation, so

$$\frac{|F(z)|}{|z|} \to 0 \quad \text{as} \quad |z| \to 0. \tag{5.14}$$

Without loss of generality we can assume that $L$ is in normal form. If $L$ has a zero eigenvalue, and so is one of the degenerate cases considered towards the end of the last section, then it is quite obvious that there exist arbitrarily small perturbations of the linear system which do not retain the property that there is an invariant line of stationary points. So we cannot hope for any robustness under perturbation in such systems. This leaves us with the six cases enumerated in the previous section: nodes, saddles, improper nodes, stars, foci and centres. We will consider each case in turn, but first suppose that

$$\dot{x} = ax + by, \quad \dot{y} = cx + dy. \tag{5.15}$$

Then in polar coordinates, with $x = r\cos\theta$ and $y = r\sin\theta$, the equation becomes

$$\dot{r} = r[a\cos^2\theta + (b+c)\cos\theta\sin\theta + d\sin^2\theta] = rR(\theta) \tag{5.16a}$$

$$\dot{\theta} = [c\cos^2\theta + (d-a)\cos\theta\sin\theta - b\sin^2\theta] = \Omega(\theta) \tag{5.16b}$$

where the functions $R$ and $\Omega$ are the obvious expressions in square brackets. Now, suppose that this system represents the linearization about a nonlinear stationary point. We want to understand the extent to which the linear description (in terms of nodes, foci and so on) provides a faithful representation of the local behaviour of the nonlinear system. It turns out that there is a subtlety that we have ignored in our linearization theorems hitherto, in that the size of the nonlinear terms as $r$ tends to 0 becomes important. Recall that in Chapter 3 we described two linearization theorems: Poincaré's Theorem, which applies to analytic equations and shows that provided certain non-resonance conditions are satisfied there is an analytic change of coordinates in a neighbourhood of the origin which brings the equation into the linear form, and Hartman's Theorem which states that provided the stationary point is hyperbolic there is a homeomorphism defined on some neighbourhood of the stationary point which takes the nonlinear flow to the linear flow. We remarked that neither of these theorems is entirely satisfactory. The conditions for Poincaré's Theorem are extremely strong: not all nonlinear systems are analytic and non-resonant; whilst the homeomorphism of Hartman's Theorem does not distinguish between nodes and foci. To find out how these results work in practice we will consider each possible case for stationary points in two-dimensional flows in turn. Note that since neither of the linearization theorems of Chapter 3 applies to non-hyperbolic cases we expect to find severe restrictions on the linearization in these cases, the centre and the degenerate cases with a zero eigenvalue, whilst we know that in the hyperbolic cases at least some features of the linearized flow are preserved. A full discussion of the non-hyperbolic cases is postponed to later chapters, where they form the basis of bifurcation theory. Here we restrict our attention to simply determining whether the linearized system is or is not a good model of a general nonlinear perturbation.

The phrase 'general nonlinear perturbation' is really the crux of the subtleties referred to above. In polar coordinates the nonlinear system with the linearization described above is

$$\dot{r} = rR(\theta) + P(r,\theta) = \rho(r,\theta) \tag{5.17a}$$

$$\dot{\theta} = \Omega(\theta) + Q(r,\theta) = \Theta(r,\theta) \tag{5.17b}$$

where the functions $P$ and $Q$ represent the nonlinear terms ignored by the linearization. Hence, for the most general nonlinear perturbations

$$\lim_{r\to 0} \frac{P(r,\theta)}{r} = 0 \quad \text{and} \quad \lim_{r\to 0} Q(r,\theta) = 0. \tag{5.18}$$

If this is not obvious see Exercise 5.1 at the end of this chapter. This information is most conveniently written using the $o$ (or 'little o') notation: $P(r,\theta) \sim o(r)$ and $Q(r,\theta) \sim o(1)$. The subtlety referred to above is that the results which can be obtained for general perturbations of this kind are different from the results which apply to the slightly less general perturbation for which

$$P(r,\theta) \sim O(r^2) \text{ and } Q(r,\theta) \sim O(r) \tag{5.19}$$

where a function $f(r,\theta)$ is $O(\alpha)$ if $\frac{|f(r,\theta)|}{\alpha}$ is bounded as $r$ tends to zero. The difference between the two possibilities is not immediately obvious, but, for example, $r^{\frac{3}{2}}$ is $o(r)$ but not $O(r^2)$ whilst any function with a valid Taylor series in a neighbourhood of the origin can be written as $f(0) + \frac{\partial f}{\partial x}x + O(|x|^2)$. So most of the functions that we have been considering fall into the 'big O' category. None the less, we shall describe the more general results here, leaving the modifications for functions which satisfy the 'big O' conditions as exercises. Now, after this somewhat lengthy technical diversion, we are in a position to go through the various cases described in the previous section.

**(i) Nodes.** Suppose that for the linearized flow the origin is a node, so the eigenvalues of $L$ are real and, without loss of generality, negative. After a linear change of coordinates we can write $L = \text{diag}(\lambda_1, \lambda_2)$ with $-\lambda_2 > -\lambda_1 > 0$. Hence the linear system $\dot{z} = Lz$ has trajectories which tend to the origin tangent to the $x$-axis (except for the $y$-axis, which is invariant). In terms of the expressions in polar coordinates derived above,

$$R(\theta) = (\lambda_2 \cos^2\theta + \lambda_1 \sin^2\theta) \tag{5.20a}$$

and

$$\Omega(\theta) = (\lambda_1 - \lambda_2)\cos\theta\sin\theta. \tag{5.20b}$$

Differentiating $R(\theta)$ with respect to $\theta$ we see that it takes its maximum and minimum values when $\cos\theta = 0$ and $\sin\theta = 0$ respectively, so $\lambda_2 < R(\theta) < \lambda_1$. Choosing $r$ sufficiently small so that $|P(r,\theta)| < \epsilon r$ for some $\epsilon > 0$ with $0 < \epsilon < |\lambda_1|$ we find that

$$\dot{r} \leq (\lambda_1 + \epsilon)r \tag{5.21}$$

and so if $-k = (\lambda_1 + \epsilon)$ then $k > 0$ and

$$r(t) \leq r_0 e^{-kt}. \tag{5.22}$$

Hence the origin is asymptotically stable. The $\dot{\theta}$ equation is

$$\dot{\theta} = (\lambda_2 - \lambda_1)\cos\theta\sin\theta + Q(r,\theta) \tag{5.23}$$

and so provided both $\cos\theta$ and $\sin\theta$ are bounded away from zero (i.e. outside some small neighbourhood of the $x$- and $y$- axes) the angular behaviour is as for the linear case (Fig. 5.9a). For trajectories which start sufficiently close to the axes we can only deduce that the $r$ coordinate tends to zero exponentially. To consider the behaviour near the axes choose some small $\delta > 0$ and define four sectors:

$$S_1 = \{(r,\theta)|\ |\theta| < \delta\},$$

$$S_2 = \left\{(r,\theta)|\ |\theta - \tfrac{1}{2}\pi| < \delta\right\},$$

$$S_3 = \{(r,\theta)|\ |\theta - \pi| < \delta\}$$

and

$$S_4 = \left\{(r,\theta)|\ |\theta - \tfrac{3}{2}\pi| < \delta\right\}.$$

Since $Q(r,\theta)$ is $o(1)$ for any small $\delta > 0$ we can choose $r$ sufficiently small so that $\dot{\theta} > \eta > 0$ for all $\theta$ between $S_2$ and $S_3$ and between $S_4$ and $S_1$ and similarly $\dot{\theta} < -\eta < 0$ for all $\theta$ between $S_1$ and $S_2$ and between $S_3$ and $S_4$. Hence all trajectories which do not stay in either $S_2$ or $S_4$ for all time tend to the origin tangential to the $x$-axis in $S_3$ or $S_1$. Now, still working in a small neighbourhood of the origin, consider a small circle of radius $p$ enclosing the origin. Let $\theta_1$ be the largest angle in $S_2$ such that the trajectory through $(p,\theta_1)$ tends to the origin in $S_2$ (tangential to the $y$-axis), and let $\theta_0$ be the smallest such angle. Then the trajectories through $(p,\theta)$ with $\theta_0 < \theta < \theta_1$ must all tend to the origin in $S_2$, tangential to the $y$-axis giving the situation shown in Figure 5.9b. However, if the nonlinear terms $P(r,\theta)$ and $Q(r,\theta)$ are (respectively) $O(r^2)$ and $O(r)$ then it is possible to prove (see the exercises at the end of this chapter) that $\theta_0 = \theta_1$ and so there is only one such trajectory as in the case of the linear node. A similar argument in $S_4$ completes the picture of the nonlinear mode, giving one of the two possibilities shown in Figures 5.9c,d.

**(ii) Saddles.** For a saddle the stable manifold theorem implies immediately that the local structure of solutions is essentially the same as the linear system with some distortion provided that the flow is sufficiently smooth (i.e. $P$ and $Q$ satisfy the 'big O' conditions above). If $P$ and $Q$ satisfy the more general nonlinear conditions then by the same arguments as used for the node there may be a set of trajectories which tend to the origin directly in forward or backwards time (see Fig 5.10).

**(iii) Stars.** If the linearized equations at the origin correspond to a star then the equations are

$$\dot{r} = \lambda r + P(r,\theta), \quad \dot{\theta} = Q(r,\theta) \tag{5.24}$$

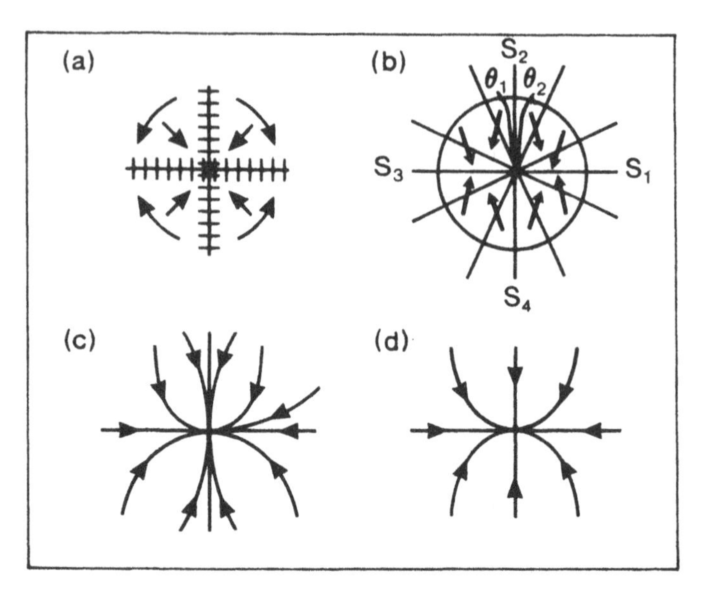


Fig. 5.9 (a) Regions of definite motion; (b) the sectors $S_i$; (c) a nonlinear node (with $o(r)$ nonlinearity); (d) a nonlinear node (with $O(r^2)$ nonlinearity).

where we can choose the direction of time so that $\lambda < 0$. As with the nonlinear node the origin is asymptotically stable, since we can choose $r$ sufficiently small so that $|P(r,\theta)| < \epsilon r$ for some $\epsilon > 0$ with $\epsilon < |\lambda|$. Then $k = -(\lambda + \epsilon) > 0$ and

$$\dot{r} \leq -kr \tag{5.25}$$

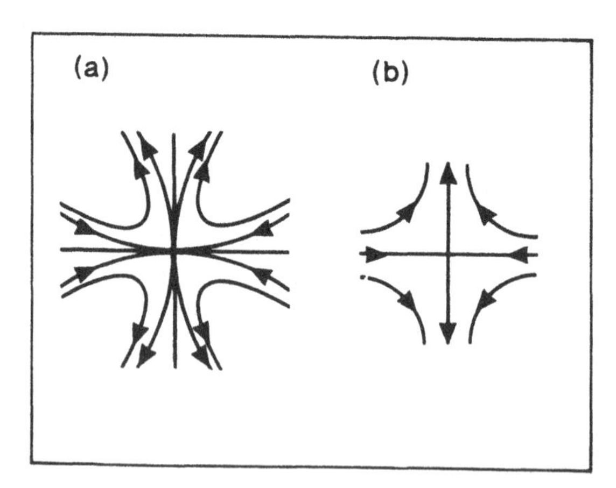


Fig. 5.10 Nonlinear saddles.

or

$$r(t) \leq r_0 e^{-kt}. \tag{5.26}$$

The angular motion is not so easy to determine here, and it is fairly easy to find functions $Q(r,\theta)$ which are $o(1)$ and for which $\theta(t) \to \infty$ as $t \to \infty$, so the star becomes a stable focus when nonlinear terms are added. It is also possible to find nonlinear perturbations under which the star retains the property that there are an infinite set of angles along which trajectories approach the origin, the trivial case $Q(r,\theta) = 0$ is an example, and also examples for which all trajectories tend to the origin tangential to one of a finite number of directions. However, if $P$ and $Q$ satisfy the 'big O' conditions then there are always an infinite number of tangent directions (see the exercises at the end of the chapter).

*Example 5.2*

Consider the system

$$\dot{x} = -x - \frac{y}{\log(x^2+y^2)}, \qquad \dot{y} = -y + \frac{x}{\log(x^2+y^2)}$$

or, in polar coordinates

$$\dot{r} = -r, \quad \dot{\theta} = (2\log r)^{-1}.$$

The linearization of this system is the star with $\lambda = -1$ and the nonlinear perturbations are $P(r,\theta) = 0$, $Q(r,\theta) = (2\log r)^{-1}$, so $P$ and $Q$ satisfy the 'little o' conditions. However, $r(t) = r_0 e^{-t}$ and so

$$\dot{\theta} = (2c - 2t)^{-1}, \quad \text{so} \quad \theta(t) = -\tfrac{1}{2}\log|c-t|,$$

where $c = \log r_0$. Hence $\theta \to -\infty$ as $t \to \infty$ and the star has become a focus.

**(iv) Improper nodes.** If the origin is an improper node then we can choose coordinates near the origin and the direction of time so that the Jacobian matrix of the differential equation evaluated at the origin is

$$\begin{pmatrix} \lambda & 1 \\ 0 & \lambda \end{pmatrix} \tag{5.27}$$

with $\lambda < 0$. In polar coordinates the nonlinear system is therefore

$$\dot{r} = r[\lambda + \cos\theta\sin\theta] + P(r,\theta), \quad \dot{\theta} = -\sin^2\theta + Q(r,\theta). \tag{5.28}$$

In this case even the $r$ equation is awkward, although we know from Hartman's Theorem that the origin remains asymptotically stable under nonlinear perturbations. To analyze this system we need to do another little change of coordinates which ensures that in the new coordinates the $r$ motion is easy to determine. In Cartesian coordinates we have the linearization near the origin

$$\dot{x} = \lambda x + y, \quad \dot{y} = \lambda y \tag{5.29}$$

so if we defined $z$ by $y = \epsilon z$ for some small $\epsilon$ with $0 < \epsilon < -\lambda$ the equation in $(x, z)$ coordinates is

$$\dot{x} = \lambda x + \epsilon z, \quad \dot{z} = \lambda z \tag{5.30}$$

and so, taking the standard polar coordinates in the $(x, z)$ plane the full nonlinear equation is

$$\dot{r} = r[\lambda + \epsilon \cos\theta \sin\theta] + P'(r, \theta), \quad \dot{\theta} = -\sin^2\theta + Q'(r, \theta) \tag{5.31}$$

where the functions $P$ and $Q$ have the same asymptotic properties as the functions $P'$ and $Q'$. In particular, $P'(r, \theta) \sim o(r)$ and $Q'(r, \theta) \sim o(1)$. Hence we can choose $r$ sufficiently small so that $|P'(r, \theta)| < \delta r$ for any $\delta > 0$ such that $\frac{1}{2}\epsilon + \delta < -\lambda$. Provided $r$ is sufficiently small (and noting that $|\cos\theta \sin\theta| < \frac{1}{2}$) we find that

$$\dot{r} \leq (\lambda + \tfrac{1}{2}\epsilon + \delta) r$$

for sufficiently small $r$ with $\lambda + \frac{1}{2}\epsilon + \delta = -k < 0$. Hence $\dot{r} \leq -kr$ and so

$$r(t) \leq r_0 e^{-kt}. \tag{5.32}$$

This shows (as we already knew) that the origin is asymptotically stable.

Now consider the $\dot{\theta}$ equation,

$$\dot{\theta} = -\sin^2\theta + Q'(r, \theta) \tag{5.33}$$

where $Q'(r, \theta) \sim o(1)$. For sufficiently small $r$ the $\sin^2\theta$ term dominates the nonlinear terms away from the $x$-axis, i.e. for all $\delta > 0$ (a different $\delta$ from the one above) there exists $\eta > 0$ and $r_1 > 0$ such that if $r < r_1$ and $\theta$ lies outside the sectors $S_1 = \{(r, \theta) | \, |\theta| < \delta\}$ and $S_3 = \{(r, \theta) | \, |\theta - \pi| < \delta\}$ then $\dot{\theta} < -\eta$. Thus the motion is inwards and clockwise outside these two sectors. Inside these sectors there are two possibilities: either solutions tend to the origin remaining in the sectors for all future time, or the nonlinear terms conspire to push the trajectories through the sectors, eventually coming out the other side. Hence we obtain the four possible cases shown in Figure 5.11. However, if the nonlinear terms satisfy the 'big O' conditions only the first case is possible, i.e. the improper node remains an improper node.

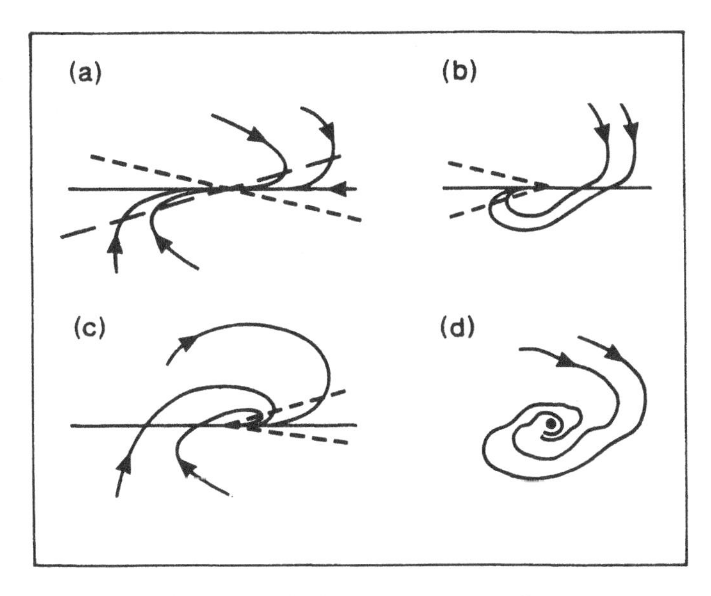


Fig. 5.11 Nonlinear improper nodes.

**(v) Foci.** The focus is the only completely unambiguous success in this program of investigating the effects of nonlinear perturbations of linear equations. The nonlinear equations are

$$\dot{r} = \rho r + P(r, \theta), \quad \dot{\theta} = \omega + Q(r, \theta) \tag{5.34}$$

where we assume without loss of generality that $\rho < 0$ and $\omega > 0$. Now, choosing $r$ small enough we can arrange that $|P(r, \theta)| < \epsilon r$ for some $\epsilon$ with $0 < \epsilon < -\rho$, and $|Q(r, \theta)| < \delta < \omega$. Hence there exist positive numbers $k_1$ and $k_2$ such that

$$\dot{r} \leq -k_1 r \text{ and } \dot{\theta} \geq k_2 \tag{5.35}$$

and so $r(t) \leq r_0 e^{-k_1 t}$, which implies that the origin is asymptotically stable and $\theta(t) \to \infty$ as $t \to \infty$. Hence the origin is a focus.

**(vi) Centres.** For a nonlinear centre the equations are

$$\dot{r} = P(r, \theta), \quad \dot{\theta} = \omega + Q(r, \theta) \tag{5.36}$$

and so, by the argument used above for the focus, the $\theta$ motion is unambiguous for sufficiently small $r$: $\theta(t) \to \infty$ as $t \to \infty$ if $\omega > 0$. Consider trajectories which start on the $y$-axis with coordinate $(x, y) = (0, y_0)$ in some small neighbourhood of the origin. There are three possibilities (see Fig 5.12). Since $\theta(t)$ is increasing to infinity there must exist some finite time after which the trajectory strikes the $y$-axis in $y > 0$ again for the first time, at $(0, y_1)$ say. If $y_1 < y_0$ (Fig 5.12a) then all trajectories which start on $(0, y)$ with $y_1 < y < y_0$ must return to the $y$-axis below $y_1$ and if $y_1 > y_0$ (Fig 5.12b) the opposite is true. Finally if $y_1 = y_0$ the trajectory is periodic. Hence the origin is either a stable focus, an unstable focus, a centre (in that all trajectories sufficiently close to the origin are periodic) or there is an infinite sequence of isolated periodic orbits which accumulate on the origin.

*Example 5.3*

Suppose

$$\dot{r} = -r^3, \quad \dot{\theta} = 1.$$

Then, integrating the $r$ equation, $r(t) = (2t + c)^{-\frac{1}{2}}$ where $c$ is a positive constant and so $r$ tends to zero as $t \to \infty$ and, of course, $\theta(t) = \theta_0 + t$. Hence the origin is a stable nonlinear focus.

*Example 5.4*

Suppose

$$\dot{x} = y, \quad \dot{y} = -x + x^3.$$

Note that if $(x(t), y(t))$ is a solution then so is $(-x(-t), y(-t))$, i.e. the equation is invariant under the (pseudo-)symmetry $(x, y, t) \to (-x, y, -t)$. Since the linearization about the origin is a centre an orbit started at $(0, y_0)$ with sufficiently small $y_0 > 0$ will strike the $y$-axis again at $(0, y_1)$ on the opposite side of the origin. Now, using the symmetry (Fig 5.13) the trajectory through $(0, y_1)$ comes back to the $y$-axis

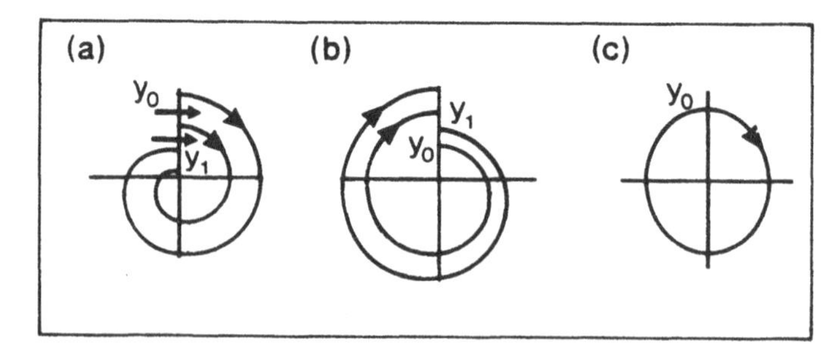


Fig. 5.12 Nonlinear centres.

at $(0, y_0)$ and so all orbits in a neighbourhood of the origin are periodic and the origin is a nonlinear centre. The existence of either a symmetry or a conserved energy-type function is frequently the only way of showing that a centre remains a centre under nonlinear perturbation.

*Example 5.5*

Suppose

$$\dot{x} = -y + x(x^2 + y^2)\sin(\log\sqrt{(x^2 + y^2)})$$

$$\dot{y} = x + y(x^2 + y^2)\sin(\log\sqrt{(x^2 + y^2)})$$

or

$$\dot{r} = r^3 \sin(\log r), \quad \dot{\theta} = 1.$$

This has periodic orbits whenever $\sin(\log r) = 0$, i.e. whenever $\log r = \pm n\pi$. Hence there is an infinite sequence of isolated periodic orbits with radii $r_n = e^{-n\pi}$, $n = 1, 2, 3, ...$ which accumulate on the origin.

## 5.3 Rabbits and sheep

Models from population dynamics provide a rich source of nonlinear behaviour. These models, introduced by Lotka and Volterra, represent the most simple models of nonlinear interaction between species. We

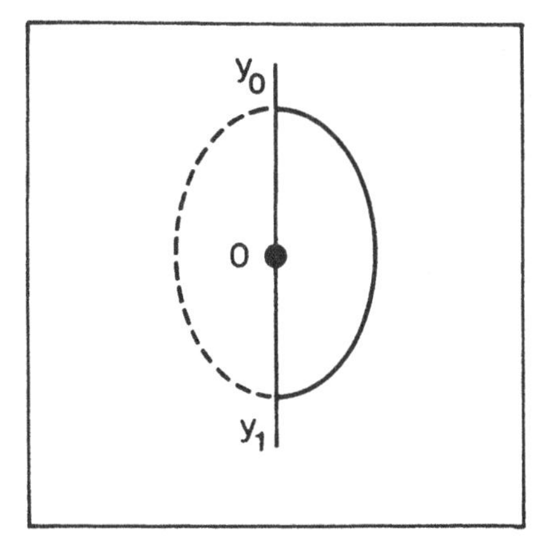


Fig. 5.13 A nonlinear centre by symmetry.

shall consider a grassy island with two species of animal, which may be in competition for the island's resources (rabbits and sheep) or one may prey on the other (wolves and sheep). For simplicity we assume that these two populations are large, so that the number of individuals of a population can be taken to be a real (positive!) number, with the understanding that if the number of sheep, $s(t)$, is small then $s(t)$ is interpreted as the number of sheep divided by some large number, so $s = 1$ might represent a population of 100,000 sheep.

Suppose that there is a grassy island supporting populations of two species, $x$ and $y$. If the populations are large then it is reasonable to let the normalized populations be continuous function of time. We propose a simple model of the change in population of the form

$$\dot{x} = x(A + a_1 x + b_1 y) \tag{5.37a}$$

$$\dot{y} = y(B + b_2 x + a_2 y) \tag{5.37b}$$

where $A$, $B$, $a_i$ and $b_i$ are constants. These equations can be interpreted as the rate of change of the population equals the present population multiplied by (the birth rate – the death rate). Consider the $x$ equation when $y = 0$: $\dot{x} = x(A+a_1 x)$. The coefficient $a_1$ describes the interaction of the species with itself and is negative since the island is finite and so large populations suffer from overcrowding. On the other hand $A > 0$ if the species eats grass (so the population increases if the initial population is small) and $A \leq 0$ if the species preys on the second species (since if $y = 0$ there is no available food and the population dies of starvation). Finally, the coefficient $b_1$ describes the effect of species $y$ on species $x$. If $b_1 > 0$ then this term increases the rate of population growth, for example if $x$ feeds upon $y$, whilst if $b_1 < 0$ this term decreases the population growth of $x$, for example if $x$ and $y$ compete for the same resources. Similar interpretations hold for $B$, $a_2$ and $b_2$. This means that there are four classes of population models, depending on the signs of $b_1$ and $b_2$: if $b_i > 0$, $i = 1, 2$, both populations benefit each other (a symbiotic relationship), if $b_i < 0$, $i = 1, 2$, both populations inhibit each other (competitive species), whilst if $b_1 < 0$ and $b_2 > 0$ we have a predator-prey model where $x$ is the predator, and if $b_1 > 0$ and $b_2 < 0$ the situation is the same but $y$ is the predator. These are two species models, and of course they can be generalized to $N$ species with populations $x_i(t)$ which satisfy

$$\dot{x}_i = x_i \left( A_i + \sum_{k=0}^{N} a_{ik} x_k \right) \tag{5.38}$$

for $i = 1, 2, ..., N$ with $a_{ii} < 0$ and the signs of the remaining coefficients depend upon the relationships between the various species.

These models provide examples which can be treated using the fairly basic knowledge about the nature of stationary points that has already been established. The strategy is first, to locate the stationary points, then determine their type and find the relevant eigenvectors of the linearization about each stationary point and finally to join together this local information into a convincing global phase diagram.

(5.1) EXERCISE

*Show that the x- and y-axes are invariant for these population models. Why should this be a necessary feature of a population model?*

We shall illustrate this technique by considering an example with two competitive species: rabbits, $r$, and sheep, $s$.

*Example 5.6*

Consider the model

$$\dot{r} = r(3 - r - 2s), \quad \dot{s} = s(2 - r - s)$$

with $r, s \geq 0$.

**Step 1.** To find the stationary points we need to solve

$$r(3 - r - 2s) = 0 \text{ and } s(2 - r - s) = 0$$

in the positive quadrant. This is a straightforward process (solving a pair of simultaneous equations) and gives four stationary points at $(r, s)$ equal to

$$(0,0) \quad (0,2) \quad (3,0) \quad \text{and} \quad (1,1).$$

**Step 2.** To determine the type of each stationary point we need the Jacobian matrix $Df(r, s)$. Differentiating the defining equations we get

$$Df(r,s) = \begin{pmatrix} 3 - 2r - 2s & -2r \\ -s & 2 - r - 2s \end{pmatrix}.$$

**Step 3.** We now need to evaluate the eigenvalues and eigenvectors of the Jacobian matrix at each stationary point. If the stationary point is a node it is important to determine the eigenvector to which trajectories are tangential as they approach or leave a neighbourhood of the stationary point (cf. Fig. 5.1).

At $(0,0)$, $Df(0,0) = \begin{pmatrix} 3 & 0 \\ 0 & 2 \end{pmatrix}$ and so the eigenvalues are $\lambda_1 = 3$ and $\lambda_2 = 2$ with eigenvectors $e_1 = \begin{pmatrix} 1 \\ 0 \end{pmatrix}$ and $e_2 = \begin{pmatrix} 0 \\ 1 \end{pmatrix}$. So $(0,0)$ is an unstable node and trajectories leave tangential to the $e_2$ eigenvector, since this corresponds to the eigenvalue of smallest modulus.

At $(0,2)$, $Df(0,2) = \begin{pmatrix} -1 & 0 \\ -2 & -2 \end{pmatrix}$ with eigenvalues $\lambda_1 = -1$ and $\lambda_2 = -2$ and corresponding eigenvectors $e_1 = (1,-2)^T$ and $e_2 = (0,1)^T$. Hence $(0,2)$ is a stable node and trajectories tend to $(0,2)$ tangential to the $e_1$ axis.

At $(3,0)$, $Df(3,0) = \begin{pmatrix} -3 & -6 \\ 0 & -1 \end{pmatrix}$ with eigenvalues $\lambda_1 = -3$ and $\lambda_2 = -1$ and corresponding eigenvectors $e_1 = (1,0)^T$ and $e_2 = (3,-1)^T$. Hence $(3,0)$ is another stable node and trajectories tend to $(3,0)$ tangential to the $e_2$ axis.

At $(1,1)$, $Df(1,1) = \begin{pmatrix} -1 & -2 \\ -1 & -1 \end{pmatrix}$ with eigenvalues $\lambda_{\pm} = -1 \pm \sqrt{2}$ and eigenvectors $e_{\pm} = (1, \mp\frac{1}{\sqrt{2}})$. Hence $(1,1)$ is a saddle, with linear stable manifold in the direction of $e_-$ and linear unstable manifold in the direction of $e_+$.

**Step 4.** Guess the global phase portrait from the local analysis. This is a little like joining up the dots in puzzle books, and gives a picture like the one shown in Figure 5.14: the stable manifold of $(1,1)$ divides the positive quadrant into two regions, everything below this curve tends to the stationary point at $(3,0)$ and everything above tends to $(0,2)$. Thus for this choice of the parameters we find that one of the species always dies out, but that the question as to which one dies is determined by the initial populations. Furthermore, very small changes of the initial condition near the stable manifold of the saddle lead to radically different asymptotic steady states.

This section has given a simple way of getting some idea of the behaviour of simple nonlinear models using the linearization results near stationary points. It is easy to implement, but has some drawbacks. The linearization determines the flow in a neighbourhood of the stationary

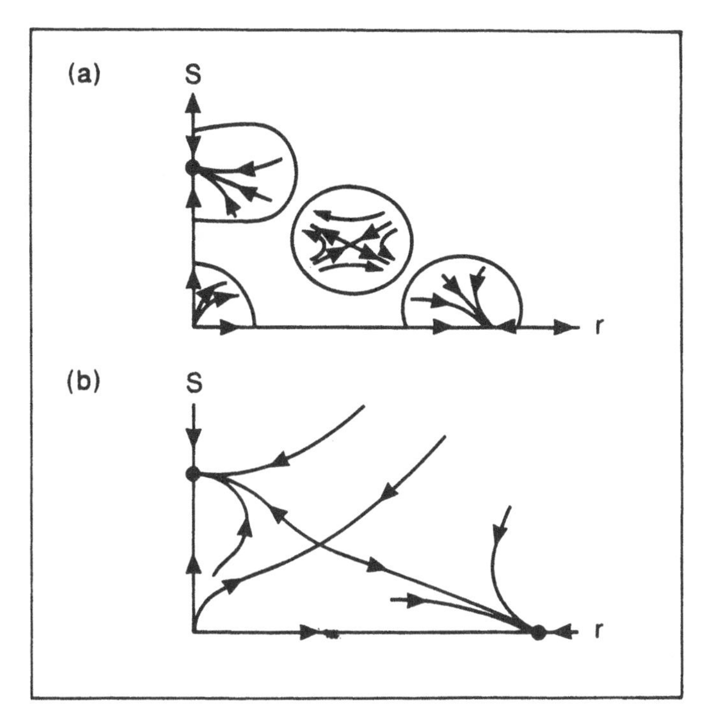


Fig. 5.14 (a) Local patches; (b) global phase portrait.

points, and away from the stationary points trajectories are locally parallel. When joining together patches there are often few topologically distinct solutions if this is done in such a way that no new stationary points are introduced. Indeed, if trajectories come in from infinity *and all stationary points are hyperbolic and there are no periodic orbits* then there must be five or more stationary points before any choice becomes possible. Figure 5.15 shows a flow for which more than one choice of patching the linear neighbourhoods together is possible. There is a symmetric solution (not drawn) and two asymmetric solutions, one of which is illustrated. Can you find other topologically distinct phase portraits for this configuration of stationary points? [Hint: the behaviour of the stable and unstable manifolds of the saddles determines the flow.]

The problems due to periodic orbits are illustrated in Figure 5.16. Suppose the flow has a stable focus, so in a neighbourhood of this stationary point trajectories are spiralling into the point. Then we cannot guarantee that there are no periodic orbits as shown in either the second or third sketch of Figure 5.16. This suggests that we need to do more

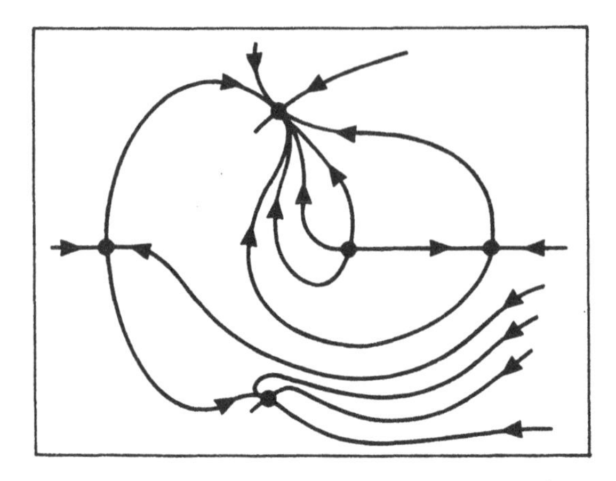

Fig. 5.15 One of the consistent non-symmetric configurations.

work on necessary conditions for the existence of periodic solutions in two dimensions.

## 5.4 Trivial linearization

If an eigenvalue of the linearized flow about a stationary point vanishes then the local dynamics is dominated by the nonlinear terms in the equation. There are no general techniques for finding the phase portrait in such cases and we need to use any bits of information which are

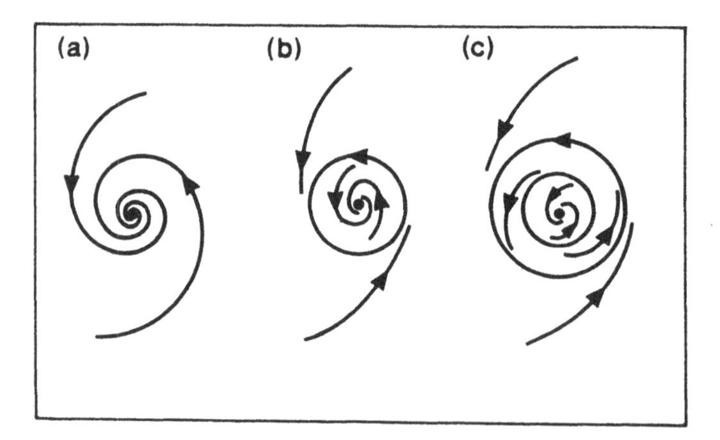


Fig. 5.16 Local behaviour and periodic orbits.

available. There are four approaches which can be used, depending upon the situation, with varying degrees of success.

1. Look at $\frac{dy}{dx}$ on trajectories. If this can be solved then we obtain the equation of trajectories in phase space which can then be sketched.
2. Find the curves in phase space on which $\dot{x} = 0$ and $\dot{y} = 0$ and hence divide phase space into regions where $\dot{x} > 0$ and $\dot{y} > 0$, $\dot{x} < 0$ and $\dot{y} > 0$ and so on. Use this information to obtain a rough picture of the behaviour of trajectories in phase space.
3. Transform into polar coordinates and determine the curves on which $\dot{r} = 0$ and $\dot{\theta} = 0$. These curves divide phase space into regions where $\dot{r} > 0$ and $\dot{\theta} > 0$, $\dot{r} < 0$ and $\dot{\theta} > 0$ and so on. Now, as in approach 2, use this information to sketch the phase portrait.
4. Look for a first integral or, more generally, some invariant curve which will help organize the sketch of phase space.

None of these approaches is truly satisfactory, but if they are used together it is often possible to get a good idea of how trajectories behave.

*Example 5.7*

Consider the equations

$$\dot{x} = x, \quad \dot{y} = y^2.$$

The linearization about the origin obviously has a zero eigenvalue and so none of the robust types of linear stationary point described in Section 5.2 are applicable here. However, on trajectories

$$\frac{dy}{dx} = \frac{y^2}{x}$$

and so we can integrate this equation to obtain

$$-\frac{1}{y} = \log x + c$$

for some constant $c$, or

$$e^{-\frac{1}{y}} = Ax$$

for some constant $A$. Noting that both the $x$- and $y$-axes are invariant (since if $x = 0$, $\dot{x} = 0$ and if $y = 0$ then $\dot{y} = 0$) we obtain the phase portrait shown in Figure 5.17a, which seems to be half saddle and half node.

*Example 5.8*

The nonlinear differential equation

$$\dot{x} = x^3 - 2xy^2, \quad \dot{y} = 2x^2y - y^3$$

can be written in polar coordinates as

$$\dot{r} = r^3 \cos 2\theta, \quad \dot{\theta} = 2r^2 \sin 2\theta.$$

Using method 3 above we note that $\dot{r} = 0$ on the diagonal and anti-diagonal lines $x = y$ and $x = -y$. Furthermore $\dot{r} > 0$ if $x^2 > y^2$ and $\dot{r} < 0$ if $x^2 < y^2$. Similarly, $\dot{\theta} = 0$ on the lines $x = 0$ and $y = 0$ with $\dot{\theta} > 0$ if $xy > 0$ and $\dot{\theta} < 0$ if $xy < 0$. These remarks are illustrated in Figure 5.17b. Now consider a trajectory starting in the positive quadrant near the $x$-axis. Initially both $r$ and $\theta$ increase, until at some time the trajectory crosses the line $x = y$. Thereafter $r$ decreases whilst $\theta$ continues to increase. Hence, asymptotically, $\theta$ must tend to $\frac{\pi}{2}$ whilst $r$ tends to the origin. A similar argument elsewhere leads to the phase portrait sketched in Figure 5.17c.

## 5.5 The Poincaré index

The Poincaré index is an invariant of closed curves (not necessarily trajectories) in phase space which can be used to rule out the possibility of periodic orbits in certain systems.

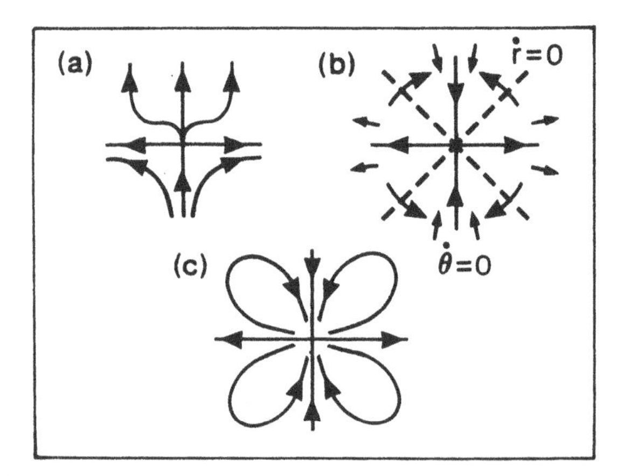


Fig. 5.17 (a) Example 5.7; (b) isoclines for Example 5.8; (c) Example 5.8.

Consider the system $(\dot{x}, \dot{y}) = (f_1(x,y), f_2(x,y))$. At each point the vector field $(f_1(x,y), f_2(x,y))$ defines a direction or angle,

$$\psi = \tan^{-1}\left(\frac{f_2(x,y)}{f_1(x,y)}\right). \tag{5.39}$$

Now let $\Gamma$ be any simple closed curve in the plane, then moving around $\Gamma$ we see that $\psi$ changes continuously and that when we return to the original starting point the value of $\psi$ has changed by a multiple of $2\pi$. This multiple, which may be positive or negative, is called the Poincaré index of $\Gamma$, $I_\Gamma$. This idea can be expressed more mathematically by noting that

$$I_\Gamma = \frac{1}{2\pi}\int_\Gamma d\psi = \frac{1}{2\pi}\int_\Gamma d\left(\tan^{-1}\left(\frac{f_2(x,y)}{f_1(x,y)}\right)\right)$$

since the change of the inverse tangent of $f_2/f_1$ about the curve is precisely how the index is defined. Now, since $\frac{d}{dx}\tan^{-1}x = 1/(1+x^2)$ we find

$$d\left(\tan^{-1}\left(\frac{f_2(x,y)}{f_1(x,y)}\right)\right) = \frac{f_1 df_2 - f_2 df_1}{f_1^2 + f_2^2}$$

and so

$$I_\Gamma = \frac{1}{2\pi}\int_\Gamma \frac{f_1 df_2 - f_2 df_1}{f_1^2 + f_2^2}, \tag{5.40}$$

which is an integer! We can, of course, go on to write

$$df_i = \frac{\partial f_i}{\partial x}dx + \frac{\partial f_i}{\partial y}dy$$

to get an expression for the index in terms of an integral with respect to $x$ and $y$; we leave this as an exercise. These two formulations of the Poincaré index (as the change in angle of the vector field around the curve and as a line integral) allow us to state various properties of the Poincaré index.

(5.2) LEMMA (PROPERTIES OF THE POINCARÉ INDEX)

i) *If $\Gamma'$ is a closed curve obtained from $\Gamma$ by a continuous deformation which does not cross any stationary points, then $I_{\Gamma'} = I_\Gamma$.*
ii) *The index of a periodic orbit is $+1$.*
iii) *The index of a node, focus or centre is $+1$.*
iv) *The index of a saddle is $-1$.*
v) *The index is additive, so if $\Gamma_3 = \Gamma_1 + \Gamma_2$ in the sense defined in Figure 5.18a then $I_{\Gamma_3} = I_{\Gamma_1} + I_{\Gamma_2}$.*

vi) *The index of a closed curve is the sum of the indices of the stationary points inside the curve.*
vii) *The index of a closed curve containing no stationary points is 0.*
viii) *The index of a curve is unchanged if $(f_1, f_2)$ is replaced by $(-f_1, -f_2)$.*

The first property follows from the continuity of the line integral provided $f_1^2 + f_2^2 \neq 0$. The line integral takes integer values and so if it is continuous it must be constant. Note that this result allows us to define the index of a stationary point as the index of any curve containing the stationary point and no other stationary points. The following three properties follow by inspection (draw a circle about a stationary point and move around the circle with a pencil showing the direction of the vector field, what angle does the pencil move through?). Property (v) is a direct consequence of the line integral formulation of the index and properties (vi) and (vii) follow from (v). Property (viii) is again a direct consequence of the line integral representation.

(5.3) COROLLARY

*If $\Gamma$ is a periodic orbit then $\Gamma$ encloses at least one stationary point. If the stationary points are hyperbolic then $\Gamma$ encloses $2n+1$ stationary points (for some $n \geq 0$) $n$ of which are saddles, the remainder are sinks or sources.*

(5.4) EXERCISE

*Prove this corollary.*

(5.5) EXERCISE

*Let the index of a manifold equal the sum of the indices of the stationary points on the manifold (including the point at infinity). Show that the index of the plane is 2. [Hint: let $\Gamma$ be a curve enclosing all the finite stationary points. To find the index at infinity work in complex variable $z$ and then invert the plane using the variable $w = z^{-1}$.]*

(5.6) EXERCISE

*Suppose that $f_1$ and $f_2$ are quadratic functions of $x$ and $y$, and that the index at infinity is $+1$. If there are no periodic orbits and all stationary points are hyperbolic, show that the nature of the flow is completely determined by the linearizations of the flow about the stationary points.*

Corollary 5.3 gives us a useful tool for showing that periodic orbits cannot exist in certain configurations (see Fig. 5.18).

## 5.6 Dulac's criterion

The divergence theorem provides another way of proving the non-existence of periodic orbits in some regions of phase space. Recall that if $\Gamma$ is a simple closed curve with outward normal $n$ enclosing a region $R$ and $f : \mathbf{R}^2 \to \mathbf{R}^2$ is a continuously differentiable vector field and $g : \mathbf{R}^2 \to \mathbf{R}$ is a continuously differentiable function then the divergence theorem states that

$$\int_\Gamma g(n.f)dl = \int\int_R \nabla.(gf)dxdy, \qquad (5.41)$$

where $gf$ is a vector, $g(x, y)f(x, y)$, and should not be confused with the composition of $g$ and $f$.

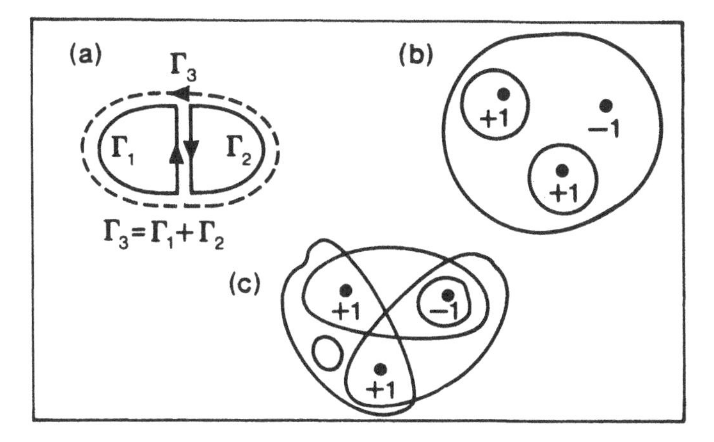


Fig. 5.18 (a) Addition of curves; (b) allowed positions; (c) closed curves which cannot be periodic orbits.

(5.7) LEMMA (DULAC'S CRITERION)

*If there exists a continuously differentiable function $g : \mathbf{R}^2 \to \mathbf{R}$ such that $\nabla.(gf)$ is continuous and non-zero on some simply connected domain $D$, then no periodic orbit can lie entirely in $D$.*

*Proof*: Suppose a periodic orbit $\Gamma$ does lie entirely in $D$. Then

$$\int\int_A \nabla.(gf)dxdy \neq 0 \tag{5.42}$$

where $A$ is the area bounded by $\Gamma$, since $\nabla.(gf)$ is either strictly greater than zero or strictly less than zero throughout $A$. But a periodic orbit is a trajectory, and hence tangential to the vector field, $f$. So $n.f = 0$, where $n$ is the outward normal to the periodic orbit. Hence

$$\int_\Gamma g(n.f)dl = 0 \tag{5.43}$$

giving a contradiction by the divergence theorem. If $g = 1$ then this result is sometimes referred to as the divergence test.

*Example 5.9*

Consider the Lotka-Volterra model.

$$\dot{x} = x(A - a_1 x + b_1 y), \quad \dot{y} = y(B - a_2 y + b_2 x)$$

where $a_i > 0$, $i = 1, 2$, since the island (or field) is finite and these terms model the effect of overcrowding. Do there exist such models which allow for periodic cycles in the populations? Consider the weighting function $g(x, y) = (xy)^{-1}$. Then

$$gf = (y^{-1}(A - a_1 x + b_1 y), x^{-1}(B - a_2 y + b_2 x))$$

and so

$$\nabla.(gf) = -a_1 y^{-1} - a_2 x^{-1},$$

which is less than zero for positive $x$ and $y$. Hence there are no periodic orbits in the positive quadrant.

## 5.7 Pike and eels

The previous example shows that there cannot be periodic variations in the populations of Lotka-Volterra models unless the coefficients $a_i$,

$i = 1, 2$ are both zero. We now want to determine whether there can be cyclic populations if this criterion is satisfied. We consider an infinite pond containing a population of pike ($x$) and eels ($y$) where we assume that the pike prey on the eels and the eels have no problem finding whatever natural resources they need to survive. Their only problem is the existence of large and hungry pike in the pond. Hence, in the absence of any pike ($x = 0$) the population of eels increases indefinitely: $\dot{y} = By$, $B > 0$. The pike need the eels to survive so in the absence of eels ($y = 0$) $\dot{x} = -Ax$, $A > 0$. We now introduce the simplest coupling between the creatures consistent with their predator-prey relationship giving the differential equation

$$\dot{x} = x(-A + b_1 y), \quad \dot{y} = y(B - b_2 x) \tag{5.44}$$

where $b_i > 0$, $i = 1, 2$. Thus if the number of pike is large the number of eels decreases (eaten alive) whilst if the number of eels is small the number of pike decreases (starved to death). This equation has only two stationary points, the origin, $(x, y) = (0, 0)$, and a non-trivial population level $(x, y) = (\frac{B}{b_2}, \frac{A}{b_1})$. The Jacobian matrix is

$$\begin{pmatrix} -A + b_1 y & b_1 x \\ -b_2 y & B - b_2 x \end{pmatrix}$$

and so the origin is a saddle (the $x$-direction is stable and the $y$-direction unstable) whilst the linearization of the flow about the non-trivial stationary point has purely imaginary eigenvalues, so the linearization gives a centre.

(5.8) EXERCISE

*Verify the linear types of the two stationary points.*

Since the linearization about the non-trivial stationary point yields a centre we need to do a little more work to determine the type of the stationary point for the full (nonlinear) equations. For general systems of nonlinear equations this can be very hard (see Chapter 8), but here we can notice that on trajectories

$$\frac{dy}{dx} = \frac{y(B - b_2 x)}{x(-A + b_1 y)} \tag{5.45}$$

and so we can separate variables to obtain

$$\int \frac{(-A + b_1 y)}{y} dy = \int \frac{(B - b_2 x)}{x} dx$$

or

$$-A \log y + b_1 y - B \log x + b_2 x = E(x, y) \tag{5.46}$$

is constant on trajectories. To sketch curves of constant $E$ we begin by finding the maxima, minima and saddle points of $E$. These are solutions of $\frac{\partial E}{\partial x} = \frac{\partial E}{\partial y} = 0$. A straightforward exercise in differentiation shows that there is only one non-trivial solution to this equation: $(x, y) = (\frac{B}{b_2}, \frac{A}{b_1})$, the stationary point of the flow. Furthermore this is a global minimum of the function $E$. Hence trajectories (i.e. level sets of $E$) are simply closed curves around the non-trivial stationary point of the flow and all trajectories in the (strictly) positive quadrant except this stationary point are periodic (see Fig. 5.19).

This result is a little surprising, and in Chapter 9 we will show that in models with three different populations it is possible to find isolated periodic orbits (which the above argument and the example at the end of the previous section show are impossible for two species population models).

## 5.8 The Poincaré-Bendixson Theorem

The results of Sections 5.5 and 5.6 have given negative criteria: ways of proving that periodic orbits do not exist. The Poincaré-Bendixson Theorem is a beautiful result which allows one to prove the existence of at least one periodic orbit under certain assumptions. The idea is to

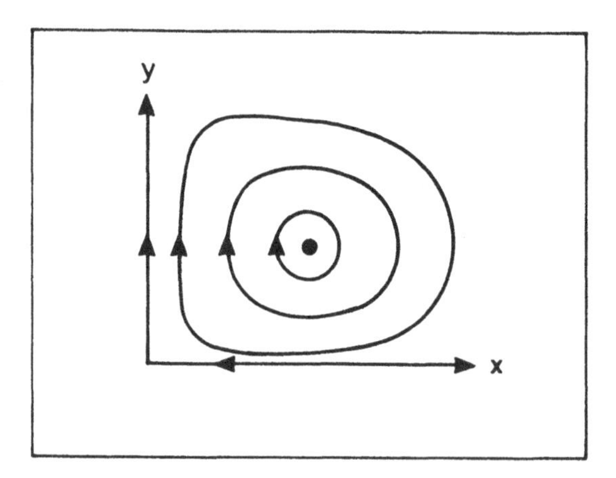


Fig. 5.19 Pike and eels.

find a 'trapping region' which does not contain any stationary points but which does contain limit points of the flow and then to show that these limit points lie on a periodic orbit. To prove this result we need a definition and a couple of (easy) technical lemmas.

(5.9) DEFINITION

*A local transversal is a line segment which all trajectories cross from the same side (see Fig. 5.20).*

(5.10) LEMMA

*If $x_0$ is not a stationary point then it is always possible to construct a local transversal in a neighbourhood of $x_0$.*

*Proof*: If $x_0$ is not a stationary point then we can choose coordinates $y$ in the direction of the flow at $x_0$ and $z$ orthogonal to $y$ such that the flow takes the form

$$\dot{y} = a + O(|(y,z) - x_0|), \quad \dot{z} = O(|(y,z) - x_0|) \tag{5.47}$$

by expanding the vector field as a Taylor series about $x_0$. Here $a$ is a non-zero constant (equal to the magnitude of the vector field at $x_0$). Now simply choose a line through $x_0$ in the $y$ direction. Since $a \neq 0$ all trajectories which cross this line sufficiently close to $x_0$ do so in the same direction (with increasing $y$ if $a > 0$ and decreasing $y$ if $a < 0$).

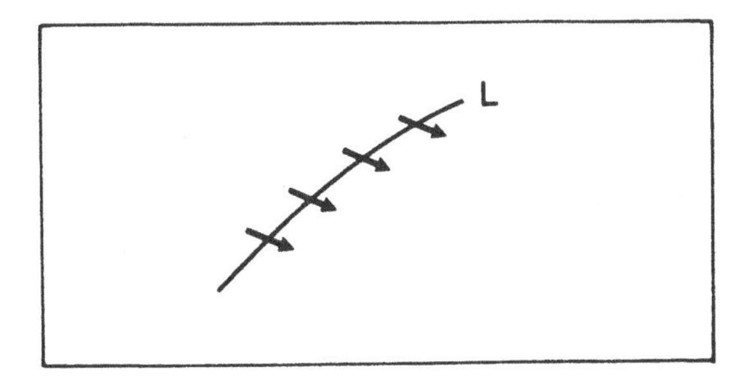


Fig. 5.20 L is a local transversal.

The important property of local transversals in $\mathbf{R}^2$, which is no longer true in higher dimensions or even for flows on the torus, is that the successive intersections of a trajectory move monotonically along the local transversal.

(5.11) LEMMA

*If a trajectory $\gamma(x)$ intersects a local transversal several times, the successive crossing points move monotonically along the local transversal.*

*Proof*: Consider two successive crossings. The geometry must be as shown in one of the sketches in Figure 5.21. In all cases $\gamma(x)$ leaves the shaded region and cannot return, so the next crossing is further away.

This is (quite literally) a sketch proof; note that it relies heavily on the Jordan curve lemma (that a closed curve in the plane separates an inside from an outside).

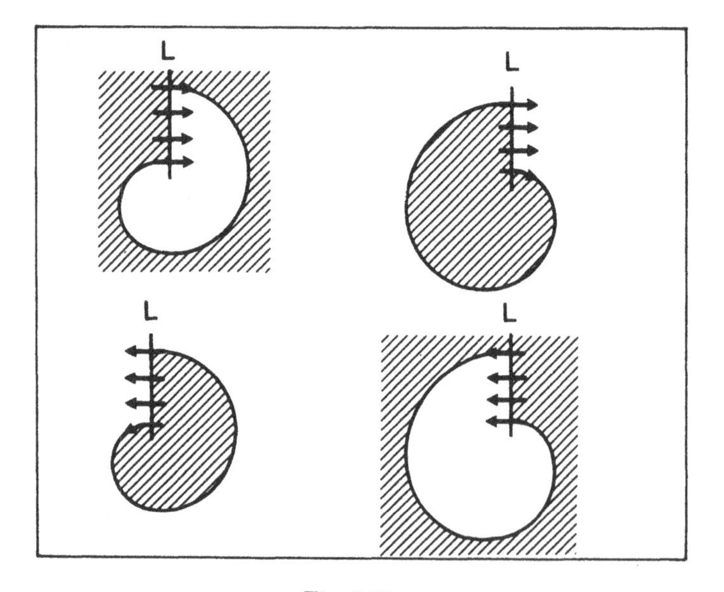


Fig. 5.21

(5.12) COROLLARY

*If $x \in \Lambda(x_0)$ is not a stationary point and $x \in \gamma(x_0)$, then $\gamma(x)$ is a closed curve.*

*Proof*: Since $x \in \gamma(x_0)$, $\Lambda(x) = \Lambda(x_0)$, and so $x \in \Lambda(x_0)$ implies that $x \in \Lambda(x)$. Choose $\Sigma$ to be a local transversal through $x$, then there is an increasing sequence $(t_i)$, $t_i > t_{i-1} > 0$, such that $\varphi(x, t_i) \to x$ as $i \to \infty$ and $\varphi(x, t_i) \in \Sigma$. Also, $\varphi(x, 0) = x$. Suppose that $\varphi(x, t_1) \neq x$, then successive intersections of $\varphi(x, t)$ with $\Sigma$ are bounded away from $x$ by Lemma 5.11, which contradicts the fact that $x \in \Lambda(x)$. Hence $\varphi(x, t_1) = x$ and $\gamma(x)$ is periodic.

We can now state and prove the Poincaré-Bendixson Theorem.

(5.13) THEOREM (POINCARÉ-BENDIXSON)

*Suppose that $\gamma(x_0)$ enters and does not leave some closed, bounded domain $D$ (so $\varphi(x_0, t) \in D$ for all $t \geq T$, some $T \geq 0$) and that there are no stationary points in $D$. Then there is at least one periodic orbit in $D$, and this orbit is in the $\omega$-limit set of $x_0$.*

The strategy of the proof is simple: the $\omega$-limit set of $x_0$ is in $D$ so take a point in this set, $x$. If $x$ is in the trajectory of $x_0$ we are finished by Corollary 5.12, otherwise consider the trajectory through $x$, which also has limit points in $D$. Construct a local transversal through such a limit point, and now show that this limit point must be in the trajectory through $x$ (and hence periodic). If it is not in the trajectory through $x$, this trajectory must move monotonically along the local transversal through its limit point. But the trajectory through $x$ is also in the limit set of $x_0$ and so the trajectory through $x_0$ must be able to move up and down the transversal: a contradiction. Let us write this down properly.

*Proof*: Since $\gamma(x_0)$ enters and does not leave the compact domain $D$, $\Lambda(x_0)$ is non-empty and is contained in $D$. Take $x \in \Lambda(x_0)$ and note that $x$ is not stationary since there are no stationary points in $D$. There are two cases:

i) $x \in \gamma(x_0)$. In this case $\gamma(x)$ is periodic by Corollary 5.12.
ii) $x \notin \gamma(x_0)$. Since $x$ is in $\Lambda(x_0)$, $\gamma^+(x) \subset \Lambda(x_0)$ and so $\gamma^+(x) \subset D$. Hence $x$ has a limit point, $x^*$, in $D$. If $x^* \in \gamma^+(x)$ then $\gamma(x^*)$ is

closed by Corollary 5.12, so we need to show that $x^* \notin \gamma^+(x)$ leads to a contradiction.

Choose a local transversal, $\Sigma$, through $x^*$. Since $x^* \in \Lambda(x)$ the trajectory through $x$ must intersect $\Sigma$ at points $p_1, p_2, \ldots$ which accumulate monotonically on $x^*$. But $p_i \in \Lambda(x_0)$ so it must be possible for $\gamma(x_0)$ to pass arbitrarily close to $p_i$, then $p_{i+1}$ and then $p_i$ again. But this implies that the intersections of $\gamma(x_0)$ with $\Sigma$ do not move monotonically along $\Sigma$, a contradiction.

To apply the Poincaré-Bendixson Theorem we need to find a region $D$ which contains no stationary points and which trajectories enter but do not leave. Typically, $D$ will be an annular region, with a source in the hole in the middle (so trajectories enter the region on the inner boundary) and the outer boundary is chosen so that the radial component of the vector field is always inwards on this boundary (see Fig. 5.22). This is illustrated in the following example.

*Example 5.10*

Consider the equation

$$\dot{x} = y + \tfrac{1}{4}x(1 - 2r^2), \quad \dot{y} = -x + \tfrac{1}{2}y(1 - r^2)$$

where $r^2 = x^2 + y^2$. To show that this has at least one periodic orbit using the Poincaré-Bendixson Theorem we need to find a region which

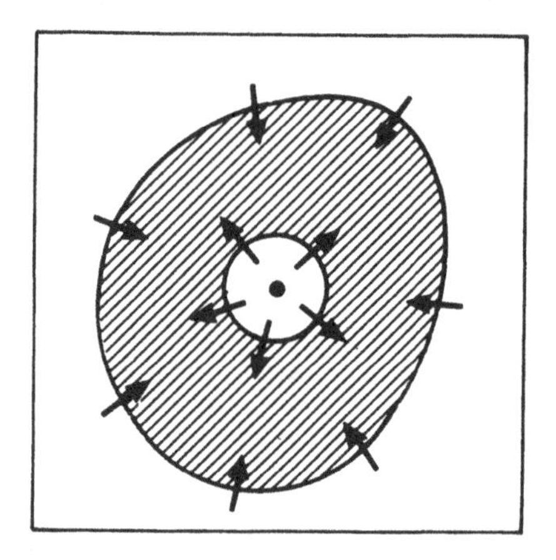

Fig. 5.22 A trapping region.

contains no stationary points of the flow and which trajectories enter but do not leave. From the $\dot{x}$ equation a stationary point must satisfy

$$y = -\tfrac{1}{4}x(1-2r^2)$$

and substituting this expression for $y$ into the equation $\dot{y}=0$ gives

$$0 = -x(1+\tfrac{1}{8}(1-2r^2)(1-r^2)).$$

Hence there is a stationary point at the origin and for any other stationary point $r^2$ must satisfy the equation

$$2r^4 - 3r^2 + 9 = 0.$$

This quadratic equation for $r^2$ has no real roots (as the descriminant is $3^2 - 4.2.9$, which is negative) and so the origin is the only stationary point of the equations. Now consider the equation for $r$. Differentiating the expression $r^2 = x^2 + y^2$ gives $r\dot{r} = x\dot{x} + y\dot{y}$ and so

$$\begin{aligned} r\dot{r} &= x(y + \tfrac{1}{4}x(1-2r^2)) + y(-x + \tfrac{1}{2}y(1-r^2)) \\ &= \tfrac{1}{4}x^2(1-2r^2) + \tfrac{1}{2}y^2(1-r^2) \\ &= \tfrac{1}{4}r^2(1+\sin^2\theta) - \tfrac{1}{2}r^4. \end{aligned}$$

Dividing through by $r$ we find that

$$\dot{r} = \tfrac{1}{4}r(1+\sin^2\theta) - \tfrac{1}{2}r^3.$$

This equation is relatively straightforward to analyze. First note that $\dot{r} > 0$ for all $\theta$ provided $\frac{1}{4}r - \frac{1}{2}r^3 > 0$, i.e. provided $r^2 < \frac{1}{2}$ and that $\dot{r} < 0$ for all $\theta$ provided $r - r^3 < 0$, i.e. provided $r > 1$. Now define the annular region $D$ on which we are going to apply the Poincaré-Bendixson Theorem by

$$D = \left\{(r,\theta) | \tfrac{1}{2} \le r \le 1\right\}.$$

Since the only stationary point is the origin there are no stationary points in $D$, and since $\dot{r} > 0$ if $r < \frac{1}{2}$ and $\dot{r} < 0$ if $r > 1$ all trajectories in $D$ remain in $D$ for all time. Hence, by the Poincaré-Bendixson Theorem there is at least one periodic orbit in $D$.

## 5.9 Decay of large amplitudes

In this section we consider a class of examples for which it is possible to prove the existence of a domain $D$ on which the conditions of the Poincaré-Bendixson Theorem hold. These are a class of nonlinear oscillators

$$\ddot{x} + f(x)\dot{x} + g(x) = 0 \tag{5.48}$$

or in Liénard coordinates

$$\dot{x} = y - F(x), \quad \dot{y} = -g(x), \quad F(x) = \int_0^x f(\xi)d\xi. \tag{5.49}$$

We shall assume that $g(0) = 0$, so $(0,0)$ is a stationary point of the flow. The stability of this stationary point is determined by the eigenvalues of the Jacobian matrix

$$\begin{pmatrix} -f(0) & 1 \\ -\frac{dg}{dx}(0) & 0 \end{pmatrix}.$$

It is a simple exercise to show that the eigenvalues of this matrix both have positive real parts (i.e. $(0,0)$ is a source) if $\frac{dg}{dx}(0) > 0$ and $f(0) < 0$. Hence, if these conditions are satisfied there all trajectories leave some small neighbourhood $U$ of the origin. $\partial U$ will be the inner boundary of the region $D$ on which we can apply the Poincaré-Bendixson Theorem. To construct the outer boundary we need a few more conditions. Let us state these conditions in a theorem.

(5.14) THEOREM

*Consider $\ddot{x} + f(x)\dot{x} + g(x) = 0$ and suppose that*

i) *$xg(x) > 0$ for $x \neq 0$, $g(0) = 0$, $f(0) < 0$ and $\frac{dg}{dx}(0) > 0$;*
ii) *$sign(x)F(x) > k > 0$ for sufficiently large $|x|$;*
iii) *$G(x) \to \infty$ as $|x| \to \infty$; where $G(x) = \int_0^x g(\xi)d\xi$. Then the system has at least one periodic orbit.*

*Proof*: By (i) $(0,0)$ is the only stationary point of the equation and it is a source by the argument above the statement of the theorem. Hence we can find some small closed curve around $(0,0)$ which trajectories cross outwards. To find a region $D$ on which to apply the Poincaré-Bendixson Theorem we have to construct an outer boundary which trajectories cannot cross in the outward direction. This is done in three parts (see Fig. 5.23).

First, we choose $x_0$ and $y_0$ large enough so that the trajectory through $(-x_0, y_0)$ strikes the line $x = x_0$ above $y = k$ with $y = y_1$, $y_0 - k \leq y_1 \leq y_0 + k$. Second, we show that there is a bounded curve in $x > x_0$

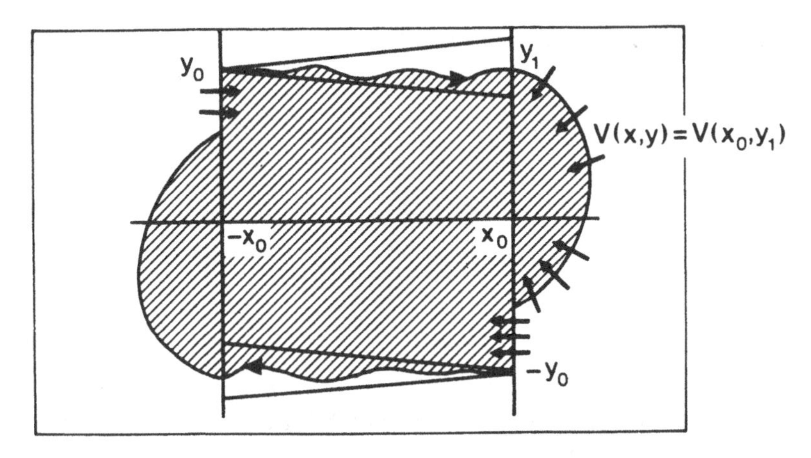


Fig. 5.23 Construction of the trapping region.

from $(x_0, y_1)$ to $(x_0, -y_2)$, $y_2 < y_0$, which trajectories cross inwards. Furthermore, the line segment from $(x_0, -y_2)$ to $(x_0, -y_0)$ is a local transversal which trajectories cross from right to left. Finally, we repeat the argument from $(x_0, -y_0)$ back to the line $x = -x_0$ to obtain the trapping region of Figure 5.23. This region contains no stationary points and trajectories enter but do not leave it, so by the Poincaré-Bendixson Theorem there is at least one periodic orbit in the region. Let us write this out properly.

**Step 1.** Note that for trajectories $\frac{dy}{dx} = \frac{-g(x)}{y-F(x)}$. Choose $x_0$ large enough so that $\text{sign}(x)F(x) > k$ for all $|x| > x_0$ (this will be needed in the next step). Since $[-x_0, x_0]$ is compact and $g$ and $F$ are continuous, they are both bounded in $[-x_0, x_0]$ so we can find a constant, $c$, such that $|g(x)|$ and $|F(x)|$ are bounded by $c$ on $[-x_0, x_0]$. Now choose $y_0 > 2k$ large enough so that

$$\left|\frac{dy}{dx}\right| = \left|\frac{-g(x)}{y - F(x)}\right| < \frac{k}{2x_0} \tag{5.50}$$

for all $y \in [y_0 - k, y_0 + k]$ and $x \in [-x_0, x_0]$. This is easy; just choose $y_0$ such that

$$0 < \frac{c}{y_0 - k - c} < \frac{k}{2x_0}$$

or

$$y_0 > k + c + \frac{2cx_0}{k}. \tag{5.51}$$

Provided this condition is satisfied, $|\frac{dy}{dx}| < \frac{k}{2x_0}$ for the trajectory through $(-x_0, y_0)$ and so it strikes $x = x_0$ at the point $(x_0, y_1)$ where

$$k < y_0 - k < y_1 < y_0 + k.$$

**Step 2.** We now want to show that the trajectory through $(x_0, y_1)$ returns to $x = x_0$ at a point $(x_0, -y_2)$. We do this by using a modified Liapounov function. Let

$$V(x,y) = \tfrac{1}{2}(y-k)^2 + G(x), \tag{5.52}$$

which has contours symmetric about $y = k$. For any given value of $y$ greater than $k$ the contour of constant $V$ through $(x_0, y)$ in $x > x_0$ cannot tend to infinity in finite $x$ (or the $(y-k)^2$ term would become unbounded) nor can it tend to infinity with finite $y$ (as $G(x) \to \infty$ as $x \to \infty$). Hence the contour must intersect the curve $y = k$ at some finite value of $x$. By the symmetry of $V$ about $y = k$ the contour can be reflected in $y = k$ and so it intersects $x = x_0$ for a second time at $-y + k$. Furthermore

$$\begin{aligned} \dot{V} &= \dot{y}(y-k) + \dot{x}g(x) = -g(x)(y-k) + g(x)(y-F) \\ &= -g(x)(F(x) - k) \end{aligned} \tag{5.53}$$

so in $x > x_0$, $F(x) > k$ and hence $\dot{V} < 0$. Now consider the part of the contour $V(x,y) = V(x_0, y_1)$ in $x > x_0$. Trajectories cross this contour inwards (since $\dot{V} < 0$) and the contour intersects $x = x_0$ at $y = y_1$ and $y = -y_2$ where

$$y_2 = y_1 - k \le y_0.$$

This implies that $(x_0, -y_2)$ lies above $(x_0, -y_0)$ and since $\dot{x}$ is less than zero for all $y$ less than $F(x_0)$, $\dot{x} < 0$ for all $(x_0, y)$ with $-y_2 > y > -y_0$. Hence the straight line from $(x_0, -y_2)$ to $(x_0, -y_0)$ is a local transversal which trajectories cross from right to left (see Fig. 5.23).

**Step 3.** We now simply repeat the argument starting from $(x_0, -y_0)$ to come back to $(-x_0, y_2')$ below $(-x_0, y_0)$.

We can now construct a region, $D$, containing no stationary points and which trajectories enter but do not leave. The inner boundary of $D$ is a small closed curve about the stationary point at the origin, which is the only stationary point of the system. The outer boundary of $D$ is the union of six curves:

i) the trajectory from $(-x_0, y_0)$ to $(x_0, y_1)$ constructed in Step 1;
ii) the contour $V(x,y) = V(x_0, y_1)$ in $x > x_0$, which intersects the line $x = x_0$ at $y = y_1$ and $y = -y_2$, trajectories cross this contour inwards;
iii) the line segment from $(x_0, -y_2)$ to $(x_0, -y_0)$, which trajectories cross from right to left;
iv) the equivalent curves defined in Step 3.

Applying the Poincaré-Bendixson Theorem (Theorem 5.13) finishes the proof.

There are many variations on this theorem, some of which are given as exercises, but they all use the same kind of argument, sometimes without needing to use all of the steps described above. With extra conditions it can be possible to prove that the periodic orbit is unique, but we will not worry about this here.

## Exercises 5

The first three exercises below are considerably harder than the rest of the questions. They deal with the effect of nonlinear terms on the local behaviour near a stationary point, following on from the discussion of Section 5.2.

1. Suppose that

$$\dot{x} = ax + by + F_1(x,y), \quad \dot{y} = cx + dy + F_2(x,y).$$

If $F_i(x,y) \sim o(\sqrt{x^2+y^2})$, $i = 1,2$, show that the differential equation can be written in polar coordinates as

$$\dot{r} = rR(\theta) + P(r,\theta), \quad \dot{\theta} = \Omega(\theta) + Q(r,\theta)$$

where $R(\theta)$ and $\Omega(\theta)$ are defined in (5.16) and

$$P(r,\theta) \sim o(r), \quad Q(r,\theta) \sim o(1).$$

If $F_i(x,y) \sim O(x^2+y^2)$, $i = 1,2$, show that

$$P(r,\theta) \sim O(r^2), \quad Q(r,\theta) \sim O(r).$$

You should find expressions for $P$ and $Q$ in terms of $F_1$ and $F_2$ explicitly. In subsequent questions the notation of this question is used.

2. (Nonlinear node). Suppose the origin is a nonlinear stable node with eigenvalues $\lambda_1$ and $\lambda_2$ with $-\lambda_1 > -\lambda_2 > 0$. Setting $a = \lambda_1$, $d = \lambda_2$ and $b = c = 0$ in Question 1, show that in the absence of nonlinear terms all trajectories tend to the origin tangential to the $y$-axis, except for trajectories which start on the $x$-axis. Now suppose that $F_i(x, y) \sim O(x^2 + y^2)$, $i = 1, 2$. Assume that two trajectories approach the origin tangential to the $x$-axis in $x > 0$,

$$y = g_1(x) \quad \text{and} \quad y = g_2(x),$$

say. Show that the functions $g_i(x)$ satisfy the differential equations

$$\frac{dg_i}{dx} = \frac{\lambda_2 g_i(x) + F_2(x, g_i(x))}{\lambda_1 x + F_1(x, g_i(x))}, \quad i = 1, 2.$$

Define $G(x) = g_1(x) - g_2(x)$, $x > 0$. Using the fact that trajectories cannot cross, argue that $G(x) > 0$ for $x > 0$ without loss of generality. Show that

$$\begin{aligned} \frac{dG}{dx} = & \frac{\lambda_2 G(x) + F_2(x, g_1(x)) - F_2(x, g_2(x))}{\lambda_1 x + F_1(x, g_1(x))} \\ & - \frac{(\lambda_2 g_2(x) + F_2(x, g_2(x))(F_1(x, g_1(x)) - F_1(x, g_2(x)))}{(\lambda_1 x + F_1(x, g_1(x)))(\lambda_1 x + F_1(x, g_2(x)))}. \end{aligned}$$

Note that

$$F_i(x, g_1(x)) - F_i(x, g_2(x)) = G(x) \frac{\partial F_i}{\partial y}(x, \xi)$$

for some $\xi$ with $g_2(x) < \xi < g_1(x)$. Hence show that

$$\frac{dG}{dx} = \frac{\lambda_2 G(x)}{\lambda_1 x}(1 + O(r)).$$

Deduce that in a sufficiently small neighbourhood of the origin there exist positive constants $C$ and $\delta$, with $\delta < 1$, such that

$$Cx^{\delta - 1} < \frac{G(x)}{x}$$

and hence obtain a contradiction.

3. (Nonlinear star). Suppose the origin is a nonlinear stable star with eigenvalue $\lambda < 0$. If

$$P(r, \theta) \sim O(r^2), \quad Q(r, \theta) \sim O(r)$$

show that for sufficiently small $r$, $\theta(t)$ is bounded. Show further that initial conditions can be chosen such that trajectories tend to the origin at an angle arbitrarily close to any given angle.

4. Consider the following Lotka-Volterra equations for the population of an island with rabbits ($r$) and sheep ($s$), $r, s \geq 0$. Find the stationary points, investigate their stability and sketch a plausible global solution.

$$\dot{s} = s(2 - r - s)$$

i) $\dot{r} = r(3 - r - s)$
ii) $\dot{r} = r(3 - 2r - s)$
iii) $\dot{r} = r(3 - r - 2s)$
iv) $\dot{r} = r(3 - 2r - 2s)$.

5. Using polar coordinates, sketch the trajectories of

$$\dot{x} = xy - x^2y + y^3, \quad \dot{y} = y^2 + x^3 - xy^2.$$

6. Show that the system

$$\dot{x} = x^2 - y - 1, \quad \dot{y} = (x - 2)y$$

has just three stationary points and determine their types. By considering the three straight lines through pairs of stationary points, show that there are no periodic orbits. Sketch the phase portrait of the system.

7. Locate the three stationary points of the system

$$\dot{x} = x(2 - y - x), \quad \dot{y} = y(4x - x^2 - 3), \quad x, y \geq 0$$

and find their linear types. Using Dulac's criterion with weighting function $(xy)^{-1}$, show that there are no periodic orbits and sketch the trajectories.

8. Consider the differential equation $\dot{x} = f(x)$ for $x \in \mathbf{R}^2$. If $\nabla \times f = 0$ in a simply connected domain $D$, prove that no periodic orbits can lie entirely in $D$. Deduce that the system

$$\dot{x} = y + 2xy, \quad \dot{y} = x + x^2 - y^2$$

has no periodic orbits.

9. Consider the differential equation

$$\dot{x} = x - y - (x^2 + \tfrac{3}{2}y^2)x, \quad \dot{y} = x + y - (x^2 + \tfrac{1}{2}y^2)y.$$

Determine the nature of the stationary point at the origin and use the Poincaré-Bendixson Theorem to show that there is at least one periodic orbit. Find bounds for the annular region in which any periodic orbit can lie inside.

10. Consider

$$\dot{x} = y + cx(1 - 2b - r^2), \quad \dot{y} = -x + cy(1 - r^2)$$

where $r^2 = x^2 + y^2$ and the constants $b$ and $c$ satisfy $0 \le b \le \frac{1}{2}$ and $0 < c \le 1$. Prove that there is at least one periodic orbit and that if there are several then they all have the same period, $P(b, c)$. Prove that for $b = 0$ there is only one periodic orbit.

11. Suppose $\ddot{x} + g(x) = 0$. If $g(0) = 0$, $xg(x) > 0$ for $x \neq 0$ and

$$\int_0^x g(u)du \to \infty \quad \text{as} \quad |x| \to \infty,$$

prove that all non-trivial trajectories are periodic.

By considering $g(x) = xe^{-x^2}$, show that if the integral condition does not hold then not all trajectories need be periodic.

12. Suppose $\ddot{x} + f(x)\dot{x} + g(x) = 0$ and let $F(x) = \int_0^x f(u)du$ and $G(x) = \int_0^x g(u)du$. Prove that if

$$g(0) = 0, \quad xg(x) > 0 \ \text{ if } \ x \neq 0, \quad f(0) < 0, \quad g'(0) > 0$$

and

$$\text{sign}(x)F(x) \to \infty \quad \text{as} \quad |x| \to \infty$$

then there is at least one periodic orbit.

13. Consider the differential equation

$$\ddot{v} + \dot{v}^3 - a\dot{v} + v = 0.$$

Show that if $a < 0$ then the origin is asymptotically stable. If $a > 0$ then use Theorem 5.14 or Exercise 12 to show that there is at least one periodic orbit. (Begin by setting $v = Y$ and $\dot{v} = X$.)

14. Let $D$ be a region on which the Poincaré-Bendixson Theorem can be applied for the system $\dot{x} = f(x)$, $x \in \mathbf{R}^2$. If $\nabla . f < 0$ in $D$ prove that there is one and only one periodic orbit in $D$. If $\dot{r} = R(r, \theta)$ and $\dot{\theta} = \Theta(r, \theta)$, show that

$$\nabla . f = \frac{1}{r}\frac{\partial}{\partial r}(rR) + \frac{\partial \Theta}{\partial \theta}.$$

(Note that $f$ is the *velocity* field.) Hence or otherwise prove that if $0 \le b < \frac{1}{2}$ and $\frac{1}{2} < c < 1$ then the system of Exercise 10 has a unique periodic orbit.

# 6

# *Periodic orbits*

The Poincaré-Bendixson Theorem can be used to prove the existence of periodic orbits for flows in $\mathbf{R}^2$, but the proof relies on the Jordan curve lemma (that a simple closed curve separates the plane into two regions, an inside and an outside) and so it cannot be applied to periodic orbits in $\mathbf{R}^n$, $n \geq 3$. The Poincaré index and Dulac's criterion suffer from similar drawbacks in higher dimension, so in order to discuss periodic orbits in a more general context we need a new idea: the study of maps induced by a flow.

Consider a periodic orbit of the autonomous differential equation

$$\dot{x} = f(x), \quad x \in \mathbf{R}^n.$$

If $x_0$ is on the periodic orbit then it is possible to find a local manifold of dimension $(n-1)$ which intersects the flow transversely (see Fig. 6.1). Given any point on this manifold we can then ask whether the trajectory through this point intersects the manifold at some later time. Since $x_0$

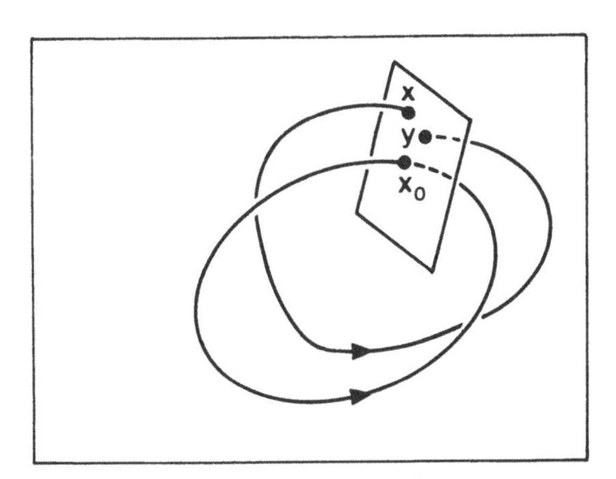


Fig. 6.1 Construction of the return map.

is on the manifold (which is called a surface of section, and is analogous to a local transversal in $\mathbf{R}^2$) some points must intersect the manifold again. We can then construct a map $R$ from the manifold to itself by setting $R(x) = y$, where $y$ is the point at which the trajectory through $x$ first intersects the manifold, assuming such a point exists. Clearly $R^k(x_0) = x_0$ for some $k \geq 1$, and the surface of section can be chosen so that $k = 1$. In this way we have reduced the problem of analyzing the flow near a periodic orbit to the problem of analyzing the return map $R$ on an $(n-1)$-dimensional manifold. The study of maps is in many ways simpler than the study of flows, and properties of the map reflect properties of the flow. In this chapter the correspondence between these induced maps and the flow will be investigated in more detail, and we will begin to outline a theory of the dynamics of maps similar to the theory of flows already undertaken.

## 6.1 Linear and nonlinear maps

A map is really just a function $f : \mathbf{R}^n \to \mathbf{R}^n$. Given such a function and a point $x \in \mathbf{R}^n$ we define the *orbit* of $x$ to be the sequence

$$(x, f(x), f(f(x)), \ldots, f^n(x), \ldots) \tag{6.1}$$

where $f^n(x) = f(f^{n-1}(x))$ is called the $n^{th}$ iterate of $f$. The orbit of a point is similar to the trajectory through a point for differential equations. The analogy between maps and flows is emphasised if the map is written as

$$x_{n+1} = f(x_n) \tag{6.2}$$

so given a point $x_0$ we form the new point, $x_1 = f(x_0)$, and then $x_2 = f(x_1) = f^2(x_0)$ and so on. Thus the iteration of maps is like the integration of a differential equation, but continuous time is replaced by discrete time. We can ask the same questions about maps as we have done about flows: what is the limit set of a point (the accumulation points of the orbit of $x$), is there an invariant set, a periodic orbit and so on? A fixed point of a map is a point $x$ such that $x = f(x)$, so $x_n = x$ for all $n \geq 0$ and a point is periodic of (minimal) period $p$ if $x = f^p(x)$ and $f^k(x) \neq x$ for $1 \leq k < p$.

*Example 6.1*

Let $f(x) = 1 - x^2$. As with nonlinear differential equations the first things we look for are fixed points: these satisfy

$$x = 1 - x^2$$

so there are two fixed points: $x_{\pm} = (-1 \pm \sqrt{5})/2$. Now note that $f^2(x) = f(1-x^2) = 1 - (1-x^2)^2$. So points of period two satisfy

$$x = 1 - (1-x^2)^2 = 2x^2 - x^4.$$

Now, it is obvious that $x = 0$ satisfies this equation and so $x = 0$ and $f(0) = 1$ are both points of period two. However, in general it is rarely so easy to factorize the fixed point equation for $f^2$ and we need to use the fact that a point of period one is also a point of period two, since if $f(x) = x$ then $f(f(x)) = f(x) = x$. This shows that the equation for the fixed points of $f$, $x^2 + x - 1$ must be a factor of $x^4 - 2x^2 + x$. Indeed it is easy to see that

$$x^4 - 2x^2 + x = (x^2 + x - 1)(x^2 - x)$$

from which we deduce that the points which are of period two but are not fixed points for $f$ are the solutions of $x^2 - x = 0$: $x = 0$ and $x = 1$.

By analogy with differential equations, the next step would be to determine the stability of these solutions: to do this we need the equivalent of the linearization of a flow near a stationary point. Suppose that $z$ is a fixed point of $x_{n+1} = f(x_n)$, so $z = f(z)$. To consider the local behaviour of the map near $z$ set $x = z + \xi$, for $\xi$ small. Then $x_{n+1} = z + \xi_{n+1}$ and expanding $f(z + \xi_n)$ as a Taylor series

$$z + \xi_{n+1} = f(z + \xi_n) = f(z) + Df(z)\xi_n + O(|\xi_n|^2). \qquad (6.3)$$

Since $z = f(z)$ this becomes

$$\xi_{n+1} = Df(z)\xi_n + O(|\xi|^2) \qquad (6.4)$$

and, again by analogy with the continuous time case, we might hope that the local behaviour is determined by the linearized system $\xi_{n+1} = Df(z)\xi_n$. This will be true for hyperbolic fixed points (which have yet to be defined), but first we need to know something about linear maps.

*Example 6.2*

Consider the map $x_{n+1} = ax_n$ for real $a$ and $x_n$. The only fixed point is at $x = 0$ and $x_1 = ax_0$, $x_2 = ax_1 = a^2x_0, \ldots$ so $x_n = a^n x_0$. Hence all orbits tend to the origin if $|a| < 1$, all orbits (except $x = 0$) diverge if $|a| > 1$, if $a = 1$ all points are fixed and if $a = -1$ the origin is fixed and all other points are of period two.

This example suggests that the hyperbolicity condition for fixed points will be that all eigenvalues of the linearized map lie off the unit circle.

The general linear map in $\mathbf{R}^n$ is

$$x_{n+1} = Ax_n \tag{6.5}$$

where $A$ is a real $n \times n$ matrix. By induction this gives $x_n = A^n x_0$, so the action of the linear map is determined by $A^n$ in the same way that a linear flow $\dot{x} = Ax$ is determined by the matrix $e^{tA}$. We leave a fuller discussion of normal forms for maps to later; here we simply observe that eigenvectors of $A$ corresponding to eigenvalues outside the unit circle give unstable directions and those which correspond to eigenvalues inside the unit circle give stable directions. That the results of Chapter 4 for hyperbolic stationary points of flows carry over to hyperbolic fixed points of nonlinear maps is confirmed by the following sequence of theorems.

(6.1) DEFINITION

*Let $z$ be a fixed point of $x_{n+1} = f(x_n)$. Then $z$ is hyperbolic iff $|\lambda| \neq 1$ for all eigenvalues, $\lambda$, of $Df(z)$.*

With this definition of hyperbolicity we can now state the results for fixed points of maps corresponding to the Stable Manifold Theorem for flows (Theorem 4.7), Poincaré's (analytic) Linearization Theorem (Theorem 4.5) and the persistence of hyperbolic stationary points (Section 4.3). The local stable and unstable manifolds of a fixed point are defined in the same way as for flows: given some small neighbourhood $U$ of a hyperbolic fixed point, $z$, of some smooth, invertible map $f$ the local stable manifold, $W^s_{loc}(z)$, of $z$ is

$$W^s_{loc}(z) = \left\{ x \in U | \lim_{n\to\infty} f^n(x) = z \text{ and } f^i(x) \in U \text{ for all } i \geq 0 \right\}$$

and the local unstable manifold, $W^u_{loc}(z)$, of $z$ is

$$W^u_{loc}(z) = \left\{ x \in U | \lim_{n\to\infty} f^{-n}(x) = z \text{ and } f^{-i}(x) \in U \text{ for all } i \geq 0 \right\}.$$

As with flows these manifolds can be extended to give the (global) stable and unstable manifolds of $z$ using

$$W^s(z) = \bigcup_{n\geq 0} f^{-n}(W^s_{loc}(z)) \quad \text{and} \quad W^u(z) = \bigcup_{n\geq 0} f^n(W^u_{loc}(z)).$$

(6.2) THEOREM (STABLE MANIFOLD THEOREM FOR MAPS)

*Suppose that the origin is a hyperbolic fixed point of the map* $x_{n+1} = f(x_n)$, *where* $f$ *is a diffeomorphism (i.e. smooth and invertible). Then the linear map* $x_{n+1} = Df(x_n)$ *has stable and unstable manifolds,* $E^s$ *and* $E^u$ *and there exist local stable and unstable manifolds* $W^s_{loc}(z)$ *and* $W^u_{loc}(z)$ *for the nonlinear system which are of the same dimension as* $E^s$ *and* $E^u$ *respectively. These manifolds are (respectively) tangential to* $E^s$ *and* $E^u$ *at the origin and are as smooth as the original map* $f$.

The equivalent of Poincaré's Linearization Theorem holds with a modified resonance condition.

(6.3) THEOREM (ANALYTIC LINEARIZATION THEOREM)

*Suppose that* $x_{n+1} = f(x_n)$, *where* $f$ *is an analytic map with a hyperbolic fixed point at the origin. Then there is a near-identity formal power series change of coordinates* $y = x + ...$ *such that in the* $y$ *coordinates* $y_{n+1} = Df(0)y_n$ *provided the eigenvalues* $\lambda_i$, $i = 1, ..., n$, *of* $Df(0)$ *do not satisfy the resonance condition*

$$\lambda_s = \prod_{k=1}^{n} \lambda_i^{m_i}, \quad s = 1, 2, ..., n,$$

*where the non-negative integers* $m_i$ *satisfy* $\sum m_i \geq 2$.

The proof of this theorem is almost exactly the same as the proof of Theorem 4.5; the only difference is that when the successive coordinate changes of higher and higher order are made the divisors of the coefficients look like $(\lambda_s - \prod_{k=1}^n \lambda_i^{m_i})$ rather than $(\lambda_s - \sum_{k=1}^n m_i\lambda_i)$. Hartman's Theorem also holds if the fixed point is hyperbolic; i.e. if $f$ is a diffeomorphism and $z$ is a hyperbolic fixed point of $f$ then there exists a homeomorphism which takes the orbits of $f$ to the orbits of $Df$. Finally, we have the persistence result.

(6.4) THEOREM (PERSISTENCE OF HYPERBOLIC POINTS)

*Suppose that $f$ is a diffeomorphism and that $z$ is a hyperbolic fixed point of $f$. Then for all diffeomorphisms $v$ there exists $\epsilon_1 > 0$ such that if $0 < \epsilon < \epsilon_1$ then $f + \epsilon v$ is a diffeomorphism with a hyperbolic fixed point $z'$ near $z$. Furthermore, the dimensions of $W^{s,u}_{loc}(z)$ and $W^{s,u}_{loc}(z')$ are the same.*

One major difference between maps and flows is that periodic orbits of maps can be treated in precisely the same way as fixed points, since a periodic point of period $p$ is a fixed point of $f^p$. Hence all the definitions and theorems for fixed points of maps can be extended trivially to periodic points by replacing $f$ by $f^p$. Stability for a fixed point depends upon the spectrum (i.e. the eigenvalues) of $Df(z)$, so for a periodic point $y$ with $f^p(y) = y$ it should depend upon $Df^p(y)$. Now, by the chain rule

$$Df^p(y) = Df(f^{p-1}(y))Df^{p-1}(y) = \prod_{k=0}^{p-1} Df(y_k) \tag{6.6}$$

where $y_0 = y$ and $y_k = f^k(y)$. This shows that the eigenvalues of $Df^p(y)$ are independent of the point $y$ on the periodic orbit, so any stability results will be properties of the periodic orbit, not just the periodic point.

## 6.2 Return maps

The idea behind return maps was sketched in the introduction to this chapter, so we will go straight into the definition of the map.

Consider a flow $\varphi(x,t)$ defined by the differential equation $\dot{x} = f(x)$, $x \in \mathbf{R}^n$, and suppose that there is a periodic orbit $\Gamma$. Then we can define a local cross-section $\Sigma \subset \mathbf{R}^n$ of dimension $n-1$ which is transverse to the flow and which therefore intersects $\Gamma$ at a single point $z$. Now for some open set $U \subset \Sigma$ with $z \in U$ define the map $R : U \to \Sigma$ by

$$R(x) = \varphi(x, \tau(x)) \tag{6.7}$$

where $\varphi(x,\tau(x)) \in \Sigma$ and $\varphi(x,t) \notin \Sigma$ for $0 < t < \tau(x)$. Note that if the period of $\Gamma$ is $T$ then $\tau(z) = T$ and that the stability of the periodic orbit is reflected in the stability of $z$, which is a fixed point of the map $R$. $R$ is called a (Poincaré) return map, and in a neighbourhood of a point on an isolated periodic orbit it is always possible to construct such a map which is as smooth as the original flow.

*Example 6.3*

Consider

$$\dot{x} = x - y - x(x^2 + y^2)$$

$$\dot{y} = x + y - y(x^2 + y^2)$$

or in polar coordinates

$$\dot{r} = r(1 - r^2), \quad \dot{\theta} = 1.$$

Choose $\Sigma = \{(r, \theta) | r > 0, \theta = 0\}$. The differential equations can be integrated to give solutions

$$r = \left[1 + \left(\frac{1}{r_0^2} - 1\right) e^{-2t}\right]^{-\frac{1}{2}}, \quad \theta = \theta_0 + t$$

so if we choose initial conditions $(r_0, \theta_0) = (x, 0)$ on $\Sigma$ we see immediately that $\tau(x) = 2\pi$ for all $x > 0$ and so

$$R(x) = \left[1 + \left(\frac{1}{x^2} - 1\right) e^{-4\pi}\right]^{-\frac{1}{2}}.$$

This has a fixed point, $x = R(x)$, when $x = 1$, so the point $(r, \theta) = (1, 0)$ lies on a periodic orbit of the flow, and

$$DR(x) = -\frac{1}{2}\left[1 + \left(\frac{1}{x^2} - 1\right) e^{-4\pi}\right]^{-\frac{3}{2}} \cdot \left(-\frac{2e^{-4\pi}}{x^3}\right).$$

Evaluating this when $x = 1$ gives $DR(1) = e^{-4\pi} < 1$, so the periodic orbit is stable.

This could, of course, have been deduced directly from the solutions; the important aspect of this section is that periodic orbits can be studied, in the abstract, by considering invertible maps of $\mathbf{R}^{n-1}$.

## 6.3 Floquet Theory revisited

There is a second way of reducing the flow near a periodic orbit to a map which is perhaps more in keeping with the linearization arguments of preceeding sections. Suppose $\Gamma$ is a periodic orbit and $x_0 \in \Gamma$, so

$\varphi(x_0, T) = x_0$, where $T$ is the period of the orbit. Then we can consider the effect of a small perturbation, $v$ to $x_0$, and try to discover whether such a perturbation grows or not. In other words we start with a small perturbation $v_0$, then calculate $v_1 = v_0(T)$ in the following sense. Consider $\varphi(x_0 + v_0, T)$, then $v_1 = \varphi(x_0 + v_0, T) - x_0$, now repeat this with $v_1$ replacing $v_0$. In this way we obtain a map $v_{n+1} = G(v_n)$ valid for small $|v|$. But using Taylor's Theorem

$$\varphi(x_0 + v_n, T) = \varphi(x_0, T) + D\varphi(x_0, T)v_n + O(|v_n|^2) \tag{6.8}$$

and so

$$v_{n+1} = \varphi(x_0 + v_n, T) - x_0 = D\varphi(x_0, T)v_n + O(|v_n|^2) \tag{6.9}$$

since $x_0 = \varphi(x_0, T)$. This gives us a linear map with a fixed point at $x_0$. Note that this map will inevitably have a neutral eigenvalue of $+1$ corresponding to a choice of $v$ such that $x_0 + v \in \Gamma$, but the remainder of the spectrum of $D\varphi(x_0, T)$ will tell us whether the fixed point is stable or unstable (the indeterminacy of stability in the direction tangential to the periodic orbit allows for phase lagging).

Unfortunately this method involves the calculation of the time evolution of the extra $n^2$ functions $\frac{\partial \varphi_i}{\partial x_j}$: the variation of the solution with respect to the initial conditions. The equations obtained are nasty (though autonomous) and this approach is natural to use in numerical computations. However, for a theoretical understanding of the problem it is much easier to consider the time evolution of a small perturbation of the periodic orbit, which gives rise to a linear equation with periodic coefficients of the kind described in Section 3.5.

Let $x = \varphi(x_0, t) + v(t)$, $|v(0)| \ll 1$, be a small perturbation of the periodic solution. Then

$$\dot{x} = f(x) = f(\varphi(x_0, t) + v) = f(\varphi(x_0, t)) + Df(\varphi(x_0, t))v + O(|v|^2).$$

But $\dot{x} = \dot{\varphi}(x_0, t) + \dot{v} = f(\varphi(x_0, t)) + \dot{v}$ and so

$$\dot{v} = Df(\varphi(x_0, t))v + O(|v|^2). \tag{6.10}$$

To obtain the map above we have simply integrated the linear part of this equation.

*Example 6.4*

In polar coordinates, Example 6.3 is

$$\dot{r} = r(1 - r^2), \quad \dot{\theta} = 1$$

and the periodic orbit can be written explicitly as $\varphi(x_0, t) = (1, t)$. If we now let $v = (\rho, \psi)$ in polar coordinates and $x = (1 + \rho, t + \psi)$ we find, substituting these perturbations into the differential equation and ignoring quadratic terms,

$$\dot{\rho} = -2\rho, \quad \dot{\psi} = 0.$$

Thus we find that the linearized system has one eigenvalue in the left half-plane (corresponding to the stable direction) and one neutral direction (corresponding to the possibility of drift around the periodic orbit).

The results for the map derived in the previous example are simply obtained by integrating the flow through one period of the oscillation.

In general, of course, $Df(\varphi(x_0, t))$ is not independent of time, but since $\varphi(x_0, t) = \varphi(x_0, t + T)$ it is a periodic $n \times n$ matrix, $A(t)$ with $A(t) = A(t + T)$. This leads us to consider linear equations

$$\dot{v} = A(t)v, \quad A(t) = A(t + T). \tag{6.11}$$

This is just a linear equation with periodic coefficients so (cf. Chapter 3) we can define a fundamental matrix $\Phi(t)$ with $\Phi(0) = I$ and such that $\Phi(t + T) = \Phi(t).C$. Setting $t = 0$ gives the monodromy matrix $C = \Phi(T)$. The eigenvalues of $\Phi(T)$ are the Floquet multipliers, $\lambda(T)$ and if $\lambda(T) = e^{\sigma T}$ then the coefficients $\sigma$ are called the Floquet exponents of $\Phi(T)$. From the definition of the fundamental matrix,

$$v(t) = \Phi(t)v_0 \tag{6.12}$$

and so $v(nT) = \Phi(nT)v_0$. Hence $|v(nT)|$ tends to zero provided all the multipliers of $\Phi(T)$ lie inside the unit circle (or the Floquet exponents all lie in the left half-plane). We have already remarked that there must be a multiplier equal to $+1$, with an eigenvector tangential to the periodic orbit at $x_0$; this neutral stability corresponds to the possibility of drift around the periodic orbit.

For periodic orbits in two dimensions there is a much simpler way of finding the Floquet multipliers, exploiting the final three exercises of Chapter 3. Basically, the idea is to use the divergence of the vector field $f$ near the periodic orbit. Suppose that the periodic orbit is $u(t)$, with $u(t) = u(t + T)$. Then the periodic matrix, $A(t)$, in equation (6.11) is

$$A(t) = Df(u(t)). \tag{6.13}$$

Since $u$ is a solution of the differential equation,

$$\dot{u} = f(u) \tag{6.14}$$

and so, differentiating with respect to time again,

$$\ddot{u} = \frac{d}{dt} f(u(t)) = Df(u)\dot{u} = A(t)\dot{u}. \tag{6.15}$$

So $\dot{u}$ satisfies the small perturbation equation, (6.11), and since $u$ is periodic $\dot{u}(t) = \dot{u}(t+T)$. But, using (6.12), this implies that

$$\dot{u}(T) = \dot{u}(0) = \Phi(T)\dot{u}(0) \tag{6.16}$$

and so one Floquet multiplier of the problem must equal unity. This result is true in arbitrary dimension, but now consider the case of differential equations on the plane.

Let $\Phi(t)$ be a fundamental matrix with initial value $\Phi(0) = I$. Then the Floquet multipliers are the eigenvalues of $\Phi(T)$. But the product of the eigenvalues of a $2 \times 2$ matrix is simply the determinant of the matrix and we have already established that one of the eigenvalues is equal to one. Hence the other eigenvalue, $\lambda$ is given by

$$\lambda = \det\Phi(T) \tag{6.17}$$

and, of course, $\lambda$ is the Floquet multiplier we are trying to calculate. This may not appear to be a great simplification of the problem, but it turns out that the equation for the time evolution of $\det\Phi(t)$ is particularly simple.

First note that since $\dot{\Phi} = A\Phi$, we can write in the usual suffix notation

$$\dot{\Phi}_{ij} = A_{i1}\Phi_{1j} + A_{i2}\Phi_{2j} \tag{6.18}$$

and

$$\det\Phi(t) = \Phi_{11}\Phi_{22} - \Phi_{12}\Phi_{21}. \tag{6.19}$$

Differentiating (6.19) using (6.18) gives

$$\begin{aligned} \frac{d}{dt}\det\Phi(t) = & (A_{11}\Phi_{11} + A_{12}\Phi_{21})\Phi_{22} \\ & + \Phi_{11}(A_{21}\Phi_{12} + A_{22}\Phi_{22}) \\ & - (A_{11}\Phi_{12} + A_{12}\Phi_{22})\Phi_{21} \\ & - \Phi_{12}(A_{21}\Phi_{11} + A_{22}\Phi_{21}). \end{aligned} \tag{6.20}$$

Tidying this up a little gives

$$\frac{d}{dt}\det\Phi(t) = (A_{11} + A_{22})\det\Phi. \tag{6.21}$$

But, by definition, $A(t) = Df(u(t))$, and so

$$A_{11} + A_{22} = \frac{\partial f_1}{\partial x_1}(u(t)) + \frac{\partial f_2}{\partial x_2}(u(t)) = \nabla . f(u(t)) \tag{6.22}$$

giving, at long last,

$$\frac{d}{dt}\det\Phi(t) = \nabla.f(u(t))\det\Phi. \tag{6.23}$$

Integrating this equation together with the initial condition $\det\Phi(0) = 1$ we obtain

$$\det\Phi(t) = \exp\left(\int_0^t \nabla.f(u(\tau))d\tau\right) \tag{6.24}$$

and so

$$\lambda = \exp\left(\int_0^T \nabla.f(u(\tau))d\tau\right) \tag{6.25}$$

and $\lambda < 1$, the condition for orbital stability, becomes

$$\int_0^T \nabla.f(u(\tau))d\tau < 0. \tag{6.26}$$

In many situations this is an easy calculation.

*Example 6.5*

In the example which runs through this section

$$\nabla.f = \frac{1}{r}\frac{\partial}{\partial r}(r[r(1-r^2)]) = 2 - 4r^2$$

and on the periodic orbit $r = 1$, so $\nabla.f = -2$. The integral is easy ($T = 2\pi$) and we find $\lambda = \exp(-4\pi) < 1$. Hence the periodic orbit is orbitally stable.

*Example 6.6*

Consider the nonlinear oscillator

$$\ddot{x} + h(x)\dot{x} + g(x) = 0.$$

In Liénard coordinates with $H(x) = \int^x h(\xi)d\xi$, the equation is

$$\dot{x} = y - H(x), \quad \dot{y} = -g(x)$$

and so $\nabla.f(x,y) = -h(x)$. Hence if $\varphi(x_0, t)$ is a periodic orbit for the system with period $T$, the orbit is stable if

$$\int_0^T h(\varphi(x_0,t))dt > 0.$$

## 6.4 Periodically forced differential equations

Another situation in which maps arise in a natural way is in the study of forced equations of the form

$$\dot{x} = f(x,t), \quad f(x,t+T) = f(x,t). \tag{6.27}$$

If we let $\theta = t$, then

$$\dot{x} = f(x,\theta), \ \dot{\theta} = 1 \text{ and } f(x,\theta+T) = f(x,\theta) \tag{6.28}$$

and we have an autonomous differential equation on $\mathbf{R}^n \times S^1$ where $S^1$ denotes the circle of length $T$, i.e. $\theta$ and $\theta + T$ are identified. Now if we choose $\Sigma = \{(x,\theta)|\theta = \theta_0\}$, $\Sigma$ is a globally transvere manifold and we can define a return map on $\Sigma$ in the standard way: $R(x) = \pi_x[\varphi(x,\theta_0 + T)]$, where $\pi_x$ is the projection operator onto the $x$-coordinate. We have therefore defined a global return map $R : \mathbf{R}^n \to \mathbf{R}^n$. This is sometimes called the time $T$ map (or Poincaré map), since it is equivalent to integrating the equation through one period of the forcing frequency. The orbital stability of a periodic orbit is reflected in the eigenvalues of the map $DR$ in the standard way. We shall not go into this in any detail, but we illustrate the idea with an example.

*Example 6.7*

Consider the equation

$$\dot{x} = -ax + \epsilon g_1(x,y,t), \quad \dot{y} = -by + \epsilon g_2(x,y,t),$$

where $a$ and $b$ are positive constants, $\epsilon > 0$ is small and the functions $g_i$ are bounded and periodic in time with period $T$, so $g_i(x,y,t) = g_i(x,y,t+T)$, $i = 1,2$. We want to show that for sufficiently small $\epsilon$ this system has a stable periodic orbit with period $T$. First, consider the case $\epsilon = 0$. Then both equations can be integrated and the time $T$ map is

$$(x,y) \to (xe^{-aT}, ye^{-bT}).$$

This is a linear map and both eigenvalues ($e^{-aT}$ and $e^{-bT}$) lie inside the unit circle and so the origin, which is a fixed point, is stable. Now consider the full equations. Multiplying the $\dot{x}$ equation by $e^{at}$ we find

$$\frac{d}{dt}xe^{at} = \dot{x}e^{at} + axe^{at} = \epsilon e^{at} g_1(x,y,t)$$

and so, integrating over the period of the forcing, $T$,

$$x(T) = x(0)e^{-aT} + \epsilon e^{-aT}\int_0^T e^{at}g_1(x,y,t)dt.$$

Similarly,

$$y(T) = y(0)e^{-bT} + \epsilon e^{-bT}\int_0^T e^{bt}g_2(x,y,t)dt.$$

Hence the time $T$ map is a small perturbation of the unperturbed time $T$ map which has a stable fixed point at the origin and so, by Theorem 6.4, the (perturbed) time $T$ map has a fixed point $\epsilon$ close to the origin for sufficiently small $\epsilon > 0$. This fixed point is stable (since the fixed point of the unperturbed map is stable) and corresponds to a stable periodic orbit of period $T$.

## 6.5 Normal forms for maps in $\mathbf{R}^2$

In Chapter 3 we showed how the linear part of a differential equation could be written in one of a number of (Jordan) normal forms by a linear change of coordinate. The same theory holds for linear maps, since if we have a linear map $y_{n+1} = Ly_n$ and choose some change of coordinates $y = Ax$, where $A$ is an invertible matrix then

$$x_{n+1} = A^{-1}y_{n+1} = A^{-1}Ly_n = A^{-1}LAx_n.$$

The invertible matrix $A$ can be chosen to make $A^{-1}LA$ take the most convenient form. For example, if $L$ has two distinct real eigenvalues, $\lambda_1$ and $\lambda_2$, $A$ can be chosen in such a way that $A^{-1}LA = \text{diag}(\lambda_1, \lambda_2)$ (choose $A$ to be the matrix which has the eigenvectors corresponding to $\lambda_1$ and $\lambda_2$ as columns). In Chapter 9 we will need the normal form for linear maps on $\mathbf{R}^2$ with a pair of complex conjugate eigenvectors, $\lambda \pm i\omega$. In this case $L$ has a (complex) eigenvector $z$ such that $Lz = (\lambda + i\omega)z$ and write $z = e_1 + ie_2$ where the vectors $e_i$ are real. Taking real and imaginary parts of $Lz = (\lambda + i\omega)z$ we find that

$$Le_1 = \lambda e_1 - \omega e_2$$

$$Le_2 = \omega e_1 + \lambda e_2$$

and so choosing $A = (e_2, e_1)$, i.e. the matrix which has $e_{3-i}$ as the $i^{th}$ column, $i = 1, 2$, we find

$$LA = A\begin{pmatrix} \lambda & -\omega \\ \omega & \lambda \end{pmatrix}$$

and hence in the new coordinates

$$x_{n+1} = \begin{pmatrix} \lambda & -\omega \\ \omega & \lambda \end{pmatrix} x_n.$$

This matrix can be written in a more geometrically appealing way by noting that $\lambda + i\omega = r\exp(i\theta)$ and so, taking real and imaginary parts again, $\lambda = r\cos\theta$ and $\omega = r\sin\theta$. Hence

$$x_{n+1} = r\begin{pmatrix} \cos\theta & -\sin\theta \\ \sin\theta & \cos\theta \end{pmatrix} x_n. \tag{6.29}$$

The effect of this transformation should now be obvious: it consists of a contraction or expansion by a factor $r$ composed with a rotation through an angle of $\theta$. We leave the description of the remaining cases for normal forms of maps in $\mathbf{R}^2$ as an exercise.

## Exercises 6

1. It is often useful to have a graphical description of the iteration of maps. Consider the map

$$x_{n+1} = 2 - x_n^2.$$

Sketch a graph of $x_{n+1}$ against $x_n$ with $x_n$ on the horizontal axis. Sketch the diagonal $x_{n+1} = x_n$ and show that fixed points of the map correspond to the intersections of your graph with the diagonal. Show that the map takes the interval $[-2, 2]$ onto itself. Choose a point $x_0$ in $(-2, 0)$. Find $x_1$ (approximately) from your graph. Find $x_2$ in the same way and thus convince yourself that iteration works by moving up or down to the graph then across to the diagonal. Repeat this construction for

$$x_{n+1} = -x_n^2$$

and hence convince yourself that all orbits in some interval tend to the point $x = 0$. Check that $x = 0$ is a fixed point of the map and that it is stable. What happens to $x = 0$ in the original map? Is the fixed point at $x = -1$ stable?

2. The Newton-Raphson method for finding the real zeros of a function $f$ can be written as the iterative procedure

$$x_{n+1} = x_n - \frac{f(x_n)}{f'(x_n)}.$$

Consider the function $x^2 - 3x + 2$. Find the real zeros of this function explicitly by solving the quadratic equation. Now consider the corresponding Newton-Raphson scheme. Determine the fixed points of this map and discuss their stability. By sketching the right hand side of the iteration scheme show that all points with $x < \frac{3}{2}$ (resp. $x > \frac{3}{2}$) tend under iteration to the fixed point in $x < \frac{3}{2}$ (resp. $x > \frac{3}{2}$).

3. Find the function $g(x)$ for which the Newton-Raphson procedure to determine solutions of $g(x) = 0$ is $x_{n+1} = 1 - x_n^2$. Find the fixed points of this map and determine their stability. Find the points of period two. Is the period two orbit stable?

4. Find the fixed points of

$$x_{n+1} = x_n + \tfrac{1}{2} \sin x_n$$

and determine their stability. Illustrate this graphically on a plot of $x_{n+1}$ against $x_n$.

5. Find the values of $\mu$ for which the map

$$x_{n+1} = \mu - x_n^2$$

has two fixed points. For what values of $\mu$ is one of these stable? Look for points of mimimal period two. For what values of $\mu$ do they exist and when is there a stable orbit of period two? Find the value of $\mu$ for which the point $x = 0$ lies on an orbit of period three.

6. Suppose $g$ maps the interval $[a, b]$ into itself. Let $h$ be a change of coordinates, so $h$ is a continuous, strictly monotonic map from $[a, b]$ onto its image. Show that in these new coordinates the map can be written as $f = hgh^{-1}$.

7. Show that the maps $g(x) = 4x(1 - x)$ on the interval $[0, 1]$ and $f(x) = 2 - x^2$ on the interval $[-2, 2]$ are related by a change of coordinates. Show that the parametrised family of maps $g_\mu(x) = \mu x(1 - x)$ maps the interval $[0, 1]$ into itself if $0 \leq \mu \leq 4$. For what values of $\mu$ is this map related by a change of coordinates to

i) $f_\lambda(x) = \lambda - x^2$;
ii) $r_\pi(x) = 1 - \pi x$?

8. Consider the linear map $x_{n+1} = Ax_n$, where $x \in \mathbf{R}^2$ and $A$ is a constant $2 \times 2$ matrix. Classify the behaviour of this map according to the nature of the eigenvalues of $A$ (cf. Section 5.1 for flows).

9. The Poincaré-Bendixson Theorem can be applied to a planar system, $\dot{x} = f(x)$, in an annular region

$$D = \{(r, \theta) |\ R_1 \leq r \leq R_2\} .$$

If $\nabla . f$ is continuous and negative in $D$ show that there is a stable periodic orbit in $D$ and deduce that it is the only periodic orbit in $D$.

10. Suppose $\dot{x} = f(x)$ for $x \in \mathbf{R}^3$. Let $D \subset \mathbf{R}^3$ be a simply connected, bounded domain with smooth boundary $\partial D$. Prove that if $\nabla . f < 0$ for all $x \in D$ then no invariant torus can lie entirely inside $D$. Can an invariant sphere lie entirely inside $D$?

# 7

## *Perturbation theory*

In this chapter we abandon, at least to some extent, the geometric approach that has been developed in the previous chapters and attempt to see how far calculations of solutions to nonlinear equations can be pushed using the techniques of perturbation theory. In all the examples considered here the nonlinearities are small (in a sense made explicit in the next section) and we will consider solutions as perturbations of the solutions to a underlying linear system. Although the approach will be analytic (approximate solutions will be constructed in closed form) we shall not abandon the qualitative approach espoused throughout this book completely in that we shall use these solutions to obtain qualitative information about the asymptotic behaviour of the nonlinear systems.

Many problems are phrased in such a way that there is a small parameter in the defining equation. When this is the case it is often possible to deduce properties of the system by treating it as a perturbation of the system when the small parameter is zero and finding solutions as series expansions in the small parameter. As a simple example consider the quadratic equation

$$x^2 + 2\epsilon x - 1 = 0 \tag{7.1}$$

where $\epsilon$ is a small parameter, $\epsilon \ll 1$. This has solutions

$$x = -\epsilon \pm \sqrt{1+\epsilon^2} \tag{7.2}$$

and the two solutions can be written, expanding the expression under the square root, as

$$x = \begin{cases} 1 - \epsilon + \frac{1}{2}\epsilon^2 - \frac{1}{8}\epsilon^4 + \dots \\ -1 - \epsilon - \frac{1}{2}\epsilon^2 + \frac{1}{8}\epsilon^4 + \dots \end{cases} . \tag{7.3}$$

The main idea in perturbation theory is to expand the solution as

$$x = x_0 + \epsilon x_1 + \epsilon^2 x_2 + \dots . \tag{7.4}$$

and then substitute into the equation to be solved to find the coefficients $x_i$ by equating terms of order $\epsilon^i$. Substituting (7.4) into (7.1) gives

$$x_0^2 - 1 + \epsilon(2x_0 + 2x_0x_1) + \epsilon^2(2x_1 + x_1^2 + 2x_0x_2) + \ldots = 0 \tag{7.5}$$

and so, looking at the terms of order $\epsilon^0$ we find

$$x_0^2 - 1 = 0 \tag{7.6a}$$

so, taking the positive root, $x_0 = 1$. The terms in $\epsilon^1$ give

$$2x_0 + 2x_0x_1 = 0 \tag{7.6b}$$

so $x_1 = -1$, and the terms in $\epsilon^2$ give

$$2x_1 + x_1^2 + 2x_0x_2 = 0 \tag{7.6c}$$

and so $x_2 = \frac{1}{2}$. Putting these results together we have obtained the leading order terms of the expansion for the positive root of the quadratic equation:

$$x = 1 - \epsilon + \tfrac{1}{2}\epsilon^2 + \ldots. \tag{7.7}$$

This chapter is devoted to developing similar techniques for differential equations, and then applying them to a number of forced oscillators to get an idea of some of the different types of nonlinear phenomena that occur in these systems. There are many textbooks devoted almost entirely to the application of perturbation methods to differential equations. For more details I would recommend either Minorsky (1974) or Nayfeh and Mook (1979).

## 7.1 Asymptotic expansions

In what follows we shall not be concerned about whether the expansions derived actually converge. We shall only ask that they are asymptotic on some domain.

(7.1) DEFINITION

*The sum $\sum_{k=0}^{\infty} f_k(\epsilon)$ is an asymptotic expansion of the function $f(\epsilon)$ iff for all $n \geq 0$*

$$\frac{f(\epsilon) - \sum_{k=0}^{n} f_k(\epsilon)}{f_n(\epsilon)} \to 0 \text{ as } \epsilon \to 0$$

*i.e. if the remainder is smaller than the last term included in the expansion.*

If $\sum_k f_k(\epsilon)$ is an asymptotic expansion of $f(\epsilon)$ then we will write

$$f(\epsilon) \sim \sum_k f_k(\epsilon) \quad (\epsilon \to 0).$$

In most of the applications of asymptotic expansions here we shall be expanding solutions of differential equations $x(t,\epsilon)$. In this case we obtain asymptotic expansions of the form

$$x(t,\epsilon) \sim \sum_k a_k(t)\delta_k(\epsilon) \tag{7.8}$$

with, typically, $\delta_k(\epsilon) = \epsilon^k$. Such an expansion is asymptotic for $t$ in some domain, $D$, if $\sum_k a_k(t)\delta_k(\epsilon)$ is an asymptotic expansion as $\epsilon \to 0$ for all $t \in D$. We shall never verify that the expansions obtained really are asymptotic expansions, but we will use the criterion that if $\sum_k a_k(t)\delta_k(\epsilon)$ is an asymptotic expansion then

$$\frac{a_{k+1}(t)\delta_{k+1}(\epsilon)}{a_k(t)\delta_k(\epsilon)} \to 0 \text{ as } \epsilon \to 0.$$

In other words, we shall simply check that successive terms in the series are small compared with the previous term!

*Example 7.1*

The van der Pol oscillator is the equation

$$\ddot{x} + \epsilon\dot{x}(x^2 - 1) + x = 0, \quad 0 < \epsilon \ll 1$$

and so for $\epsilon$ small it is a small perturbation of the standard equation of simple harmonic motion. If we set $x(t,\epsilon) \sim x_0(t) + \epsilon x_1(t) + \ldots$ then the equations for $x_0$ and $x_1$, obtained by equating terms of order $\epsilon^0$ and $\epsilon^1$, are

$$\ddot{x}_0 + x_0 = 0, \quad \ddot{x}_1 + x_1 = -\dot{x}_0(x_0^2 - 1).$$

Hence $x_0 = A\cos(t - t_0)$, and choosing an initial condition such that $t_0 = 0$ the equation for $x_1$ becomes

$$\ddot{x}_1 + x_1 = A\sin t(A^2\cos^2 t - 1).$$

Tidying the right hand side of this equation using $\sin^3 t = (3\sin t - \sin 3t)/4$ we obtain

$$\ddot{x}_1 + x_1 = \tfrac{1}{4}A(A^2 - 4)\sin t + \tfrac{1}{4}A^3 \sin 3t.$$

This is a linear differential equation which can easily be solved to give

$$x_1 = B\cos(t - t_1) - \tfrac{1}{8}A(A^2 - 4)t\cos t - \tfrac{1}{32}A^3 \sin 3t.$$

The second term in this solution grows like $t \cos t$ and comes from the resonance of the forcing terms in $\sin t$. In other words, the particular integral part of the general solution is resonant: the forcing term on the right hand side of the $x_1$ equation contains the complementary function of the linear differential equation on the left hand side of the $x_1$ equation. Hence, if we want an asymptotic expansion for $x(t)$ valid over times of order $\epsilon^{-1}$ this term must vanish, i.e. we must take $A = 2$ to give the leading order periodic solution

$$x(t) \sim 2\cos t + O(\epsilon).$$

Thus the limit cycle for the van der Pol equation (which can be shown to exist using the decay of large amplitudes result of the previous chapter) has amplitude $2 + O(\epsilon)$ and approximate period $2\pi$.

If all perturbation theory was as straightforward as this example, this chapter would be shorter! In fact, we have been extremely lucky to be able to carry through the perturbation analysis this far, as the next example shows.

*Example 7.2*

The Duffing equation is

$$\ddot{x} + x + \epsilon x^3 = 0, \quad 0 < \epsilon \ll 1$$

and, as in the van der Pol equation, the unperturbed equation is the simple harmonic oscillator. As before we pose the asymptotic expansion $x(t) \sim x_0(t) + \epsilon x_1(t) + \dots$ and find the leading order equations

$$\ddot{x}_0 + x_0 = 0, \quad \ddot{x}_1 + x_1 = -x_0^3.$$

Choose initial conditions so that $x_0 = A\sin t$ so the second equation becomes

$$\ddot{x}_1 + x_1 = -A^3 \sin^3 t = \tfrac{1}{4}A^3(\sin 3t - 3\sin t).$$

Once again we have a resonant forcing term $(-\frac{3}{4}A^3 \sin t)$ which will give a solution proportional to $t\sin t$, which breaks the assumption that the series $x_0 + \epsilon x_1 + \dots$ is asymptotic on time scales of order $\epsilon^{-1}$. The big difference between this example and the previous one is that the only way of making the series asymptotic is to set $A = 0$. Does this imply that the only periodic solution is the trivial stationary point at the origin? The answer to this is no, as can easily be seen by noting that there is a simple first integral for the system: if $y = \dot{x}$ then $\frac{1}{2}y^2 + \frac{1}{2}x^2 + \frac{1}{4}\epsilon x^4$ is constant on trajectories, and all contours on which this function is constant are closed curves, representing periodic orbits in phase space. The point is that we have been sloppy about constructing the asymptotic series in the first place. The solution to the full equation for $x_1$ is

$$x_1 = B\sin(t - t_1) - \tfrac{1}{32}A^3 \sin 3t + \tfrac{3}{8}A^3 t\cos t$$

and so

$$x(t) \sim A\sin t + \epsilon(B\sin t - \tfrac{1}{32}A^3\sin 3t + \tfrac{3}{8}A^3 t\cos t) + \dots$$

where the dots denote terms of order $\epsilon^2$ and higher. This is *not* an asymptotic expansion for times of order $\epsilon^{-1}$ (unless $A = 0$), since the $\frac{3}{8}\epsilon t A^3 \cos t$ term is of order 1 when $t = O(\epsilon^{-1})$. But

$$A\sin t + \epsilon\tfrac{3}{8}A^3 t\cos t = A\sin(t + \tfrac{3}{8}\epsilon A^2 t) + \dots$$

and so we could equally well have rearranged terms in the solution of $x(t)$ and written

$$x(t) \sim A\sin(t + \tfrac{3}{8}\epsilon A^2 t) + \epsilon(B\sin t - \tfrac{1}{32}A^3 \sin 3t) + \dots,$$

which is now an asymptotic series (for any finite choice of $A$). This suggests that we should think a little harder about the way in which we develop the asymptotic series, taking into account the drift in phase (and possibly the amplitude) of the leading order solution which may destroy the asymptotic nature of a naively constructed series.

## 7.2 The method of multiple scales

The method of multiple scales (sometimes known as two-timing for reasons which will become obvious in a minute) is a method of dealing with the amplitude and phase drift described above in a systematic way. The important observation, made in the previous example, is that solutions oscillate on a time scale of order $t$ with an amplitude and phase which drifts on a time scale of order $\epsilon t$. This suggests that we should seek solutions of the form

$$R(\epsilon t)\sin(t + \theta(\epsilon t)) \tag{7.9}$$

or equivalently

$$A(\epsilon t)\sin t + B(\epsilon t)\cos t. \tag{7.10}$$

Thus there are two time scales, a fast time scale on which solutions oscillate, $\tau = t$, and a slow time scale, $T = \epsilon t$, on which the amplitude and phase evolve. More generally we pose an asymptotic expansion solution

$$x(t) = x(\tau, T, \epsilon) \sim x_0(\tau, T) + \epsilon x_1(\tau, T) + \dots \tag{7.11}$$

which is asymptotic for times up to order $\epsilon^{-1}$. In this expression we treat $\tau$ and $T$ as independent variables, so

$$\frac{d}{dt} = \frac{\partial}{\partial \tau} + \epsilon \frac{\partial}{\partial T}. \tag{7.12}$$

Using this last expression and the definition of the asymptotic series we find that

$$x \sim x_0(\tau, T) + \epsilon x_1(\tau, T) + \dots, \tag{7.13a}$$

$$\dot{x} \sim x_{0\tau} + \epsilon(x_{1\tau} + x_{0T}) + \dots \tag{7.13b}$$

and

$$\ddot{x} \sim x_{0\tau\tau} + \epsilon(x_{1\tau\tau} + 2x_{0T\tau}) + \dots \tag{7.13c}$$

where the subscripts denote partial differentiation. We can now treat the two examples of the previous section properly.

*Example 7.3*

Recall that the van der Pol oscillator is

$$\ddot{x} + \epsilon \dot{x}(x^2 - 1) + x = 0, \quad 0 < \epsilon \ll 1$$

and choose initial conditions $x(0) = 1$, $\dot{x}(0) = 0$. We pose the asymptotic series solution

$$x(t) \sim x_0(\tau, T) + \epsilon x_1(\tau, T) + \dots$$

valid for time scales of order $\epsilon^{-1}$. The equation at order $\epsilon^0$ is

$$x_{0\tau\tau} + x_0 = 0, \; x_0(0,0) = 1, \; x_{0\tau}(0,0) = 0$$

with solutions

$$x_0 = R(T)\cos(\tau + \Theta(T)), \; R(0) = 1, \; \Theta(0) = 0.$$

The terms of order $\epsilon^1$ give

$$x_{1\tau\tau} + x_1 = -2x_{0\tau T} - x_{0\tau}(x_0^2 - 1)$$

and the initial conditions become $x_1 = 0$ and $x_{1\tau} + x_{0T} = 0$ when $\tau = T = 0$. From the solution to the $x_0$ equation we see that

$$x_{0\tau} = -R(T)\sin(\tau + \Theta(T))$$

and

$$x_{0\tau T} = -R_T \sin(\tau + \Theta(T)) - R\Theta_T \cos(\tau + \Theta(T)).$$

Hence the right hand side of the $x_1$ equation is

$$2(R_T \sin(\tau + \Theta(T)) + R\Theta_T \cos(\tau + \Theta(T)))$$

$$+R\sin(\tau + \Theta(T))(R^2\cos^2(\tau + \Theta(T)) - 1),$$

which can be rewritten using the identity $\sin^3 t = \frac{1}{4}(3\sin t - \sin 3t)$ as

$$(2R_T + \tfrac{1}{4}R(4 - R^2))\sin(\tau + \Theta) + R\Theta_T \cos(\tau + \Theta)$$

$$+\tfrac{1}{4}R^3 \sin 3(\tau + \Theta).$$

The first two terms in this sum are resonant, in that they will cause $x_1$ to have solutions which are proportional to $\tau\cos(\tau+\Theta)$ and $\tau\sin(\tau+\Theta)$, which will break the assumption that the series is asymptotic for times of order $\epsilon^{-1}$. Hence, in order for the series to be asymptotic we must set

$$2R_T + \tfrac{1}{4}R(4 - R^2) = 0 = R\Theta_T,$$

(these are called *nonresonance* or *secularity* conditions). Together with the initial conditions from the $x_0$ equation this implies that we must solve

$$\Theta_T = 0,\ \Theta(0) = 0$$

and

$$R_T = \tfrac{1}{8}R(4 - R^2),\ R(0) = 1.$$

The first equation gives $\Theta(T) = 0$, which explains why we were apparently able to solve the problem without the method of multiple scales at the end of the previous chapter, and the $R$ equation is easily solved using partial fractions to give

$$R(T) = \frac{2}{(1 + 3e^{-T})^{\frac{1}{2}}}.$$

So for large $T$ solutions tend to $R = 2$ and we obtain the periodic orbit

$$x(t) \sim 2\cos t + O(\epsilon)$$

but now we know how the periodic orbit is approached and that it is stable. Note that we could have deduced that the periodic orbit is stable directly from the $R_T$ equation by looking at the graph of $R_T$ against $R$ as shown in Figure 7.1. This shows that $R_T > 0$ if $R < 2$ and $R_T < 0$ if $R > 2$, which implies that solutions tend to $R = 2$.

*Example 7.4*

The Duffing equation

$$\ddot{x} + x = -\epsilon x^3, \quad 0 < \epsilon \ll 1$$

can be treated in the same way. We look for a series solution $x(t) \sim x_0(\tau, T) + \epsilon x_1(\tau, T) + \ldots$, which is asymptotic for times of order $\epsilon^{-1}$, where $\tau = t$ and $T = \epsilon t$. At order $\epsilon^0$ we obtain the equation

$$x_{0\tau\tau} + x_0 = 0$$

and so

$$x_0 = R(T)\cos(\tau + \Theta(T))$$

where the functions $R$ and $\Theta$ are determined by the secularity or non-resonance condition at order $\epsilon$. The $x_1$ equation at order $\epsilon$ is

$$\begin{aligned} x_{1\tau\tau} + x_1 &= -2x_{0\tau T} - x_0^3 \\ &= 2(R\sin(\tau + \Theta))_T - R^3\cos^3(\tau + \Theta). \end{aligned}$$

So, differentiating the first term on the right hand side of this equation and using the identity $\cos^3 t = \frac{1}{4}(3\cos t + \cos 3t)$ we have

$$\begin{aligned} x_{1\tau\tau} + x_1 = 2R_T\sin(\tau + \Theta) &+ 2R\Theta_T\cos(\tau + \Theta) \\ &- \tfrac{1}{4}R^3(3\cos(\tau + \Theta) + \cos 3(\tau + \Theta)). \end{aligned}$$

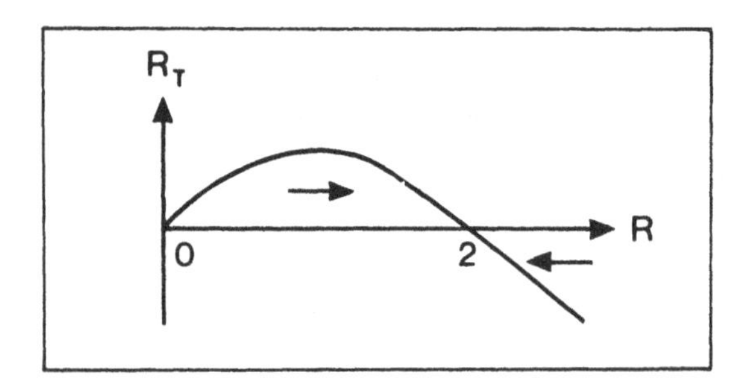


Fig. 7.1 Graph of $R_T$ against $R$.

The $\cos 3(\tau + \Theta)$ is harmless since it leads to a particular integral which is of order one, but the $\cos(\tau+\Theta)$ and $\sin(\tau+\Theta)$ terms on the right hand side are resonant and so if the series solution is to remain asymptotic over time scales of order $\epsilon^{-1}$ they must vanish, giving the secularity conditions

$$R_T = 0 \quad \text{and} \quad \Theta_T = \tfrac{3}{8}R^2.$$

Solving the first equation gives $R = R_0$, where $R_0$ is a constant, and so the second equation can be solved to give $\Theta = \frac{3}{8}R_0^2 T + t_0$, where $t_0$ is a constant (determined by the initial conditions). Hence the leading order solution is

$$x(t) \sim R_0 \cos(t + t_0 + \tfrac{3}{8}R_0^2 \epsilon t) + O(\epsilon).$$

(7.2) EXERCISE

*Find the first integral of Duffing's equation and hence sketch solution curves. How do these solutions compare with the asymptotic solution described above? [Hint: Think about $\Theta_T$ for the large amplitude solutions.]*

It is worth pointing out that this method is by no means limited to the analysis of perturbed oscillators; it can be applied to any system which is a perturbation of the linear differential equation $L(x, \dot{x}, \ddot{x}) = 0$ and can be extended to systems involving higher derivatives.

The method of multiple scales therefore involves a number of relatively straightforward steps. Given the equation

$$L(x, \dot{x}, \ddot{x}) = \epsilon f(x, \dot{x})$$

with $\epsilon \ll 1$ and where $L(x, \dot{x}, \ddot{x}) = 0$ is a linear differential equation and both $L$ and $f$ can depend explicitly on $t$ as described later in this chapter, we pose a series solution

$$x(t) \sim x_0(\tau, T) + \epsilon x_1(\tau, T) + \ldots$$

in terms of a fast time $\tau = t$ and a slow time $T = \epsilon t$, which is an asymptotic series on timescales of order $\epsilon^{-1}$. Using (7.13) the equation at order $\epsilon^0$ is

$$L(x_0, x_{0\tau}, x_{0\tau\tau}) = 0,$$

which can be solved, but the 'constants' are now functions of $T$. At order $\epsilon$ the equation is

$$L(x_1, x_{1\tau}, x_{1\tau\tau}) = g(x_0, x_{0\tau}, x_{0T}, x_{0\tau T})$$

where, for example, if $L(x, \dot{x}, \ddot{x}) = \ddot{x} + a\dot{x} + bx$ then

$$g(x_0, x_{0,\tau}, x_{0T}, x_{0\tau T}) = -2x_{0\tau T} - ax_{0T} + f(x_0, x_{0\tau}).$$

The first two terms on the right hand side of this equation come from the extra terms in (7.13b,c). We now identify the resonant terms in $g$, i.e. those which involve the complementary function of $L$ *and* which will give a term in the particular integral which breaks the assumption that the solution is asymptotic on timescales of order $\epsilon^{-1}$. The coefficients of these terms are then set to zero (the nonresonance condition), providing equations for the $T$ evolution of the 'constants' in the expressions for $x_0$. Solving these equations gives the lowest order approximation to the solution.

## 7.3 Multiple scales for forced oscillators: complex notation

The major difficulty in finding the drift equations for the amplitude and phase of the zeroth order solution lies in the identification of the resonant terms at order $\epsilon$. When the linear part of the zeroth order approximation is the equation of simple harmonic motion, $\ddot{x} + x = 0$, it is often more convenient to work in complex notation rather than using sines and cosines. This obviates the need to use nasty formulae for $\sin^3\theta$ and $\cos^3\theta$ and so on, since we need only look for the coefficients of $e^{\pm it}$ in the first order equation, and these can usually be read off without too much trouble.

To illustrate this point we shall go through the calculation for the van der Pol oscillator once again. Looking for a solution $x(t) \sim x_0(\tau, T) + \epsilon x_1(\tau, T) + \ldots$ (with $\tau = t$ and $T = \epsilon t$ as before) which is valid for times of order $\epsilon^{-1}$ we obtain the now familiar equations

$$x_{0\tau\tau} + x_0 = 0 \tag{7.14a}$$

$$x_{1\tau\tau} + x_1 = -2x_{0\tau T} - x_{0\tau}(x_0^2 - 1). \tag{7.14b}$$

The solution to the first of these equations can be written as

$$x_0(\tau, T) = A(T)e^{i\tau} + A^*(T)e^{-i\tau} \tag{7.15}$$

where $A$ is a complex function of the real variable $T$ and $A^*$ is the complex conjugate of $A$. This implies that

$$x_{0\tau} = iAe^{i\tau} - iA^*e^{-i\tau} \tag{7.16a}$$

and

$$x_{0\tau T} = iA_Te^{i\tau} - iA_T^*e^{-i\tau}. \tag{7.16b}$$

The right hand side of the equation for $x_1$ is therefore

$$-2(iA_Te^{i\tau} - iA_T^*e^{-i\tau}) - (iAe^{i\tau} - iA^*e^{-i\tau})([Ae^{i\tau} + A^*e^{-i\tau}]^2 - 1)$$

and it is now a formality to read off the terms which resonate with the left hand side of the equation (i.e. the terms in $e^{i\tau}$ and $e^{-i\tau}$). The coefficient of $e^{i\tau}$ is

$$-2iA_T - 2iA^2A^* + iA^*A^2 + iA$$

and the coefficient of $e^{-i\tau}$ is

$$2iA_T^* + 2iA^{*2}A - iAA^{*2} - iA^*,$$

which is, not surprisingly, the complex conjugate of the coefficient of $e^{i\tau}$. Hence the nonresonance or secularity condition is

$$2iA_T + iA^*A^2 - iA = 0 \tag{7.17}$$

or

$$2A_T = A(1 - |A|^2). \tag{7.18}$$

We can now solve for $A$ in two different ways. Setting $A = u + iv$ gives

$$u_T = \tfrac{1}{2}u(1 - (u^2 + v^2)), \quad v_T = \tfrac{1}{2}v(1 - (u^2 + v^2)) \tag{7.19}$$

(taking real and imaginary parts of (6.18)) or alternatively, setting $A = re^{i\theta}$ gives

$$A_T = r_Te^{i\theta} + ir\theta_Te^{i\theta}. \tag{7.20}$$

Now, since $2A_T = A(1 - |A|^2) = re^{i\theta}(1 - r^2)$, this implies that

$$r_T = r(1 - r^2), \quad r\theta_T = 0. \tag{7.21}$$

Hence $r \to 1$ and $\theta = \theta_0$ and so

$$x_0(\tau, T) \sim r(T)e^{i(\tau+\theta_0)} + r(T)e^{-i(\tau+\theta_0)} = 2r(T)\cos(\tau + \theta_0). \tag{7.22}$$

Since $r \to 1$ as $T \to \infty$ we find a periodic solution of period $2\pi + O(\epsilon^2)$ at amplitude $2 + O(\epsilon)$, as above.

## 7.4 Higher order terms

The method outlined above provides us with an asymptotic expansion which is valid on times of order $\epsilon^{-1}$. If we want to construct expansions which are valid on longer time scales we need to introduce further slow (or slower) times. For example, to obtain an expansion valid to times of order $\epsilon^{-2}$ we define three times,

$$\tau = t, \quad T_1 = \epsilon t \quad \text{and} \quad T_2 = \epsilon^2 t$$

and look for an asymptotic series

$$x(t) \sim x_0(\tau, T_1, T_2) + \epsilon x_1(\tau, T_1, T_2) + \epsilon^2 x_2(\tau, T_1, T_2) + \dots$$

treating $\tau$, $T_1$ and $T_2$ as independent variables. Hence

$$\frac{d}{dt} = \frac{\partial}{\partial \tau} + \epsilon \frac{\partial}{\partial T_1} + \epsilon^2 \frac{\partial}{\partial T_2}$$

and so

$$\dot{x} \sim x_{0\tau} + \epsilon(x_{1\tau} + x_{0T_1}) + \epsilon^2(x_{2\tau} + x_{1T_1} + x_{0T_2}) + \dots$$

and

$$\ddot{x} \sim x_{0\tau\tau} + \epsilon(x_{1\tau\tau} + 2x_{0\tau T_1}) + \epsilon^2(x_{2\tau\tau} + 2x_{0\tau T_2} + x_{0T_1T_1} + 2x_{1\tau T_1}) + \dots.$$

These equations can then be substituted into the perturbed equation of the form

$$L[x, \dot{x}, \ddot{x}] + \epsilon f(x, \dot{x}, t) + \epsilon^2 g(x, \dot{x}, t) = 0$$

where $L$ is a linear function, to obtain the desired equations to second order. The secularity conditions at first and second order then give the desired asymptotic solution. This is extremely messy and tedious. We will not go into further detail here.

## 7.5 Interlude

In the previous two sections we have described a method for treating differential equations which are perturbations of differential equations which we are able to solve. In the following sections we shall use these techniques to explore some of the nonlinear effects which can occur in perturbations of simple harmonic oscillators. First, though, we want to think a little about the type of behaviour that we might expect to see.

Consider the differential equation

$$\ddot{x} + x = \epsilon f(x, \dot{x}), \quad 0 < \epsilon \ll 1. \tag{7.23}$$

Solutions to the unperturbed equation, $\ddot{x} + x = 0$ are all periodic apart from the stationary point at the origin, which is a centre. We have already seen that centres are not persistent, small perturbations tend to make them either stable or unstable foci, so there are three possible types of behaviour that we might expect to observe in such systems. The origin might become stable, and all solutions tend to the origin (this is called quenching) or the origin might become unstable, and all solutions diverge from it for times of order $\epsilon^{-1}$ (this is called resonance again). A further possibility is that one or more of the periodic solutions of the unperturbed equation survives and so some combination of stable periodic orbits and, perhaps, stable stationary points could be observed. This is the case in the van der Pol and Duffing equations. The unperturbed equation, $\ddot{x}+x=0$ is Hamiltonian: it can be rewritten as $\dot{x} = y$, $\dot{y} = -x$, and so if $H(x,y) = \frac{1}{2}(x^2 + y^2)$ we have

$$\dot{x} = \frac{\partial H}{\partial y}, \ \dot{y} = -\frac{\partial H}{\partial x} \tag{7.24}$$

and $H$ is constant on trajectories. In Section 7.6 we outline a method which exploits this structure in order to determine which periodic orbits of the unperturbed system are picked out by the perturbation.

The equations become more interesting if the perturbing function is a function of $t$ as well as $x$ and $\dot{x}$. If $f$ is a periodic function of $t$, of period $2\pi/\omega$ say, then there are two competing periods in the problem: the natural period of the unperturbed equation and the forcing period. In Sections 7.7 and 7.8 we consider parametrically forced equations:

$$\ddot{x} + \omega(t)x = \epsilon f(x, \dot{x}), \quad 0 < \epsilon \ll 1 \tag{7.25}$$

where $\omega = 1 + \sigma(t)$ and $\sigma$ is a small, periodic function. This gives rise to a phenomenon known as resonance. For particular values of the period of $\sigma$ the amplitude of solutions can grow exponentially. This is called *parametric excitation* since resonance is due to the periodic variation in $\omega$, which can be considered as a parameter in the problem rather than external forcing. In Sections 7.9 and 7.10 we consider more general forcing

$$\ddot{x} + x = \epsilon f(x, \dot{x}, t), \quad 0 < \epsilon \ll 1 \tag{7.26}$$

where $f$ is periodic in the time variable, $t$. More precisely, we consider equations of the form

$$\ddot{x} + x = a(1 - \omega^2)\cos\omega t + \epsilon f(x, \dot{x}), \quad 0 < \epsilon \ll 1 \qquad (7.27)$$

where $\omega$ is a constant close to 1. The unperturbed equation ($\epsilon = 0$) has solutions

$$x = A\cos(t + \psi) + a\cos\omega t \qquad (7.28)$$

and so solutions are periodic if $\omega$ is rational and quasi-periodic if $\omega$ is irrational. For $\omega$ close to 1 and $\epsilon$ small but non-zero we find that solutions tend to be periodic: one of the two competing (but close) frequencies dominates and all solutions are 'locked' onto periodic orbits of approximate period $2\pi$. For some choices of $f$ there can be more than one stable periodic solution leading to *hysteresis*. As $\epsilon$ is increased slowly we follow one periodic orbit which becomes unstable, so solutions jump to another stable periodic orbit, whilst if $\epsilon$ is decreased slowly we follow the latter stable periodic orbit until it loses stability and solutions jump back to the original stable solution (see Fig. 7.2). In Sections 7.11 and 7.12 we consider cases in which $\omega$ is not close to the natural frequency of the oscillator. For typical frequencies we find that the amplitude $A$ of the unperturbed solution can decay, leaving only the $a\cos\omega t$ term (*asynchronous quenching*), but for particular values of $\omega$ more complicated and suprising behaviour is possible. In particular there can be periodic solutions whose frequency is a fraction of the forcing frequency, $\omega$. This is called *subharmonic resonance*.

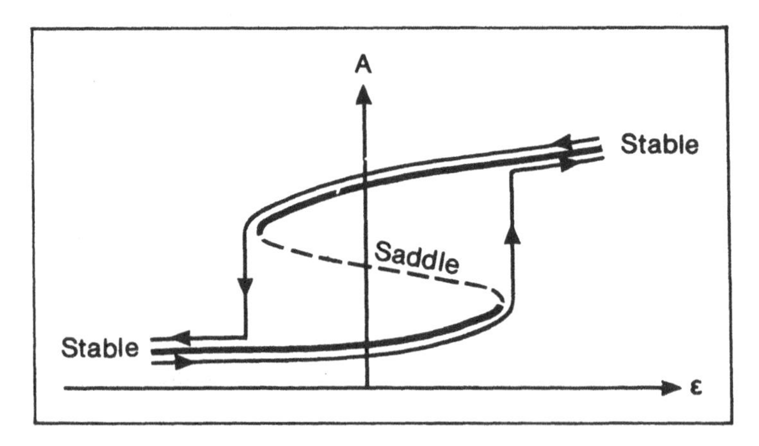


Fig. 7.2 Amplitude ($A$) vs. parameter ($\epsilon$) diagram showing hysteresis.

Finally, we look briefly at relaxation oscillators. If $\epsilon$ is very large the van der Pol oscillator has a stable periodic orbit which lies close to a simple cubic graph in phase space for nearly all the period. The solutions 'relax' onto this curve until they fall off it (at its turning points) where they zip across (very quickly) to another part of the curve and the process is repeated (Fig. 7.3).

Throughout this chapter we consider perturbations of oscillators with frequency 1 (and so with period $2\pi$). There is no loss of generality here since by rescaling time the oscillator $\ddot{x} + \omega_0^2 x = 0$ can be brought into the standard form $\ddot{x} + x = 0$. For example, in Section 7.11 we consider the equation

$$\ddot{x} + \epsilon\dot{x}(x^2 - 1) + x = a(1 - \omega^2)\cos\omega t \qquad (7.29)$$

with $\omega \approx 3$. All our results apply to the more general oscillator

$$\ddot{x} + \epsilon\dot{x}(x^2 - 1) + \omega_0^2 x = a(\omega_0^2 - \Omega^2)\cos\Omega t \qquad (7.30)$$

since if we define a rescaled time $s = \omega_0 t$ then $\frac{d}{dt} = \omega_0\frac{d}{ds}$ this equation becomes

$$\omega_0^2 x_{ss} + \epsilon\omega_0 x_s(x^2 - 1) + \omega_0^2 x = a(\omega_0^2 - \Omega^2)\cos\left(\frac{\Omega}{\omega_0}s\right).$$

Now setting $\epsilon' = \frac{\epsilon}{\omega_0}$ and $\omega = \frac{\Omega}{\omega_0}$ this equation becomes

$$x_{ss} + \epsilon' x_s(x^2 - 1) + x = a(1 - \omega^2)\cos\omega s.$$

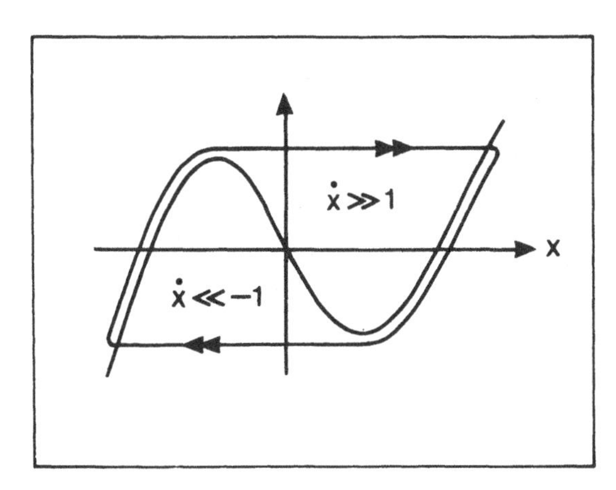


Fig. 7.3 A relaxation oscillator.

The analysis of (7.29) and (7.30) is therefore equivalent provided $\epsilon \ll \omega_0$ and the case $\omega \approx 3$ corresponds to $\Omega \approx 3\omega_0$.

## 7.6 Nearly Hamiltonian systems

The method of multiple scales requires a great deal of manipulation in all but the most simple of examples. If we are only interested in finding the approximate amplitudes of *autonomous* perturbations there is a much easier approach based on the fact that the unperturbed equations in all the examples considered so far are Hamiltonian, that is, there exists a function $H(x,y)$ such that

$$\dot{x} = \frac{\partial H}{\partial y}(x,y), \ \dot{y} = -\frac{\partial H}{\partial x}(x,y). \tag{7.31}$$

*Example 7.5*

Consider the equation of simple harmonic motion, $\ddot{x} + x = 0$. This can be written in two dimensions as $\dot{x} = y$, $\dot{y} = -x$. So if we set $H(x,y) = \frac{1}{2}(x^2 + y^2)$ then $\dot{x} = y = \frac{\partial H}{\partial y}$ and $\dot{y} = -x = -\frac{\partial H}{\partial x}$. $H$ is simply the first integral of the equation (see Chapter 1).

The phase space of Hamiltonian systems in the plane is particularly easy to sketch, since trajectories are simply curves with $H(x,y) = c$ for some constant, $c$. To see that $H$ is constant on trajectories consider $\frac{dH}{dt}$.

$$\frac{dH}{dt} = \dot{x}\frac{\partial H}{\partial x} + \dot{y}\frac{\partial H}{\partial y} = \frac{\partial H}{\partial y}\frac{\partial H}{\partial x} + (-\frac{\partial H}{\partial x})\frac{\partial H}{\partial y} = 0. \tag{7.32}$$

Now consider any small perturbation of a Hamiltonian system in the plane,

$$\dot{x} = f_1(x,y) + \epsilon g_1(x,y), \ f_1(x,y) = \frac{\partial H}{\partial y}(x,y) \tag{7.33a}$$

$$\dot{y} = f_2(x,y) + \epsilon g_2(x,y), \ f_2(x,y) = -\frac{\partial H}{\partial x}(x,y). \tag{7.33b}$$

Then if $\Gamma_\epsilon$ is a periodic orbit for this perturbed equation we must have

$$\int_{\Gamma_\epsilon} dH = 0 \tag{7.34}$$

since $\Gamma_\epsilon$ is a closed curve. But $dH = (\frac{\partial H}{\partial x})dx + (\frac{\partial H}{\partial y})dy$ and so

$$\int_{\Gamma_\epsilon} (-\dot{y} + \epsilon g_2(x,y))dx + (\dot{x} - \epsilon g_1(x,y))dy = 0. \tag{7.35}$$

Now, since $\Gamma_\epsilon$ is a trajectory

$$\int_{\Gamma_\epsilon} (-\dot{y}dx + \dot{x}dy) = \int_{\Gamma_\epsilon} (-\dot{y}\dot{x} + \dot{x}\dot{y})dt = 0$$

and so

$$\int_{\Gamma_\epsilon} (g_2(x,y)dx - g_1(x,y)dy) = 0. \tag{7.36}$$

But now, if we assume that $\Gamma_\epsilon$ is a small perturbation of a solution $\Gamma$ of the unperturbed equation we find that the periodic orbit for the perturbed equation is close to a periodic orbit for the unperturbed equation, $\Gamma$, such that

$$\int_{\Gamma} (g_2(x,y)dx - g_1(x,y)dy) = 0. \tag{7.37}$$

If we know the closed solutions of the unperturbed equation this integral can be calculated. By Green's theorem in the plane this expression equals

$$\int\int_{int(\Gamma)} \left(\frac{\partial g_1}{\partial x} + \frac{\partial g_2}{\partial y}\right) dxdy = 0. \tag{7.38}$$

A further way of expressing this condition is to note that on trajectories $dx \approx f_1(x,y)dt$ and $dy \approx f_2(x,y)dt$ and so if the unperturbed closed orbit has period T and is given by $(x(t), y(t))$ we have

$$\int_0^T (f_1 g_2 - f_2 g_1)dt = 0. \tag{7.39}$$

*Example 7.6*

The van der Pol oscillator

$$\dot{x} = y, \quad \dot{y} = -x - \epsilon y(x^2 - 1)$$

is a perturbation of a Hamiltonian system with $H = \frac{1}{2}(x^2 + y^2)$ and $g_1 = 0$, $g_2(x,y) = -y(x^2 - 1)$. Unperturbed periodic solutions have $x^2 + y^2 = A$ for any $A > 0$ and so the perturbed equation will have a periodic solution which satisfies (7.38):

$$\int\int_{x^2+y^2<A} (x^2 - 1)dxdy = 0$$

i.e.

$$\int_{r=0}^{A}\int_{\theta=0}^{2\pi}(r^2\cos^2\theta - 1)r dr d\theta = 0.$$

Evaluating the double integral we find

$$\frac{\pi A^4}{4} - \pi A^2 = 0$$

and so $A = 2$, as we had already deduced.

Alternatively we could use (7.39). Note that the unperturbed system has solutions

$$x = A\cos t, \quad y = -A\sin t$$

and that periodic solutions of the perturbed equation satisfy (7.39) with $f_1(x,y) = y$, $f_2(x,y) = -x$, $g_1(x,y) = 0$ and $g_2(x,y) = -y(x^2-1)$. Hence the approximate amplitude, $A$, of the perturbed periodic solution is given by

$$-\int_0^{2\pi} A^2\sin^2 t(A^2\cos^2 t - 1)dt = 0$$

or

$$-\tfrac{1}{4}\pi A^4 + \pi A^2 = 0.$$

As before, this little calculation gives $A = 2$.

The stability of the periodic orbit can also be determined using this method. Recall from Section 6.3 that a periodic solution, $u(t)$, is stable if

$$\int_0^T \nabla.f(u(\tau))d\tau < 0,$$

where $T$ is the period of the solution and $f$ is the defining vector field. From (7.33)

$$\nabla.f = \frac{\partial f_1}{\partial x} + \frac{\partial f_2}{\partial y} + \epsilon\left(\frac{\partial g_1}{\partial x} + \frac{\partial g_2}{\partial y}\right),$$

but since

$$f_1 = \frac{\partial H}{\partial y}, \quad \text{and} \quad f_2 = -\frac{\partial H}{\partial x}$$

the first two terms vanish to leave

$$\nabla.f = \epsilon\left(\frac{\partial g_1}{\partial x} + \frac{\partial g_2}{\partial y}\right).$$

The condition for stability is therefore

$$\epsilon \int_0^T \left( \frac{\partial g_1}{\partial x} + \frac{\partial g_2}{\partial y} \right) d\tau < 0 \tag{7.40}$$

where the partial derivatives are evaluated on the approximate periodic solutions.

*Example 7.6 continued*

Evaluating (7.40) for the van der Pol oscillator we find that the periodic orbit is stable if

$$\int_0^{2\pi} (x^2 - 1) d\tau > 0.$$

The approximate solution has $x = 2\cos t$ so this condition is

$$\int_0^{2\pi} (4\cos^2\tau - 1) d\tau > 0.$$

Evaluating the integral we obtain $2\pi$ which is positive, hence the periodic orbit is stable.

## 7.7 Resonance in the Mathieu equation

The Mathieu equation is $\ddot{x} + (1 + 4\epsilon\cos 2t)x = 0$, which we will write as

$$\ddot{x} + x = -4\epsilon x \cos 2t, \quad 0 < \epsilon \ll 1. \tag{7.41}$$

To apply the method of Section 7.3 we look for a solution

$$x(t) \sim x_0(\tau, T) + \epsilon x_1(\tau, T) + \ldots \tag{7.42}$$

with $\tau = t$, $T = \epsilon t$, which is asymptotic for $t = O(\epsilon^{-1})$. This gives the leading order solution

$$x_0 = A(T)e^{i\tau} + A^*(T)e^{-i\tau}. \tag{7.43}$$

Hence the equation at order $\epsilon$ is

$$x_{1\tau\tau} + x_1 = -2x_{0\tau T} - 4x_0 \cos 2\tau$$

$$= -2(iA_T e^{i\tau} - iA_T^* e^{-i\tau}) - 2(Ae^{i\tau} + A^* e^{-i\tau})(e^{2i\tau} + e^{-2i\tau}). \tag{7.44}$$

Reading off the resonant $e^{i\tau}$ terms in this latter expression and setting them to zero gives

$$iA_T + A^* = 0 \tag{7.45}$$

or, if $A = u(T) + iv(T)$,

$$-v_T + u = 0 \text{ and } u_T - v = 0. \tag{7.46}$$

Hence $u_{TT} = u$, so $u(T) = c_1 e^T + c_2 e^{-T}$ and $v(T) = c_1 e^T - c_2 e^{-T}$. The solution, asymptotic for time of order $\epsilon^{-1}$, has an amplitude which grows exponentially like $e^{\epsilon t}$ and we obtain solutions with increasingly large amplitudes. This phenomenon is called resonance.

## 7.8 Parametric excitation in Mathieu equations

We might now ask how robust a phenomenon the resonance described in the previous section is. To this end we investigate a modification of the Mathieu equation

$$\ddot{x} + (\Delta + 4\epsilon \cos 2t)x = 0, \quad 0 < \epsilon \ll 1 \tag{7.47}$$

for values of $\Delta$ close to unity. To do this in such a way as to keep track of small terms we expand $\Delta$ as a power series in $\epsilon$, $\Delta = 1 + \epsilon\delta + \ldots$ to obtain

$$\ddot{x} + x = -\epsilon x(\delta + 4\cos 2t). \tag{7.48}$$

Once again we pose the asymptotic expansion, valid for times of order $\epsilon^{-1}$, $x(t) \sim x_0(\tau, T) + \epsilon x_1(\tau, T) + \ldots$ in terms of the fast time, $\tau = t$ and the slow time $T = \epsilon t$ which gives the leading order solution

$$x_0 = A(T)e^{i\tau} + A^*(T)e^{-i\tau} \tag{7.49}$$

with

$$\begin{aligned} x_{1\tau\tau} + x_1 = & -2(iAe^{i\tau} - iA^*e^{-i\tau}) - \delta(Ae^{i\tau} + A^*e^{-i\tau}) \\ & -2(Ae^{i\tau} + A^*e^{-i\tau})(e^{2i\tau} + e^{-2i\tau}). \end{aligned} \tag{7.50}$$

Comparing this with the equation derived in the previous section we see that the only terms that we have not considered already are those involving $\delta$ which are easy, so the non-resonance condition at $O(\epsilon)$ is

$$2iA_T + \delta A + 2A^* = 0. \tag{7.51}$$

Setting $A = u + iv$ as before we find that

$$v_T = (1 + \tfrac{1}{2}\delta)u, \text{ and } u_T = (1 - \tfrac{1}{2}\delta)v \tag{7.52}$$

or $u_{TT} = (1 - \frac{1}{2}\delta)(1 + \frac{1}{2}\delta)u = (1 - \frac{1}{4}\delta^2)u$. Thus solutions are stable (on timescales of order $\epsilon^{-1}$) provided $\delta^2 > 4$. In terms of the original parameters $\epsilon$ and $\Delta$ this implies that the system is resonant if

$$|\Delta - 1| < 2\epsilon. \tag{7.53}$$

## 7.9 Frequency locking

In this section we want to consider oscillators which are forced by an almost resonant term. Consider the modified van der Pol oscillator

$$\ddot{x} + \epsilon\dot{x}(x^2 - 1) + \Delta x = A\cos t, \quad 0 < \epsilon \ll 1 \tag{7.54}$$

where $\Delta$ is close to 1 and $A$ is small (of the same order as $\epsilon$). Here, as described in the interlude, we are interested in the way that the natural frequency of the system ($\Delta$) and the forcing frequency (1) compete with each other. In order to obtain an asymptotic series we must ensure that terms of order $\epsilon$ are identified, so set $\Delta = 1 + \delta\epsilon$ and $A = \alpha\epsilon$ to get

$$\ddot{x} + x = -\epsilon(\dot{x}(x^2 - 1) + \delta x - \alpha\cos t). \tag{7.55}$$

As ever, we look for a solution in terms of a slow and a fast time of the form $x(t) = x_0(\tau, T) + \epsilon x_1(\tau, T) + \ldots$ which is asymptotic for $t = O(\epsilon^{-1})$. This gives $x_0 = A(T)e^{i\tau} + A^*(T)e^{-i\tau}$ and

$$x_{1\tau\tau} + x_1 = -2x_{0\tau T} - x_{0\tau}(x_0^2 - 1) - \delta x_0 + \alpha\cos\tau. \tag{7.56}$$

Substituting the expression for $x_0$ into the right hand side of this equation we find

$$\begin{aligned}
&-2[iA_Te^{i\tau} - iA_T^*e^{-i\tau}] \\
&-[iAe^{i\tau} - iA^*e^{-i\tau}][A^2e^{2i\tau} + 2AA^* + A^{*2}e^{-2i\tau} - 1] \\
&-\delta[Ae^{i\tau} + A^*e^{-i\tau}] + \frac{\alpha}{2}[e^{i\tau} + e^{-i\tau}].
\end{aligned}$$

Collecting up the resonant $e^{i\tau}$ terms and setting them to zero we obtain the equation

$$-2iA_T - iA^*A^2 + iA - \delta A + \frac{\alpha}{2} = 0$$

or

$$2A_T = (1 + i\delta)A - |A|^2A - \frac{i}{2}\alpha. \tag{7.57}$$

Setting $A(T) = r(T)e^{i\theta(T)}$ we obtain the two real differential equations

$$r_T = \tfrac{1}{2}\left(r - r^3 - \frac{\alpha}{2}\sin\theta\right) \tag{7.58a}$$

$$\theta_T = \tfrac{1}{2}\left(\delta - \frac{\alpha}{2r}\cos\theta\right). \tag{7.58b}$$

To investigate these equations in the $(r, \theta)$ plane we use some of the techniques for planar equations described in Chapter 5. We begin by looking for stationary solutions. These satisfy $r_T = \theta_T = 0$, i.e.

$$\sin\theta = \frac{2}{\alpha}r(1 - r^2) \tag{7.59}$$

and

$$r = \frac{\alpha}{2\delta}\cos\theta. \tag{7.60}$$

Equation (7.60) describes a circle of radius $\alpha/4\delta$ centred on the point $(r, \theta) = (\alpha/4\delta, 0)$ but (7.59) is a little more difficult. Figure 7.4 shows the right hand side of (7.59) as a function of $r$. It has a maximum when $2 - 6r^2 = 0$ at which value the function $\frac{2}{\alpha}r(1 - r^2)$ takes the value $\frac{4}{3\sqrt{3}\alpha}$. Hence if $\frac{4}{3\sqrt{3}\alpha} < 1$ there is a set of $\theta$ values which are not realized by (7.59) whilst if $\frac{4}{3\sqrt{3}\alpha} > 1$ all $\theta$ values between 0 and $\pi$ have two solutions for $r$ and all $\theta$ values between $\pi$ and $2\pi$ have a single solution for $r$. Hence the contours of $r_T = 0$ are as shown in Figure 7.5. We will concentrate on the case $\frac{4}{3\sqrt{3}\alpha} > 1$; the other case is left as an exercise.

Putting the contour of $\theta_T = 0$ and the contours of $r_T = 0$ together we see that there are two possibilities depending on the relative values of $\alpha$ and $\delta$ (both of which are assumed positive). There is a critical value of $\frac{\alpha}{\delta}$, $\mu_c$ say, such that if $\frac{\alpha}{\delta} > \mu_c$ then the circle $r = \frac{\alpha}{2\delta}\cos\theta$ intersects the contour of $r_T = 0$ at four points as shown in Figure 7.6a whilst if $\frac{\alpha}{\delta} < \mu_c$ the circle intersects the contours of $r_T = 0$ in two points (Fig. 7.6c). Figure 7.6b shows the limiting situation when $\frac{\alpha}{\delta} = \mu_c$.

In principle we should now find the spectrum of the linear flow near each of these points in order to determine their type, but the information

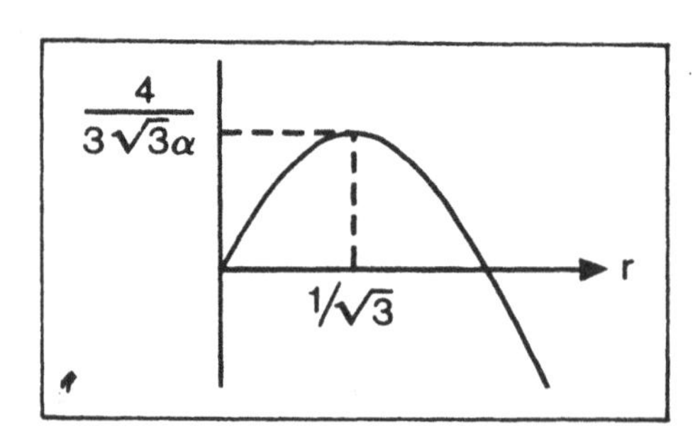


Fig. 7.4

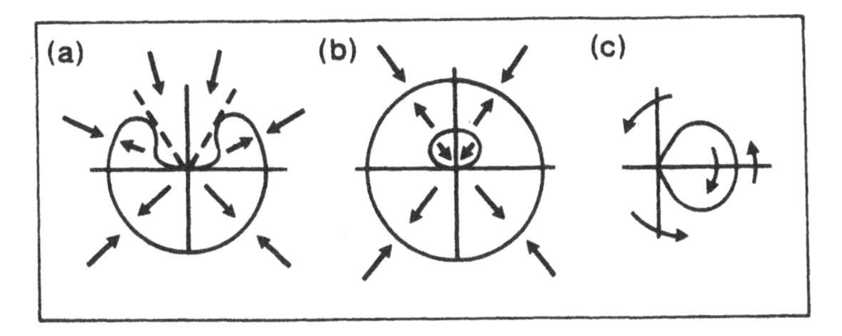


Fig. 7.5 (a) $r$ motion, $\alpha > \frac{4}{3\sqrt{3}}$; (b) $r$ motion, $\alpha < \frac{4}{3\sqrt{3}}$; (c) $\theta$ motion.

in Figures 7.5 and 7.6 of the direction of $\theta_T$ and $r_T$ is sufficient for us to be able to deduce the type of these points without going into this detail (see Fig. 7.7). In Figure 7.8 we show the flows which we can obtain in the three cases. Note that in Figure 7.8a there is an invariant curve made up of a stable sink, a saddle and the unstable manifold of the saddle (there is nowhere else for this unstable manifold to go!). This invariant curve becomes a stable periodic orbit when these two stationary points annihilate at $\frac{\alpha}{\delta} = \mu_c$. Note also that there is no stationary point at the origin (the intersection of the curves $r_T = 0$ and $\theta_0 = 0$ is a coordinate singularity). We conclude that when $\frac{\alpha}{\delta} > \mu_c$ solutions are attracted to a (non-trivial) stationary point, whilst if $\frac{\alpha}{\delta} < \mu_c$ solutions are attracted to a stable periodic orbit. In the former case we obtain solutions $x(t)$ which are periodic. This is called frequency locking: instead of two frequencies we have only one and solutions are periodic, whilst in the latter case the asymptotic solution has a modulus and phase which oscillate on a timescale of order $\epsilon t$ corresponding to beating between the two frequencies.

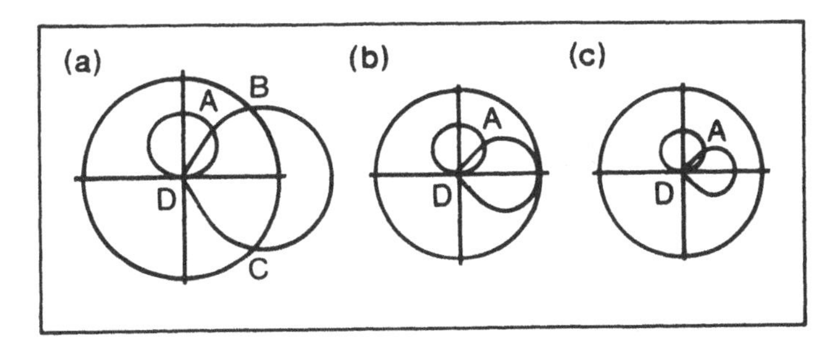


Fig. 7.6 Isoclines. (a) $\frac{\alpha}{\delta} > \mu_c$; (b) $\frac{\alpha}{\delta} = \mu_c$; and (c) $\frac{\alpha}{\delta} < \mu_c$.

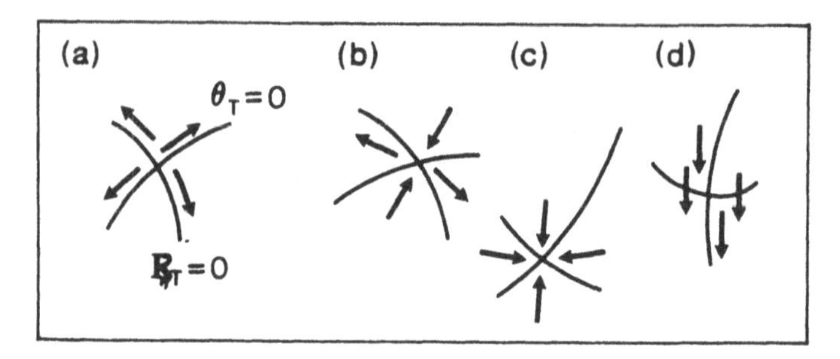


Fig. 7.7 Local pictures at (a) $A$; (b) $B$; (c) $C$; and (d) $D$.

When $\frac{\alpha}{\delta}$ passes through the critical value $\mu_c$ there is a qualitative change in the behaviour of the system as the two stationary points on the invariant closed curve come together and annihilate, leaving the periodic orbit as the attracting object. This is an example of a bifurcation, and we will return to a more general study of such phenomena in the next chapter.

## 7.10 Hysteresis

Consider the equation

$$\ddot{x} + B\dot{x} + (1+\Delta)x + \epsilon x^3 = A\cos t, \quad 0 < \epsilon \ll 1 \tag{7.61}$$

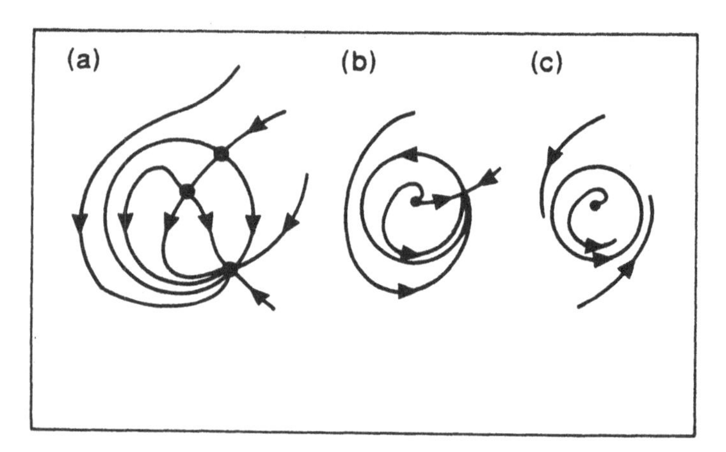


Fig. 7.8 Phase portraits. (a) $\frac{\alpha}{\delta} > \mu_c$; (b) $\frac{\alpha}{\delta} = \mu_c$; and (c) $\frac{\alpha}{\delta} < \mu_c$.

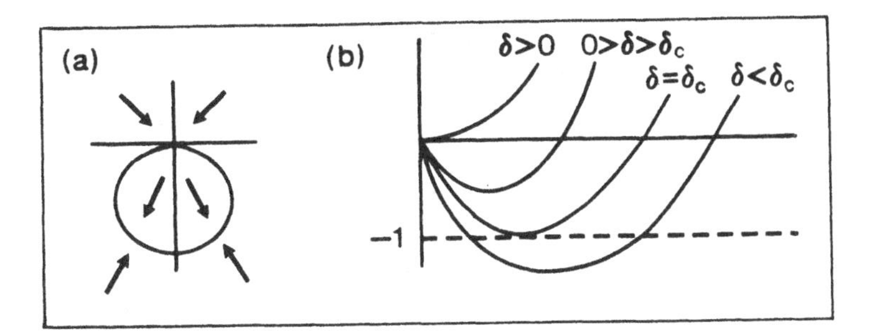


Fig. 7.9 (a) $r$ motion; (b) right hand side of (7.67) at different values of $\delta$.

where $B > 0$ and $B$, $\Delta$, $\epsilon$ and $A$ are all small parameters. This is just a mixture of systems which we have already looked at with the addition of a frictional term $B\dot{x}$. To bring it into standard form we shall assume that $\epsilon > 0$ and set $B = \beta\epsilon$, $\beta > 0$, $\Delta = \delta\epsilon$ and $A = \alpha\epsilon$ so

$$\ddot{x} + x = -\epsilon(\beta\dot{x} + \delta x + x^3 - \alpha\cos t). \tag{7.62}$$

Now posing the expansion in terms of a fast time $\tau = t$ and a slow time $T = \epsilon t$, $x(t) \sim x_0(\tau, T) + x_1(\tau, T) + \ldots$, asymptotic for $t = O(\epsilon^{-1})$, we obtain the solution

$$x_0 = A(T)e^{i\tau} + A^*(T)e^{-i\tau} \tag{7.63}$$

and the non-resonance condition at $O(\epsilon)$, which we leave as an exercise (apart from the $\beta\dot{x}$ term we have covered each case earlier), is

$$-2iA_T - \beta iA - \delta A - 3|A|^2A + \frac{\alpha}{2} = 0$$

i.e.

$$2A_T = -(\beta - i\delta)A + 3i|A|^2A - \tfrac{i}{2}\alpha. \tag{7.64}$$

In polar coordinates, $A = re^{i\theta}$, this implies that

$$2r_T = -\beta r - \frac{\alpha}{2}\sin\theta \tag{7.65a}$$

$$2\theta_T = \delta + 3r^2 - \frac{\alpha}{2r}\cos\theta. \tag{7.65b}$$

We now proceed as with the previous example: $r_T = 0$ when

$$r = -\frac{\alpha}{2\beta}\sin\theta, \tag{7.66}$$

which is the circle shown in Figure 7.9a, and $\theta_T = 0$ when

$$\cos\theta = \frac{2r}{\alpha}(\delta + 3r^2). \tag{7.67}$$

Figure 7.9b shows the right hand side of (7.67) for different values of $\delta$. If $\delta > 0$ each value of $\cos\theta$ between 0 and 1 has a unique value of $r$ associated to it, and so the contour looks like the sketch in Figure 7.10a. For $0 > \delta > \delta_c$ there are two values of $r$ for each value of $\cos\theta$ in some interval $(-z, 0)$, $0 < z < 1$, and one value of $r$ for each value of $\cos\theta$ between 0 and 1, giving the contour shown in Figure 7.10b. If $\delta = \delta_c$ the minimum of the right hand side of the equation for $\cos\theta$ equals $-1$ and so for $\delta < \delta_c$ there are two values of $r$ for each value of $\cos\theta$ between $-1$ and 0 and one value for $\cos\theta$ between 0 and 1 and so the contour $\theta_T = 0$ is as shown in Figure 7.10c. Note that the value of $\delta_c$ is easy to calculate since the turning point of the function of $r$ which determines $\cos\theta$ is $\frac{\sqrt{-\delta}}{9}$ and so

$$\delta_c^3 = -\frac{9^2}{8^2}\alpha^2. \tag{7.68}$$

Before putting the contours of $r_T = 0$ and $\theta_T = 0$ together in order to deduce the nature of the stationary points of the non-resonant equation we observe that there are two important features of the flow which are easy to see if the equations are written in Cartesian coordinates. Writing $A = u + iv$ the non-resonance condition becomes

$$2u_T = -\beta u - \delta v - 3(u^2 + v^2)v \tag{7.69a}$$

$$2v_T = \delta u - \beta v + 3(u^2 + v^2)u - \frac{\alpha}{2} \tag{7.69b}$$

from which we see that the origin is not a stationary point of the equations unless $\alpha = 0$. Furthermore, the divergence of the right hand side of the equations is $-2\beta$ and so, using Dulac's criterion with weighting function equal to unity, we see that if $\beta \neq 0$ there can be no periodic orbits in the system.

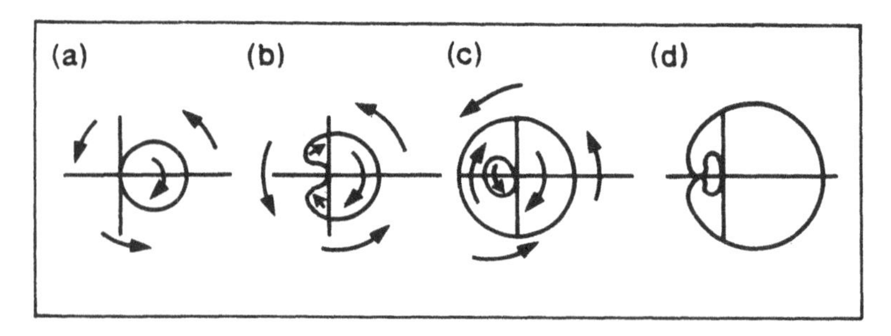


Fig. 7.10 $\theta$ motion. (a) $\delta > 0$; (b) $0 > \delta > \delta_c$; (c) $\delta_c > \delta$; and (d) as in (b) with $\delta$ near $\delta_c$.

Rather than give a complete description of all the possible behaviour that can be observed in this system we shall restrict our attention to one possible sequence. Fix $\alpha > 0$ and $\delta \in (\delta_c, 0)$ but close to $\delta_c$ so that the contour $\theta_T = 0$ is as shown in Figure 7.10d.

For sufficiently small (but positive) $\beta$ the circle $r_T = 0$ comes out of the mouth of $\theta_T = 0$ (Fig. 7.11, A) and so there is only one stationary point, and it lies in the negative right half-plane. It is not hard to see that it is stable (Fig. 7.12a). As $\beta$ increases there is a critical value $\beta_1$ at which the two curves $r_T = 0$ and $\theta_T = 0$ are tangent (B) and as $\beta$ increases further there are two new stationary points (C). The one inside the mouth of $\theta_T = 0$ is stable and the other is a saddle. As $\beta$ increases the saddle moves around the circle towards the original stationary point and when $\beta = \beta_2$ (D) there is a second tangency which removes these two stationary points. For $\beta > \beta_2$ (E) the stable stationary point in the mouth of $\theta_T = 0$ is the only stationary point of the system.

Consider increasing $\beta$ slowly from a small positive value. Initially solutions are attracted to the upper branch of solutions and provided $\beta$ is changed slowly will remain on the upper branch until it folds over, when the solution will jump (apparently discontinuously!) to the lower branch. If we now decrease $\beta$ slowly, solutions will stay on the lower branch until it folds, where solutions jump back up to the upper branch

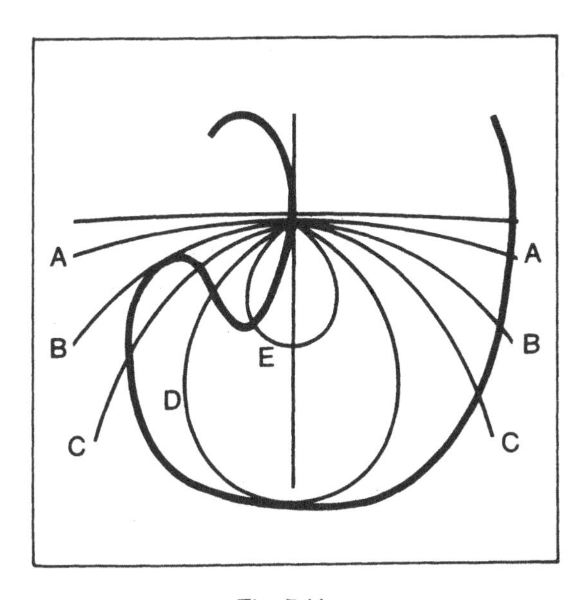


Fig. 7.11

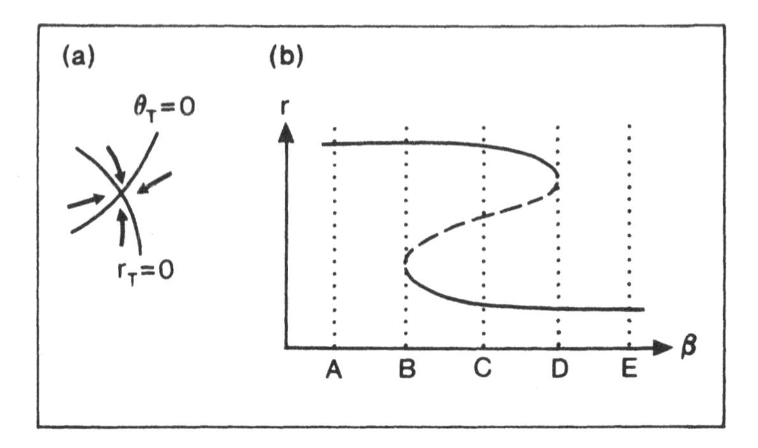


Fig. 7.12 (a) Close up of the right hand stationary point for $A$; (b) hysteresis.

(see Fig. 7.12b). This effect is called hysteresis; the selection of which attractor the solution chooses in the overlap region depends upon where solutions begin. In terms of the original problem each stationary point corresponds to a periodic orbit, so at the boundaries of the overlap region two periodic orbits come together and annihilate. Once again, these are bifurcations (but there is no invariant curve this time).

## 7.11 Asynchronous quenching

We now move on to consider differential equations of the form

$$\ddot{x} + x + a(\omega^2 - 1)\cos\omega t = \epsilon f(x, \dot{x}), \quad 0 < \epsilon \ll 1 \tag{7.70}$$

where $|\omega - 1|$ is not of order $\epsilon$ (because in the case that it is we get back to the analysis of Section 7.9). Without loss of generality we will always assume that $\omega > 0$. As before, we pose the asymptotic expansion, valid for $t = O(\epsilon^{-1})$, in terms of the fast time $\tau = t$ and the slow time $T = \epsilon t$, $x(t) \sim x_0(\tau, T) + \epsilon x_1(\tau, T) + \ldots$. For the case

$$\ddot{x} + \epsilon\dot{x}(x^2 - 1) + x + a(\omega^2 - 1)\cos\omega t = 0, \quad 0 < \epsilon \ll 1 \tag{7.71}$$

the $x_0$ and $x_1$ equations are

$$x_{0\tau\tau} + x_0 + a(\omega^2 - 1)\cos\omega\tau = 0 \tag{7.72}$$

and

$$x_{1\tau\tau} + x_1 = -2x_{0\tau T} - x_{0\tau}(x_0^2 - 1). \tag{7.73}$$

so the $x_0$ solution is

$$x_0 = A(T)e^{i\tau} + A^*(T)e^{-i\tau} + a\cos\omega\tau. \tag{7.74}$$

Recall that in the method of multiple scales we now substitute this expression into the $x_1$ equation and then choose $A$ such that there are no resonant (or secular) terms on the right hand side of the equation, that is, we choose $A$ so that the coefficient of any term with $e^{\pm i\tau}$ vanishes in order to avoid responses like $\tau e^{\pm i\tau}$ in $x_1$ which break the assumption that the solution is asymptotic for $t = O(\epsilon^{-1})$. The right hand side of the $x_1$ equation is

$$\begin{aligned}&-2(iA_Te^{i\tau} - iA_T^*e^{-i\tau})\\ &-\left(iAe^{i\tau} - iA^*e^{-i\tau} + \tfrac{1}{2}ia\omega(e^{i\omega\tau} - e^{-i\omega\tau})\right)\\ &\times\left([Ae^{i\tau} + A^*e^{-i\tau} + \tfrac{1}{2}ae^{i\omega\tau} + \tfrac{1}{2}ae^{-i\omega\tau}]^2 - 1\right).\end{aligned}$$

Some of these terms are familiar from the treatment of the van der Pol equation in previous sections. These old terms (which do not involve $e^{\pm i\omega\tau}$) can be treated as before, whilst looking at the expression for the right hand side we see that we have new terms like:

$e^{\pm i\tau(2+\omega)}$ and $e^{\pm i\tau(2-\omega)}$, which are resonant if $\omega = -1, -3$ or $\omega = 1, 3$ respectively;
$e^{\pm i\tau(1+2\omega)}$ and $e^{\pm i\tau(1-2\omega)}$, which are resonant if $\omega = -1$ or $\omega = 1$ respectively and both are resonant if $\omega = 0$;
$e^{\pm 3i\omega\tau}$, which are resonant if $\omega = \pm\frac{1}{3}$; and
$e^{\pm i\omega\tau}$, which are resonant if $\omega = \pm 1$.

Now, by assumption, $\omega > 0$ and looking at each expression we find that provided $\omega \neq 0, \frac{1}{3}, 1, 3$ the non-resonance condition becomes

$$-2iA_T + iA - i|A|^2A - \frac{a^2}{2}iA = 0 \tag{7.75}$$

or, setting $A = re^{i\theta}$,

$$\theta_T = 0 \tag{7.76a}$$

$$r_T = \tfrac{1}{2}r\left(1 - \frac{a^2}{2} - r^2\right). \tag{7.76b}$$

Hence if $a^2 > 2$, $r_T < 0$ for all $r > 0$ and the origin is asymptotically stable. In other words if $a^2 > 2$ the 'free mode' dies away and we are left only with the forced response $a\cos\omega t$. This is called asynchronous quenching, where the term 'asynchronous' refers to the fact that we have avoided the resonant frequencies $\omega = 0, \frac{1}{3}, 1, 3$.

(7.3) EXERCISE

*Describe the behaviour if $a^2 < 2$. This is called soft excitation.*

## 7.12 Subharmonic resonance

The unperturbed solution to the forced van der Pol oscillator of the previous example has solutions

$$R\cos(t-\psi) + a\cos\omega t$$

which, at the resonant frequencies $\omega = 0, \frac{1}{3}, 1, 3$ identified in the previous section, give solutions which are periodic of period $2\pi$ if $\omega = 0, 1, 3$ or $6\pi$ if $\omega = \frac{1}{3}$. A closer inspection of the resonance at $\omega = 3$ will show that although quenching can occur, giving solutions with frequency $\omega = 3$, it is also possible to have non-trivial stable periodic solutions of frequency $\frac{\omega}{3}$. This is an example of *subharmonic resonance*; the response frequency is a fraction of the frequency of the external forcing. To investigate this possibility further we consider the same equation as before,

$$\ddot{x} + \epsilon\dot{x}(x^2-1) + x + a(\omega^2-1)\cos\omega t = 0, \tag{7.77}$$

for $\omega$ close to 3, i.e. $\omega = 3(1+\delta\epsilon)$. Rather than work with a complicated $\epsilon$ dependence in the cosine it is more convenient to rescale time before beginning the analysis (we need to be absolutely certain that all terms of order $\epsilon$ are isolated). Set $s = (1+\delta\epsilon)t$, so $\frac{d}{dt} = (1+\delta\epsilon)\frac{d}{ds}$ and (7.77) becomes

$$(1+\delta\epsilon)^2\frac{d^2x}{ds^2} + \epsilon(1+\delta\epsilon)\frac{dx}{ds}(x^2-1) + x + a(9(1+\delta\epsilon)^2-1)\cos 3s = 0.$$

Dividing through by $(1+\delta\epsilon)^2$ gives

$$\ddot{x} + \epsilon(1+\delta\epsilon)^{-1}\dot{x}(x^2-1) + (1+\delta\epsilon)^{-2}x + a(9-(1+\delta\epsilon)^{-2})\cos 3s = 0$$

where the dots now represent differentiation with respect to $s$. Now expanding expressions $(1+\delta\epsilon)^{-1}$ and $(1+\delta\epsilon)^{-2}$ as power series in $\epsilon$ and simplifying matters by redefining $\delta$ and $a$ a little and renaming $s$ by $t$ we obtain the equation

$$\ddot{x} + \epsilon\dot{x}(x^2-1) + (1+\delta\epsilon)x + 8a\cos 3t = O(\epsilon^2). \tag{7.78}$$

Going through the same rigmarole as in the previous section for this modified equation (so $x(t) \sim x_0(\tau, T) + \epsilon x_1(\tau, T) + \ldots$ with $x_0 =$

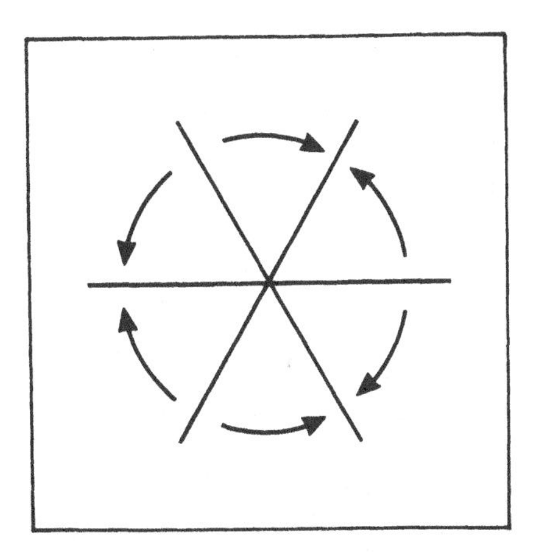

Fig. 7.13 $\theta$ motion ($\delta = 0$).

$A(T)e^{i\tau} + A^*(T)e^{-i\tau} + a\cos 3\tau$ and so on) we find that the non-resonance condition at order $\epsilon$ is

$$-2iA_T + i\left(1 - \frac{a^2}{2} + i\delta\right)A - i|A|^2A - \frac{ia\omega}{2}A^{*2} = 0 \qquad (7.79)$$

or, in polar coordinates

$$r_T = \tfrac{1}{4}r([2 - a^2] - ar\cos 3\theta - 2r^2) \qquad (7.80)$$

$$\theta_T = \tfrac{1}{4}(2\delta + ar\sin 3\theta). \qquad (7.81)$$

We shall consider two cases separately: first $\delta = 0$, i.e. the situation when $\omega = 3$, and second the more general case, $\omega \approx 3$, where we will consider only the case $\delta > 0$.

(i) $\delta = 0$. If $\delta = 0$ the $\theta$ equation becomes

$$\theta_T = \tfrac{1}{4}ar\sin 3\theta \qquad (7.82)$$

so (see Fig. 7.13) provided $a > 0$ and $r \neq 0$ we see that $\theta \to \frac{1}{3}(\pi + 2n\pi)$ and hence $\cos 3\theta \to -1$.

Substituting this value of $\cos 3\theta$ into the $r$ equation we find

$$r_T = \tfrac{1}{4}r([2 - a^2] + ar - 2r^2). \qquad (7.83)$$

The expression in round brackets on the right hand side of (7.83) is sketched in Figure 7.14 for the three qualitatively different regions. If $a^2 < 2$ then all solutions tend to a non-trivial stationary point (soft excitation) whilst if $2 < a^2 < \frac{16}{7}$ solutions tend to the non-trivial stationary

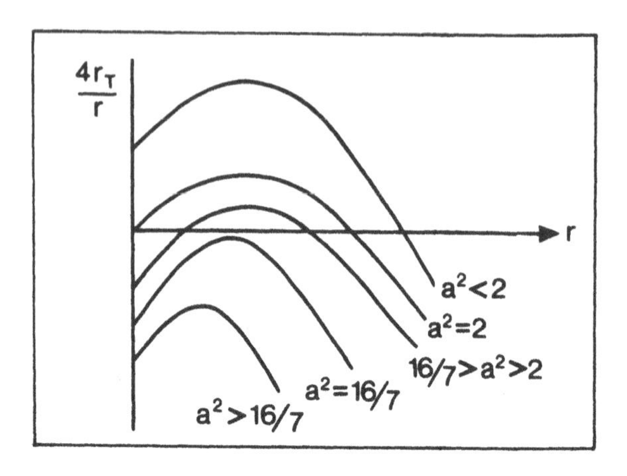


Fig. 7.14 The graph of $4r_T/r$ for different values of $a^2$.

point if the initial amplitude is sufficiently large and to the origin otherwise (this is called hard excitation) whilst if $a^2 > \frac{16}{7}$ all solutions are quenched.

(ii) $\delta > 0$. To determine the behaviour of the general equations we need to superpose the diagrams obtained by solving for $\theta_T = 0$ and $r_T = 0$ as we did in Section 7.10. First $\theta_T = 0$ if

$$\sin 3\theta = -\frac{2\delta}{ar}, \tag{7.84}$$

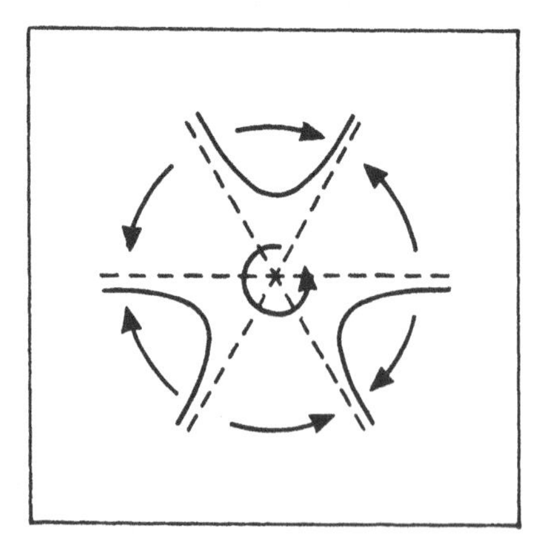

Fig. 7.15 $\theta$ motion ($\delta > 0$).

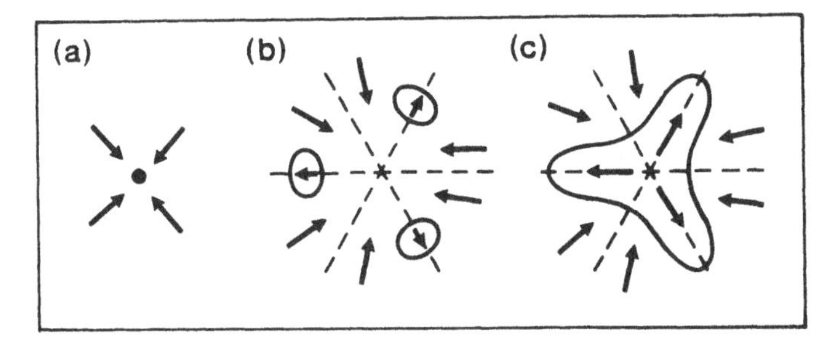


Fig. 7.16 $r$ motion: (a) $a^2 > \frac{16}{7}$; (b) $2 < a^2 < \frac{16}{7}$; and (c) $a^2 < 2$.

which gives the contours of Figure 7.15, which also shows the direction of $\theta_T$ in the different regions. Note that for sufficiently small $r$, $\theta_T < 0$ whilst for large $r$ the contours of $\theta_T = 0$ tend to the straight lines of case (i).

The contours for $r_T = 0$ are given by

$$\cos 3\theta = \frac{2 - a^2 - 2r^2}{ar}, \tag{7.85}$$

which is harder to sketch. There are three cases as shown in Figure 7.16.

Putting together Figures 7.15 and 7.16 we see that if $a^2 > \frac{16}{7}$ the free mode is quenched as before, whilst if $2 < a^2 < \frac{16}{7}$ there are two possibilities depending upon the size of $\delta$. If $\delta$ is small (Fig. 7.17a) there is hard excitation whilst if $\delta$ is large the solutions are quenched (Fig. 7.17b).

Finally, if $a^2 < 2$ there are two possibilities: if $\delta$ is small we have soft excitation (Fig. 7.18a) and if $\delta$ is large there is an attracting periodic

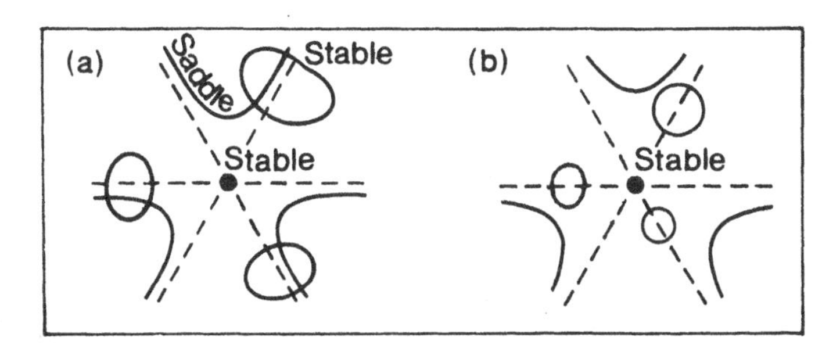


Fig. 7.17 Stationary points: (a) $\delta$ small (hard excitation); and (b) $\delta$ large (quenching).

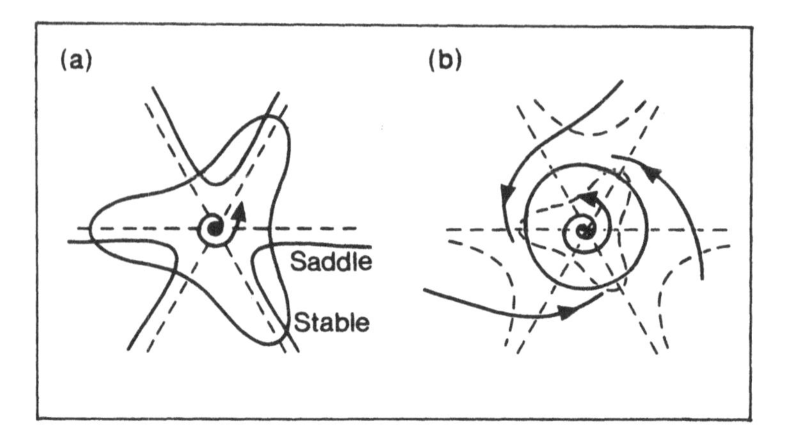


Fig. 7.18 Phase portraits, ($a^2 < 2$): (a) $\delta$ small; and (b) $\delta$ large.

orbit (by the Poincaré-Bendixson Theorem) as shown in Figure 7.18b. This corresponds to beating between the two frequencies.

This description is not entirely satisfactory: where did the periodic orbit come from? What do small and large mean? We shall return to these questions in Chapter 10 where we will be able to apply the techniques of bifurcation theory to this system as a means of getting more precise answers.

## 7.13 Relaxation oscillators

In all the previous examples we had a small parameter; in this section we investigate the effect of a large parameter (although, of course, given a large parameter then its inverse is a small parameter, but let's not quibble ...). The van der Pol oscillator can be written as

$$\dot{x} = y - \epsilon x(\tfrac{1}{3}x^2 - 1), \;\; \dot{y} = -x \tag{7.86}$$

where $\epsilon$ is now assumed to be much greater than 1. Let

$$F(x) = \epsilon x(\tfrac{1}{3}x^2 - 1),$$

then $|\dot{x}|$ is very large (i.e. $O(\epsilon)$) unless $y \approx F(x)$. Hence if a point $(x, y)$ lies off the curve $y = F(x)$ it relaxes back onto the curve very quickly (see Fig. 7.19). This suggests that trajectories are attracted to a periodic orbit (ABCD) made up of two parts of the curve $y = F(x)$ as shown in Figure 7.19, and two almost horizontal pieces where the orbit falls off the curve. More precisely, note that $y = F(x)$ has turning points at $x = \pm 1$ giving the two points $B = (-1, \frac{2}{3}\epsilon)$ and $D = (1, -\frac{2}{3}\epsilon)$. Now let $C$ be the other point on $y = F(x)$ with $y = \frac{2}{3}\epsilon$, i.e. $C = (2, \frac{2}{3}\epsilon)$ and A be the corresponding point with $y = -\frac{2}{3}\epsilon$, i.e. $A = (-2, -\frac{2}{3}\epsilon)$. Starting at $D$ the orbit shoots across to $A$, with $\dot{x} = O(\epsilon)$ and so taking a time proportional to $\epsilon^{-1}$, and then climbs slowly up the curve to $B$ taking a time proportional to $(\Delta y/\dot{y}) \sim O(\epsilon)$ and then shoots across to $C$ in time of order $\epsilon^{-1}$ and then back down to $D$ in time of order $\epsilon$. Hence the total period of the oscillation, $T$, is given approximately by twice the time taken from $A$ to $B$, i.e.

$$\tfrac{1}{2}T \approx \int_{-\frac{2}{3}\epsilon}^{\frac{2}{3}\epsilon} \frac{dy}{\dot{y}}\Big|_{y=F(x)}$$

but $\dot{y} = -x$ and on the curve $dy = \epsilon(x^2 - 1)$, so

$$\tfrac{1}{2}T \approx \int_{-2}^{-1} \frac{\epsilon(x^2-1)}{-x} dx = \tfrac{1}{2}\epsilon(3 - 2\log 2). \tag{7.87}$$

Hence the period of the oscillation is $T = \epsilon(3 - 2\log 2) + O(1)$.

(7.4) EXERCISE

*Consider all possible differential equations*

$$\dot{x} = y - F(x), \quad \dot{y} = -x$$

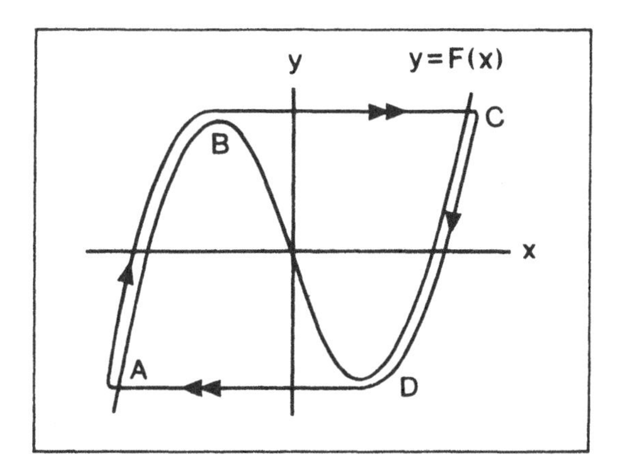


Fig. 7.19 Relaxation oscillator.

*when* $F(x) = \epsilon(x^5 + a_4x^4 + a_3x^3 + a_2x^2 + a_1x + a_0)$ *and* $\epsilon$ *is a large parameter. Describe the different types of periodic orbits which can exist in such equations; can more than one stable periodic orbit exist?*

## Exercises 7

1. Prove that there is a periodic orbit for the system

$$\ddot{x} + \mu(x^2 - 4)\dot{x} + x = 1.$$

Describe the periodic solution and find an approximation to its period for $0 < \mu \ll 1$ and for $\mu \gg 1$.

2. Discuss the nature of the stationary points of the equation

$$\ddot{x} + \epsilon(\dot{x}^2 - \alpha + 2E(x))\dot{x} + \frac{dE}{dx} = 0$$

where $0 < \epsilon \ll 1$ and $E$ is a polynomial of the form $\beta x^2 + x^4$. Find the number of periodic solutions as a function of $\alpha$ and $\beta$.

3. Use the methods of Section 7.6 to find the amplitude of periodic solutions to

$$\ddot{x} + \epsilon(x^2\dot{x}^2 - 2)\dot{x} + 4x = 0,$$

$0 < \epsilon \ll 1$. Find the Floquet multiplier for this orbit and deduce that it is stable.

4. Investigate the behaviour of

$$\ddot{x} + \epsilon(x^2 - x)\dot{x} + x = \alpha(1 - \omega^2)\cos\omega t$$

$(0 < \epsilon \ll 1)$. For what values of $\omega$ can resonance occur? Discuss the evolution of solutions to $O(\epsilon)$ in the three cases

i) no resonance;
ii) $\omega \approx 2$;
iii) $\omega \approx \frac{1}{2}$.

In cases (ii) and (iii) you need to set (e.g.) $\omega = 2(1 + \mu\epsilon)$ and then rescale time, setting $s = (1 + \mu\epsilon)t$ so that the cosine term becomes $\cos 2s$. This gives extra terms on the left hand side of the equation. [Warning: cases (ii) and (iii) are long.]

5. Find the leading order solution as $\epsilon \to 0$ of the amplitude (or resonance) equations for

$$\ddot{x} + \epsilon(x^2 - 1)\dot{x} + x + \epsilon x^3 = 0$$

with $x(0) = 1$, and $\dot{x}(0) = 0$.

6. Investigate the stability of periodic solutions to the equation

$$\ddot{x} + \epsilon(x^4 - 1)\dot{x} + x = -\epsilon b \sin t,$$

for $b$ constant and $0 < \epsilon \ll 1$. Show that a periodic orbit of the form $A \cos t + O(\epsilon)$ is stable if $|A|$ is sufficiently large.

7. Find the resonance equations at $O(\epsilon)$ for

$$\ddot{x} + x = \epsilon(-\dot{x} + x^3 + 2a \cos t)$$

and determine the location and stability of any stationary points.

8. Investigate the evolution of

$$\ddot{x} + x = -\epsilon(\dot{x} + x^3 + ax \cos \omega t)$$

where $\omega = 2(1 + \sigma\epsilon)$ and $a$ and $\sigma$ are $O(1)$. Determine the boundary of the $(a, \sigma)$ plane outside which the solution $x = \dot{x} = 0$ is stable. If this solution is unstable is there a stable periodic solution?

9. Consider the differential equation

$$\ddot{x} + x = -\epsilon f(x, \dot{x})$$

with $|\epsilon| \ll 1$. Let $y = \dot{x}$. Show that if $E(x, y) = \frac{1}{2}(x^2 + y^2)$ then

$$\dot{E} = -\epsilon f(x, y)y.$$

Hence show that an approximate periodic solution of the form $x = A \cos t + O(\epsilon)$ exists if

$$\int_0^{2\pi} f(A \cos t, -A \sin t) \sin t \, dt = 0.$$

Let $E_n = E(x(2\pi n), y(2\pi n))$ and $E_0 = E(x(0), y(0))$. Show that to lowest order $E_n$ satisfies a difference equation of the form

$$E_{n+1} = E_n + \epsilon F(E_n)$$

and write down $F(E_n)$ explicitly as an integral. Hence deduce that a periodic orbit with approximate amplitude $A^* = \sqrt{2E^*}$ exists if $F(E^*) = 0$ and that this orbit is stable if

$$\epsilon \frac{dF}{dE}(E^*) < 0.$$

10. Using the result of the previous question, find the approximate amplitude of the periodic orbits for the equations

$$\ddot{x} + \epsilon(x^2 - 1)\dot{x} + x = 0$$

and

$$\ddot{x} + \epsilon(x^2\dot{x}^2 - 2) + 4x = 0,$$

$0 < \epsilon \ll 1$, and verify that these periodic orbits are stable.

11. Consider the equation

$$\ddot{x} + \dot{x} = -\epsilon(x^2 - x), \ 0 < \epsilon \ll 1.$$

(The dot on the second term is not a misprint.) Using the method of multiple scales show that

$$x_0(\tau, T) = A(T) + B(T)\mathrm{e}^{-\tau}$$

and identify any resonant terms at order $\epsilon$. Show that the non-resonance condition is

$$A_T = A - A^2$$

and describe the asymptotic behaviour of solutions.

# 8

# *Bifurcation theory I: stationary points*

Bifurcation theory describes the way that topological features of a flow (properties such as the number of stationary points and periodic orbits) vary as one or more parameters are varied. There are many approaches to the problem of understanding the possible changes which occur in differential equations, ranging from a straightforward analytic description to a topological classification of all possible behaviours which may occur under an arbitrary, but small, perturbation of the system. In this chapter we aim to describe some of the simple techniques used to describe bifurcations of stationary points, adopting a heuristic approach rather than a rigorous mathematical treatment. Section 8.1 on the Centre Manifold Theorem follows the graduate textbook by Guckenheimer and Holmes (1983) quite closely. This book has become one of the standard introductions to nonlinear equations, and the next five chapters are littered with references to it. A great deal more of the detailed mathematical justification for the results described here and their implications can be found there. Our approach remains at the level of understanding systems and being able to deduce properties of examples.

The fundamental observation for stationary points of flows is that if the stationary point is hyperbolic, i.e. the eigenvalues of the linearized flow at the stationary point all have non-zero real parts, then the local behaviour of the flow is completely determined by the linearized flow (at least up to homeomorphism; see Chapter 4 and the Stable Manifold Theorem). Furthermore, small perturbations of the equation will also have a hyperbolic stationary point of the same type (Section 4.3). Hence bifurcations of stationary points can only occur at parameter values for which a stationary point is non-hyperbolic. This gives an easy criterion for detecting bifurcations: simply find parameter values for which the linearized flow near a stationary point has a zero or purely imaginary eigenvalue. One of the most important techniques for studying such bifurcations is based on the non-hyperbolic equivalent of the Stable Manifold Theorem, called the Centre Manifold Theorem. This generalizes the idea of the centre manifold for linear systems to nonlinear systems.

In this chapter we begin by outlining how to find the centre manifold for nonlinear systems and then go on to show how this can be used to derive the dynamics of systems near a system with a non-hyperbolic stationary point. We then describe the simple bifurcations that can arise in systems depending upon a single real parameter. Before describing this process in detail we can give a flavour of the type of manipulation that will be involved. Suppose that we have a system of equations on the real line ($x \in \mathbf{R}$) which depend upon a real parameter, $\mu$. Thus each value of $\mu$ defines a differential equation and we are interested in the way that qualitative features of the solutions vary as $\mu$ takes different values. We have already determined that problems will arise when the system has a non-hyperbolic stationary point, so suppose that

$$\dot{x} = f(x, \mu) \tag{8.1}$$

where $f(0,0) = \frac{\partial f}{\partial x}(0,0) = 0$. Then the origin is a stationary point if $\mu = 0$ and since the Jacobian matrix vanishes it is non-hyperbolic. Now, expanding $f(x, \mu)$ as a Taylor series in some neighbourhood of $(x, \mu) = (0,0)$ we obtain

$$\dot{x} = A(\mu) + B(\mu)x + C(\mu)x^2 + \dots \tag{8.2}$$

where

$$A(\mu) = f_\mu \mu + \tfrac{1}{2} f_{\mu\mu}\mu^2 + \dots, \tag{8.3}$$

$$B(\mu) = f_{\mu x}\mu + \tfrac{1}{2} f_{\mu\mu x}\mu^2 + \dots \tag{8.4}$$

and

$$C(\mu) = f_{xx} + \tfrac{1}{2} f_{\mu xx}\mu + \dots. \tag{8.5}$$

Subscripts in these equations denote partial differentiation with respect to the relevant variable and all derivatives are evaluated at $(x, \mu) = (0,0)$. Much of bifurcation theory is simply about determining the stationary points of such systems (as functions of $\mu$), i.e. looking for solutions of

$$0 = A(\mu) + B(\mu)x + C(\mu)x^2 + \dots \tag{8.6}$$

in a neighbourhood of $(x, \mu) = (0,0)$. The number of solutions and their stability varies according to whether certain partial derivatives or combinations of partial derivatives vanish and their sign if they do not vanish. Hence bifurcation theory (in one dimension) is really about being able to solve (8.6).

## 8.1 Centre manifolds

In some of the examples in the previous chapter we have seen that pairs of stationary points can come together and disappear as a parameter is varied. This is an example of a bifurcation. More precisely, a bifurcation value of a parameter $\mu$ is a value at which the qualitative nature of the flow changes.

*Example 8.1*

Consider the simple equation

$$\dot{x} = \mu - x^2.$$

For $\mu < 0$ there are no stationary points, whereas for $\mu > 0$ there are two, one at $x_+ = \sqrt{\mu}$ and the other at $x_- = -\sqrt{\mu}$. The linearized flow is given by the $(1 \times 1)$ Jacobian matrix, $-2x$, so the stationary point in $x < 0$ is unstable and the stationary point in $x > 0$ is stable. An important change in the behaviour of the system clearly occurs as $\mu$ passes through zero: two stationary points are created, one stable and the other unstable. This is an example of a saddlenode bifurcation, and the bifurcation value of $\mu$ is $\mu = 0$. At this value of the parameter, $\dot{x} = -x^2$, so $x = 0$ is a stationary point of the flow, but the linear flow vanishes. In other words, if $\mu = 0$ there is a non-hyperbolic stationary point at $x = 0$. It is often useful to illustrate bifurcations by plotting the position of stationary points as a function of parameter as shown in Figure 8.1. The stability of solutions is indicated by solid lines for stable solutions and dotted lines for unstable solutions. Such pictures are referred to as *bifurcation diagrams*.

Recall that in Chapter 4 we defined the stable and unstable manifolds of a hyperbolic stationary point and, in Chapter 3, the linear centre manifold for the linear system $\dot{z} = Lz$, $z \in \mathbf{R}^n$ to be the space spanned by the generalized eigenvectors of $L$ corresponding to eigenvalues $\lambda$ with $\text{Re}(\lambda) = 0$. To understand local bifurcations of stationary points we need the nonlinear equivalent of $E^c(0)$. This is given by the following theorem (Carr, 1981, Hirsch, Pugh and Shub, 1977, Kelley, 1967, ...).

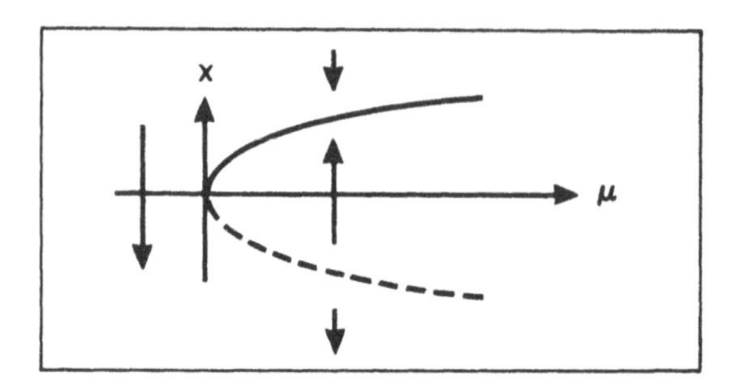


Fig. 8.1 Position ($x$) against parameter ($\mu$) for the saddlenode bifurcation.

(8.1) THEOREM (CENTRE MANIFOLD THEOREM)

*Let $f \in C^r(\mathbf{R}^n)$ with $f(0) = 0$. Divide the eigenvalues, $\lambda$, of $Df(0)$ into three sets, $\sigma_u$, $\sigma_s$ and $\sigma_c$, where $\lambda \in \sigma_u$ if $\mathrm{Re}(\lambda) > 0$, $\lambda \in \sigma_s$ if $\mathrm{Re}(\lambda) < 0$ and $\lambda \in \sigma_c$ if $\mathrm{Re}(\lambda) = 0$. Let $E^u$, $E^s$ and $E^c$ be the corresponding generalized eigenspaces. Then there exist $C^r$ unstable and stable manifolds ($W^u$ and $W^s$) tangential to $E^u$ and $E^s$ respectively at $x = 0$ and a $C^{r-1}$ centre manifold, $W^c$, tangential to $E^c$ at $x = 0$. All are invariant, but $W^c$ is not necessarily unique.*

The proof of the theorem is similar in style to the proof of the Stable Manifold Theorem, but we will not give it here. For our purposes it is enough to know that the centre manifold exists and in the rest of this section we shall see how it can be used and constructed for simple examples. The non-uniqueness of the centre manifold may appear to be a little bizarre at first sight. It is, however, not as dangerous as it sounds, and reflects the possibility of adding exponentially small terms without changing the property of the manifold. Intuitively, it reflects the fact that a number of trajectories can do essentially the same thing tangential to $E^c$, and any one of these is a suitable choice for 'the' centre manifold. This is shown explicitly in Example 8.2.

The Centre Manifold Theorem implies that at a bifurcation point, where a stationary point is non-hyperbolic ($\sigma_c \neq \emptyset$), the system can be written locally in coordinates $(x, y, z) \in W^c \times W^s \times W^u$ on the invariant manifolds as

$$\dot{x} = g(x) \tag{8.7a}$$

$$\dot{y} = -By \tag{8.7b}$$

$$\dot{z} = Cz \tag{8.7c}$$

where $B$ and $C$ are positive definite matrices. The motion on $W^s$ is unequivocally towards the stationary point and the motion on $W^u$ is unequivocally away from the stationary point, so the local behaviour can be understood by solving (8.7a). In order to do this we must find a way to calculate the function $g(x)$.

For simplicity we shall consider the situation when $\sigma_u = \emptyset$, so the equations can be written (in coordinates $x$ in the direction of $E^c$ and $y$ in the direction of $E^s$) as

$$\dot{x} = Ax + f_1(x, y) \tag{8.8a}$$

$$\dot{y} = -By + f_2(x, y) \tag{8.8b}$$

where the eigenvalues of $A$ all have zero real parts, the eigenvalues of $B$ all have strictly positive real parts and the functions $f_i$, $i = 1, 2$, represent nonlinear terms, so both $f_i$ and their first derivatives with respect to the $x$ and $y$ variables vanish at $(x, y) = (0, 0)$. Since $W^c$ is tangential to $E^c = \{(x, y)|y = 0\}$ it can be represented locally as the graph of a function of $x$, so

$$W^c = \{(x, y)|y = h(x),\ h(0) = 0,\ Dh(0) = 0\}$$

where $h : U \to \mathbf{R}^s$ is defined on some neighbourhood $U$ of the origin in $\mathbf{R}^c$ as illustrated in Figure 8.2 and $Dh$ is the Jacobian matrix of $h$. Thus, on $W^c$ the flow is approximated (projecting onto the $x$ directions) by

$$\dot{x} = Ax + f_1(x, h(x)). \tag{8.9}$$

This, then, is the equation for $g(x)$ we have been looking for. All that remains to do is to find the function $h(x)$, which can be done in precisely the same way as the stable and unstable manifolds were approached in Chapter 4. Assume that $h(x)$ can be expanded locally as a power series in $x$, building in the conditions that $h(0) = 0$ and $Dh(0) = 0$, i.e. that the manifold passes through the origin tangential to the linear centre manifold, $E^c$, at $x = 0$. Then equate powers of $x$ from the two forms of $\dot{y}$ obtained by differentiating the equation $y = h(x)$ and from the definition of $\dot{y}$ in (8.8b).

On the centre manifold $y = h(x)$ and so $\dot{y} = Dh(x)\dot{x}$, where $Dh(x)$ is the $s \times c$ Jacobian matrix of $h$ and $\dot{x}$ is a $c$ dimensional vector. Substituting for $\dot{x}$ from (8.8a) gives

$$\dot{y} = Dh(x)[Ax + f_1(x, h(x))]. \tag{8.10}$$

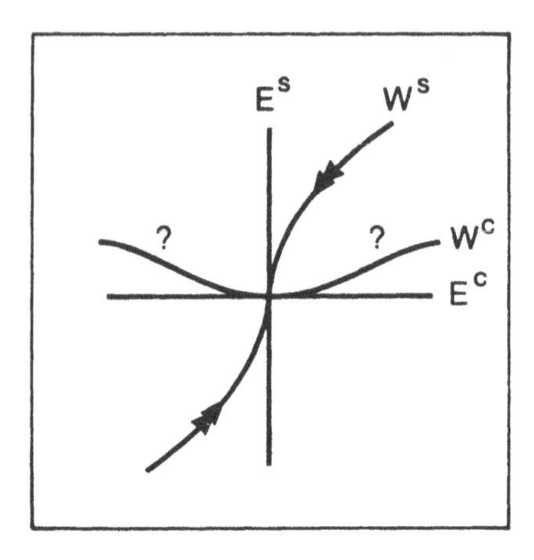


Fig. 8.2 Linear and nonlinear stable and centre manifolds.

But on the centre manifold (8.8b) implies that

$$\dot{y} = -Bh(x) + f_2(x, h(x)). \tag{8.11}$$

Comparing these two equations and including the conditions on $h(x)$ yields the problem

$$Dh(x)[Ax + f_1(x, h(x))] = -Bh(x) + f_2(x, h(x)) \tag{8.12}$$

$$h(0) = 0, \quad Dh(0) = 0. \tag{8.13}$$

These equations can be solved, at least in principle, to arbitrary order in $x$ by posing a series solution and equating coefficients at each order. This is probably best illustrated by a simple example.

*Example 8.2*

Consider the equations

$$\dot{x} = xy$$
$$\dot{y} = -y - x^2.$$

The origin is a non-hyperbolic stationary point, and the linearization about the origin is already in normal form, so $E^c(0) = \{(x, y)|y = 0\}$ and $E^s(0) = \{(x, y)|x = 0\}$. The nonlinear centre manifold will be (at least locally) the graph of a function $y = h(x)$ which passes through the origin tangential to the linear centre manifold (so $h(0) = h'(0) = 0$). This suggests that we try a solution of the form

$$y = h(x) = ax^2 + bx^3 + cx^4 + dx^5 + O(x^6).$$

Substituting for $y$ this gives

$$\dot{y} = h'(x)\dot{x} = h'(x)xy = xh(x)h'(x),$$

and using the trial solution we find

$$\begin{aligned}\dot{y} &= x(ax^2 + bx^3 + \ldots)(2ax + 3bx^2 + \ldots)\\ &= 2a^2x^4 + 5abx^5 + O(x^6).\end{aligned}$$

On the other hand, $\dot{y} = -y - x^2$ so

$$\begin{aligned}\dot{y} &= -h(x) - x^2\\ &= -(a+1)x^2 - bx^3 - cx^4 - dx^5 + O(x^6).\end{aligned}$$

Now, equating coefficients at order $x^2$ gives $(a+1) = 0$; the cubic terms obviously give $b = 0$ and the quartic terms give $2a^2 = -c$, so

$$a = -1; \quad b = 0; \quad c = -2; \quad d = 0;$$

giving an approximation to the nonlinear centre manifold of the form

$$y = h(x) = -x^2 - 2x^4 + O(x^6).$$

The motion on the centre manifold (which is tangential to the $x$-axis at $x = 0$) is therefore given by the equation

$$\dot{x} = xh(x) = -x(x^2 + 2x^4 + O(x^6))$$

and so this shows that the motion on the centre manifold is (at least locally) towards the origin. Since the $y$-axis is (approximately) the stable manifold of the origin we see that the non-hyperbolic stationary point is in fact a nonlinear sink (Fig. 8.3). Figure 8.3 also illustrates the point about the non-uniqueness of the centre manifold made earlier. Any one of the trajectories which tends to the origin tangential to

$$y = -x^2 - 2x^4 + O(x^6)$$

can be chosen to be 'the' centre manifold. For each such trajectory, the the distance between the derived manifold (defined by a power series) in a suffiiently small neighbourhood of the origin is exponentially small, and so they all have the same power series expansion about the origin. Hence any of them can be chosen as a centre manifold.

*Remark:* The lowest order approximation to the centre manifold for this example could have been obtained by observing that on the centre manifold $\dot{y}$ is approximately zero and so $y$ is approximately $-x^2$. This sort of observation can often act as a useful way of checking that an algebraic slip has not been made in the calculation of the centre manifold.

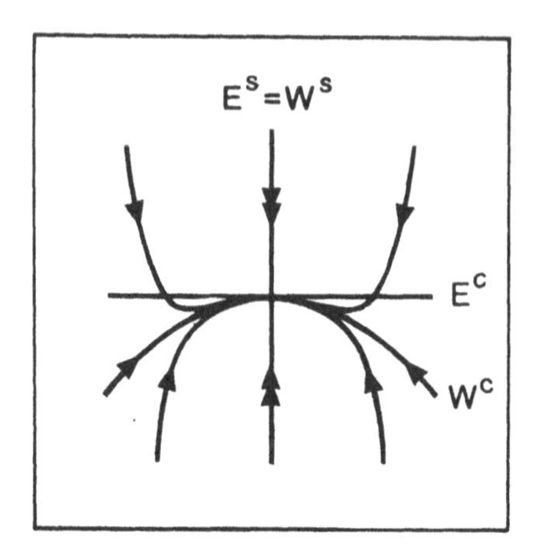


Fig. 8.3 Phase portrait for Example 8.2.

## 8.2 Local bifurcations

The analysis of the preceding section allows us to determine the nature of the non-hyperbolic stationary point by looking at the motion on the centre manifold where the linear behaviour does not completely determine the flow in a neighbourhood of the stationary point. The next step towards understanding local bifurcations is to introduce parameters, and extend the idea of the centre manifold into parameter space in such a way as to capture the behaviour of families of systems near bifurcation values of the parameter. To this end we consider differential equations which depend on one or more parameter, $\mu \in \mathbf{R}^m$, so the differential equation is $\dot{w} = f(w, \mu)$, where $w \in \mathbf{R}^n$ and $f : \mathbf{R}^n \times \mathbf{R}^m \to \mathbf{R}^n$ is a smooth function with $f(0,0) = 0$, so that the origin (in $\mathbf{R}^n$) is a stationary point of the flow. Now suppose that if $\mu = 0$ the linearized flow, $\dot{w} = Lw$ near $w = 0$ has some eigenvalues with zero real part. In this case we can choose coordinates $(x, y, z)$ in the linear eigenspaces of $L$ as in the previous section such that

$$E^c(0) = \{(x, y, z) | y = z = 0\},$$
$$E^u(0) = \{(x, y, z) | x = y = 0\},$$

and

$$E^s(0) = \{(x, y, z) | x = z = 0\}.$$

In these coordinates the differential equation $\dot{w} = f(w, \mu)$ becomes

$$\dot{x} = A(\mu)x + f_1(x, y, z; \mu) \tag{8.14a}$$

$$\dot{y} = -B(\mu)y + f_2(x, y, z; \mu) \tag{8.14b}$$

$$\dot{z} = C(\mu)z + f_3(x, y, z; \mu) \tag{8.14c}$$

where all the eigenvalues of $A(0)$ have zero real parts and there is some open neighbourhood of $\mu = 0$ for which the eigenvalues of both $B(\mu)$ and $C(\mu)$ have strictly positive real parts. (This means that we have assumed implicitly that the eigenvalues of $B(0)$ and $C(0)$ have real parts which are bounded away from zero. If $n$ is finite this is not a problem, but in infinite dimensional problems this has to be excluded explicitly.) The functions $f_i(x, y, z; \mu)$ contain the nonlinear terms in $x$, $y$ and $z$, so they vanish, together with their first derivatives with respect to the variables $x$, $y$ and $z$ at the bifurcation point $(x, y, z; \mu) = (0, 0, 0; 0)$.

To describe the dynamics in a neighbourhood of $(0, 0, 0; 0)$ we use the centre manifold on an extended system of equations: we want to include the $\mu$ variables as part of the centre manifold. To do this, add the trivial equations

$$\dot{\mu} = 0. \tag{8.14d}$$

On the face of it, this may appear to be a pretty useless thing to do. After all, this trivial equation has a very easy solution, $\mu$ equals a constant. The point is that by including these equations we obtain a centre manifold which stretches into our parameter space, so the centre manifold can be used on the extended system in such a way as to be valid for small $|x|$ *and* $|\mu|$. We can then 'solve' the trivial equation ($\dot{\mu} = 0$) without too much effort to get a simplified equation for the evolution of $x$ which involves the parameters.

The extended system has a centre manifold of dimension dim $E^c(0) + m$ which is tangential to $E^c(0)$ and $\mu = 0$ at $(0, 0, 0; 0)$. Thus, using precisely the same arguments as in the previous section, we can try to find an approximation to the centre manifold by solving for $y$ and $z$ as a graph over $x$ and $\mu$. So set

$$y = h_s(x, \mu) \tag{8.15}$$

and

$$z = h_u(x, \mu) \tag{8.16}$$

as the equation of the graph of the centre manifold and proceed as before. Substituting (8.15) and (8.16) into (8.14a) and (8.14d) we find

$$\dot{x} = A(\mu)x + f_1(x, h_s(x, \mu), h_u(x, \mu); \mu) \tag{8.17a}$$

$$\dot{\mu} = 0, \tag{8.17b}$$

valid for sufficiently small $|x|$ and $|\mu|$. As we have already remarked, the second of these two equations is not hard to solve, and so we are left with the ($\dim E^c(0)$) equation

$$\dot{x} = A(\mu)x + f_1(x, h_s(x, \mu), h_u(x, \mu); \mu) = G(x, \mu) \tag{8.18}$$

which describes the local dynamics on the centre manifold for $|x|$ and $|\mu|$ sufficiently small. The remainder of this chapter is devoted to the study of such equations for particular choices of $\dim E^c(0)$ and restrictions on higher derivatives of the function $G(x, \mu)$. This process will result in a sequence of bifurcation theorems, each saying that if $A(0)$ has a particular form and certain genericity (or non-degeneracy) conditions hold for $G(x, \mu)$ then particular changes in the dynamics of the family must occur as $\mu$ passes through zero.

## 8.3 The saddlenode bifurcation

Suppose that the equation $\dot{w} = f(w, \mu)$ has a non-hyperbolic stationary point (which we can take to be at the origin, $w = 0$) if $\mu = 0$. If the Jacobian matrix of the linear flow at $w = 0$ has a simple zero eigenvalue for $\mu = 0$ and all other eigenvalues lie off the imaginary axis, then $\dim E^c(0) = 1$ and the equation on the centre manifold is

$$\dot{x} = G(x, \mu)$$

where $x \in \mathbf{R}$ and we can take $\mu \in \mathbf{R}$ so $G : \mathbf{R} \times \mathbf{R} \to \mathbf{R}$. From the definition of $G$ above we see that $G(0,0) = 0$ and $A(0) = G_x(0,0) = 0$, where the subscript denotes partial differentiation with respect to $x$. To understand the local behaviour of the flow for $(x, \mu)$ near $(0,0)$, expand $G$ as a Taylor series in both the variables about $(x, \mu) = (0,0)$. Then, since $G(0,0) = G_x(0,0) = 0$,

$$\dot{x} = G_\mu \mu + \tfrac{1}{2}(G_{xx}x^2 + 2G_{x\mu}\mu x + G_{\mu\mu}\mu^2) + O(3). \tag{8.19}$$

All the partial derivatives are, of course, evaluated at $(x, \mu) = (0,0)$. This is a simple differential equation in one dimension which can be analyzed using the standard techniques developed over previous chapters, valid for $|x|$ and $|\mu|$ sufficiently small. The first step in any such analysis is to determine the locus of stationary points of the equation. In this case it is quite possible to do this rigorously without any serious problem (use the Implicit Function Theorem). However, we shall leave this more mathematical approach until the end of this chapter and concentrate on developing asymptotic expansions for the locus of stationary points in the $(x, \mu)$ plane. We begin by rewriting the equation on the centre manifold as

$$\dot{x} = \sum_{k\geq 0} A_k(\mu)x^k \tag{8.20}$$

where

$$A_0(\mu) = G_\mu \mu + \tfrac{1}{2}G_{\mu\mu}\mu^2 + O(\mu^3), \quad A_1(\mu) = G_{\mu x}\mu + O(\mu^2),$$

$$A_2(\mu) = \tfrac{1}{2}G_{xx} + O(\mu)$$

and so on. Trying to solve the lowest order approximation, $A_0 + A_1 x \sim 0$ gives solutions $x \sim -A_0/A_1$ which is $O(1)$ or larger if $G_\mu \neq 0$, so this will not give us a local solution. Including the quadratic term, $A_0 + A_1 x + A_2 x^2 \sim 0$ gives solutions

$$x \sim \frac{-A_1 \pm \sqrt{A_1^2 - 4A_0 A_2}}{2A_2} \tag{8.21}$$

and on substituting for the functions $A_i(\mu)$ we find the leading order solutions

$$x \sim \pm\sqrt{\frac{-2G_\mu \mu}{G_{xx}}}. \tag{8.22}$$

Hence if $G_\mu/G_{xx} > 0$ there is a pair of solutions near the origin if $\mu < 0$ and there are no solutions if $\mu > 0$. On the other hand, if $G_\mu/G_{xx} < 0$ there is a pair of solutions near the origin if $\mu > 0$, and no solutions exist if $\mu < 0$.

Suppose that $G_\mu/G_{xx} < 0$, so solutions bifurcate into $\mu > 0$. We pose the asymptotic series suggested by (8.22)

$$x \sim \sum_{n\geq 1} \alpha_n \mu^{\frac{n}{2}}$$

which we can now substitute for the full equation, (8.19), to determine the coefficients $\alpha_n$ to whatever order of accuracy is desired. For example, at order $\mu$ the equations are

$$G_\mu + \tfrac{1}{2}G_{xx}\alpha_1^2 = 0 \tag{8.23}$$

giving $\alpha_1^2 = -2G_\mu/G_{xx}$, or $\alpha_1 = \pm\sqrt{-2G_\mu/G_{xx}}$, which we already knew. At order $\mu^{\frac{3}{2}}$ we find that we need to include cubic terms from (8.19) to obtain

$$G_{\mu x}\alpha_1 + G_{xx}\alpha_1\alpha_2 + \tfrac{1}{6}G_{xxx}\alpha_1^3 = 0 \qquad (8.24)$$

and so

$$\alpha_2 = -\frac{1}{G_{xx}}(G_{\mu x} + \tfrac{1}{6}G_{xxx}\alpha_1^2) = \frac{1}{3G_{xx}^2}\left(G_\mu G_{xxx} - 3G_{\mu x}G_{xx}\right). \qquad (8.25)$$

Hence, provided both $G_\mu/G_{xx} < 0$ (and, in particular, $G_\mu \neq 0$ and $G_{xx} \neq 0$) we obtain a pair of stationary points if $\mu > 0$ at

$$x \sim \pm\sqrt{-\frac{2G_\mu}{G_{xx}}}\mu^{\frac{1}{2}} + \frac{1}{3G_{xx}^2}\left(G_\mu G_{xxx} - 3G_{\mu x}G_{xx}\right)\mu + O(\mu^{\frac{3}{2}}). \qquad (8.26)$$

In principle we could continue to find $\alpha_3$, $\alpha_4$ and so on. Note that if $G_\mu/G_{xx} > 0$ then we would find a pair of solutions if $\mu < 0$ which could be represented as an asymptotic series in powers of $(-\mu)^{\frac{1}{2}}$.

Finally we should investigate the stability of these stationary points (for small $|\mu|$). This is determined by the sign of the Jacobian matrix which, for one-dimensional systems, is simply the function $A_1 + 2A_2x + \ldots$ evaluated at the stationary points. From this it is clear that the dominant term is $A_2x$ (which is of order $|\mu|^{\frac{1}{2}}$) and so the positive (resp. negative) stationary point is stable if $G_{xx} < 0$ (resp. $G_{xx} > 0$) whilst the stationary point in $x < 0$ (resp. $x > 0$) is unstable.

This argument shows (informally) that if $x = 0$ is a non-hyperbolic stationary point of the family of differential equations $\dot{x} = G(x,\mu)$ on the real line when $\mu = 0$ then provided $G_\mu(0,0)$ and $G_{xx}(0,0)$ are non-zero a curve of stationary states bifurcates from $(x,\mu) = (0,0)$, tangential to the $x$-axis. The direction of the bifurcation is determined by the sign of $(G_\mu/G_{xx})$ and the stability of the bifurcation solutions is determined by the sign of $G_{xx}$. This set of results can be expressed as the following theorem.

(8.2) THEOREM (SADDLENODE BIFURCATION)

*Suppose that* $\dot{x} = G(x,\mu)$ *with* $G(0,0) = G_x(0,0) = 0$. *Then provided*

$$G_\mu(0,0) \neq 0 \quad \textit{and} \quad G_{xx}(0,0) \neq 0$$

*there is a continuous curve of stationary points in a neighbourhood of* $(x,\mu) = (0,0)$ *which is tangent to* $\mu = 0$ *at* $(0,0)$. *If* $G_\mu G_{xx} < 0$ *(resp.*

*$G_\mu G_{xx} > 0$) there are no stationary points near $(0,0)$ if $\mu < 0$ (resp. $\mu > 0$) whilst for each value of $\mu > 0$ (resp. $\mu < 0$) in some sufficiently small neighbourhood of $\mu = 0$ there are two stationary points near $x = 0$. For $\mu \neq 0$ both stationary points are hyperbolic and the upper one is stable and the lower unstable if $G_{xx} < 0$. The stability properties are reversed if $G_{xx} > 0$.*

This is an example of a bifurcation theorem (this bifurcation is called the saddlenode bifurcation). Typically such theorems state that provided some genericity conditions hold (in this case the non-vanishing of two partial derivatives of the equations at the bifurcation point) then, locally, certain changes in the flow will arise. In Section 8.7 we will outline a different approach to such changes, but the next few sections will be concerned with what happens if one or other of the genericity conditions for the saddlenode bifurcation fails to be satisfied.

*Example 8.3*

Consider the equations

$$\dot{x} = G(x,\mu) = x^3 + x^2 - (2+\mu)x + \mu.$$

We want to locate bifurcations of stationary points for this system, so we begin by looking for solutions to $x^3 + x^2 - (2+\mu)x + \mu = 0$. This is made easier by noticing that the cubic factorizes, so

$$(x-1)(x^2 + 2x - \mu) = 0.$$

Hence $x = 1$ is a stationary point for all values of $\mu$ whilst the quadratic terms give real solutions

$$x_\pm = -1 \pm \sqrt{1+\mu}$$

provided $\mu > -1$. This is strong evidence that there is a saddlenode bifurcation from $x = -1$ when $\mu = -1$. To confirm this we note that

$$G_x(x,\mu) = 3x^2 + 2x - (2+\mu), \quad G_\mu(x,\mu) = -x + 1$$

$$\text{and} \quad G_{xx}(x,\mu) = 6x + 2$$

and so, evaluating these quantities at the bifurcation value $(x,\mu) = (-1,-1)$ we find

$$G(-1,-1) = G_x(-1,-1) = 0, \quad G_\mu(-1,-1) = 2$$

$$\text{and} \quad G_{xx}(-1,-1) = -4.$$

Hence there is indeed a saddlenode bifurcation from $(-1,-1)$ and since $G_\mu G_{xx} < 0$ the pair of solutions bifurcates into $\mu > -1$, as we knew already from the form of the solutions $x_\pm$. Furthermore, since $G_{xx} < 0$ $x_+$ is stable and $x_-$ is unstable for $\mu+1 > 0$ sufficiently small. This is all we can determine from the local analysis, but note that if $\mu = 3$, $x_+ = 1$ and so the stationary point $x_+$ coincides with the stationary point at $x = 1$. Evaluating $G_x$ at $(x,\mu) = (1,3)$ we find that this stationary point is non-hyperbolic. Is there a further bifurcation? The answer to this rhetorical question is, of course, yes, but since $G_\mu(1,3) = 0$ it cannot be a saddlenode bifurcation. Figure 8.4 shows what happens here. The full analysis of this second bifurcation is left until the next section.

*Remark:* This example shows that a much more sensible way of determining the nature and existence of saddlenode bifurcations is simply to look at the equations for a stationary point and find places where a pair of zeros disappear (typically this will be when some discriminant vanishes). In effect, this is precisely what we have done above, but rather than work with a particular example we have used the general form of the (local) Taylor series of the system.

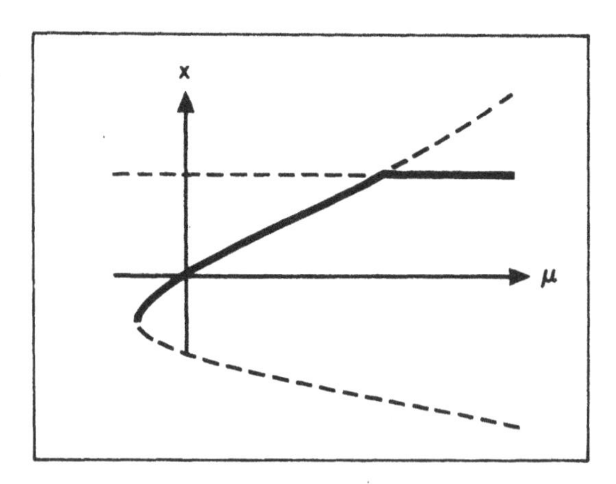


Fig. 8.4 Bifurcation diagram for Example 8.3.

## 8.4 The transcritical bifurcation

The transcritical bifurcation is another bifurcation which occurs if dim $E^c(0) = 1$ for a non-hyperbolic stationary point at the origin, but in this case $G_\mu(0,0) = 0$, and so the genericity conditions for the saddle-node bifurcation do not apply.

*Example 8.4*

Consider the differential equation

$$\dot{x} = \mu x - x^2.$$

Looking for stationary points (solutions of $\dot{x} = 0$) we find that there are two solutions, one at $x = 0$ and another at $x = \mu$. So for $\mu < 0$ and for $\mu > 0$ there are two stationary points and if $\mu = 0$ there is only one stationary point (at $x = 0$). The $(1 \times 1)$ Jacobian matrix of the equation is $\mu - 2x$, which is zero if $x = \mu = 0$, so the stationary point for $\mu = 0$ is not hyperbolic. From the Jacobian we can also see that the stationary point at $x = 0$ is stable if $\mu < 0$ and unstable if $\mu > 0$, whilst the stationary point at $x = \mu$ is unstable if $\mu < 0$ and stable if $\mu > 0$. The bifurcation diagram is shown in Figure 8.5. This bifurcation is usually referred to as a transcritical bifurcation, but (for obvious reasons) it is sometimes called an exchange of stability.

The important feature of this example is that the origin, $x = 0$, is constrained to be a stationary point both before and after the bifurcation

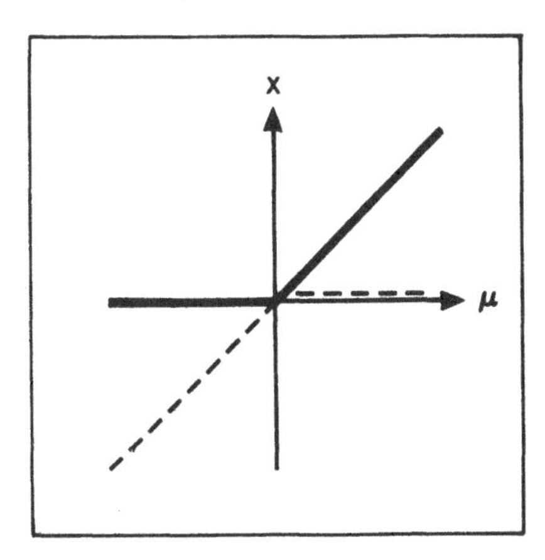


Fig. 8.5 Bifurcation diagram for Example 8.4.

(at $\mu = 0$) and it is easy to verify that if $G(x,\mu) = \mu x - x^2$ then $G(0,0) = G_x(0,0) = G_\mu(0,0) = 0$ and $G_{xx}(0,0) = 2$. This suggests that the first attempt to generalize this example should be by considering differential equations on the centre manifold,

$$\dot{x} = G(x,\mu) = \tfrac{1}{2}(G_{xx}x^2 + 2G_{\mu x}\mu x + G_{\mu\mu}\mu^2) + O(3), \tag{8.27}$$

i.e. assuming that the first derivative of $G$ with respect to each variable vanishes. As in the previous section we attempt to find the leading order expressions for stationary points of this system near $(x,\mu) = (0,0)$ by solving the quadratic equation obtained by setting the right hand side of the $\dot{x}$ equation to zero and ignoring terms of order $x^3$ and higher. This gives

$$x \sim \frac{-G_{\mu x}\mu \pm \mu\sqrt{G_{\mu x}^2 - G_{xx}G_{\mu\mu}}}{G_{xx}} \tag{8.28}$$

provided $G_{xx} \neq 0$. Set $\Delta^2 = G_{\mu x}^2 - G_{xx}G_{\mu\mu}$, then provided $\Delta^2 > 0$ there are two curves of stationary points near $(0,0)$ given by

$$x \sim -\frac{G_{\mu x} \pm \Delta}{G_{xx}}\mu + O(\mu^2). \tag{8.29}$$

These two curves intersect transversely at $(x,\mu) = (0,0)$. As with the saddlenode bifurcation the stability of these stationary points is easy to determine: the upper branch of solutions is stable if $G_{xx} < 0$ (with the lower branch being unstable) and stability properties are reversed if $G_{xx} > 0$ (see Fig. 8.6).

Taking a deep breath, it is now possible to give the conditions for a transcritical bifurcation to occur at $(x,\mu) = (0,0)$ for the system $\dot{x} = G(x,\mu)$.

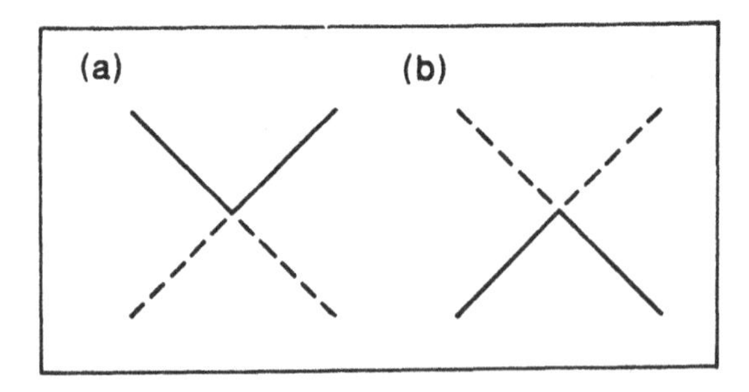


Fig. 8.6 Transcritical bifurcations ($x$ against $\mu$): (a) $G_{xx} < 0$; (b) $G_{xx} > 0$.

(8.3) THEOREM (TRANSCRITICAL BIFURCATION)

*Suppose* $\dot{x} = G(x,\mu)$ *and* $G(0,0) = G_x(0,0) = 0$. *Then if*

$$G_\mu = 0, \quad G_{xx} \neq 0 \ \text{ and } G_{\mu x}^2 - G_{xx}G_{\mu\mu} > 0$$

*there are two curves of stationary points in a neighbourhood of* $(x,\mu) = (0,0)$. *These curves intersect transversely at* $(0,0)$ *and for each* $\mu \neq 0$ *sufficiently small there are two hyperbolic stationary points near* $x = 0$. *The upper stationary point is stable (resp. unstable) and the lower stationary point is unstable (resp. stable) if* $G_{xx} < 0$ *(resp.* $G_{xx} > 0$*).*

*Example 8.5*

Consider Example 8.3:

$$\dot{x} = G(x,\mu) = x^3 + x^2 - (2+\mu)x + \mu.$$

We have already shown that there is a saddlenode bifurcation when $\mu = -1$ and that if $\mu = 3$ there is a non-hyperbolic stationary point at $x = 1$ with $G_\mu(1,3) = 0$. From the expression in Example 3 we see that $G_{xx}(1,3) = 8$ and also

$$G_{\mu x}(x,\mu) = -1 \quad \text{and} \quad G_{\mu\mu} = 0.$$

Hence $G_{xx}(1,3) \neq 0$ and $G_{\mu x}^2 - G_{xx}G_{\mu\mu} = 1 > 0$, so there is a transcritical bifurcation for $\mu = 3$ (this should be obvious from the sketch in Fig. 8.4). Recall that the two stationary points involved in this bifurcation are $x = 1$ and $x_+ = -1 + \sqrt{1+\mu}$ and so $x_+ < 1$ if $\mu < 3$ and $x_+ > 1$ if $\mu > 3$. Since $G_{xx} > 0$ this implies that $x_+$ is stable and $x = 1$ is unstable if $\mu < 3$ and the stability properties are exchanged if $\mu > 3$. This completes the full analysis of the bifurcations in this example.

## 8.5 The pitchfork bifurcation

The final example of bifurcations involving only stationary points of a flow is often found in systems which are invariant under the transformation $x \to -x$. It occurs if both $G_\mu$ and $G_{xx}$ vanish at the origin (together with $G$ and $G_x$, the standard conditions for existence of a bifurcation which we assume all along).

*Example 8.6*

Consider the differential equation

$$\dot{x} = \mu x - x^3.$$

The origin, $x = 0$, is always a fixed point, and the Jacobian is $\mu - 3x^2$. Thus $x = 0$ is non-hyperbolic when $\mu = 0$ and is stable if $\mu < 0$ and unstable if $\mu > 0$. Provided $\mu > 0$ there are two other stationary points: $x_{\pm} = \pm\sqrt{\mu}$, both of which are stable. This situation is shown schematically in Figure 8.7, which also explains the name *pitchfork*.

The analytic approach to this bifurcation is similar to the saddlenode bifurcation, in that the only thing we need to worry about is the correct asymptotic expansion for the stationary points. Recall that we are considering bifurcations on the centre manifold of the system

$$\dot{x} = G(x, \mu)$$

where $G(0,0) = G_x(0,0) = 0$, so that $x = 0$ is a non-hyperbolic stationary point when $\mu = 0$, together with the extra conditions $G_\mu(0,0) = G_{xx}(0,0) = 0$. Hence the Taylor expansion of $G$ in a neighbourhood of $(x, \mu) = (0,0)$ can be written as

$$\begin{aligned}\dot{x} = G(x,\mu) = {} & \tfrac{1}{2}(2G_{\mu x}\mu x + G_{\mu\mu}\mu^2) \\ & + \tfrac{1}{3!}(G_{xxx}x^3 + 3G_{xx\mu}x^2\mu + 3G_{x\mu\mu}x\mu^2 + G_{\mu\mu\mu}\mu^3) \\ & + O((|x| + |\mu|)^4).\end{aligned} \tag{8.30}$$

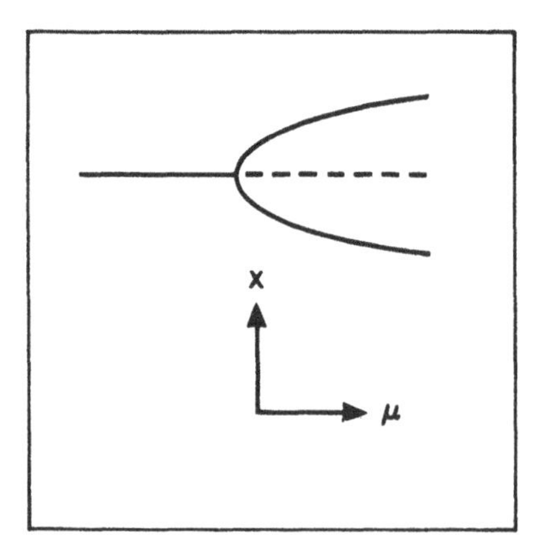


Fig. 8.7 Bifurcation diagram for Example 8.6.

In the two previous sections there has been only one scaling of $x$ which leads to a natural balance in setting the right hand side of the $\dot{x}$ equation to zero: $x \sim O(|\mu|^{\frac{1}{2}})$ for the saddlenode bifurcation and $x \sim O(|\mu|)$ for the transcritical bifurcation. Here, however, both of these possibilities lead to coherent asymptotic expansions for curves of stationary points. If we set $x \sim O(|\mu|^{\frac{1}{2}})$ then we get

$$G_{\mu x}\mu x + \tfrac{1}{3!}G_{xxx}x^3 \approx 0 \tag{8.31}$$

at leading order, whilst if we set $x \sim O(|\mu|)$ we find

$$2G_{\mu x}\mu x + G_{\mu\mu}\mu^2 \approx 0 \tag{8.32}$$

at leading order. This suggests that we need to investigate both possibilities in order to determine the nature of stationary points near $(x,\mu) = (0,0)$. The latter scaling of $x$ is clearly easier, so pose the solution

$$x \sim \sum_{n \geq 1} \alpha_n \mu^n \tag{8.33}$$

for stationary points. Substituting into the right hand side of (8.30) gives

$$\alpha_1 = -\frac{G_{\mu\mu}}{2G_{\mu x}} \tag{8.34}$$

at order $\mu^2$ provided $G_{\mu x} \neq 0$. At order $\mu^3$ the equation is

$$G_{\mu x}\alpha_2 + \frac{1}{6}(G_{xxx}\alpha_1^3 + 3G_{xx\mu}\alpha_1^2 + 3G_{x\mu\mu}\alpha_1 + G_{\mu\mu\mu}) = 0 \tag{8.35}$$

which, provided $G_{\mu x} \neq 0$ can easily be used to find $\alpha_2$. Hence there is a curve of stationary points through $(x,\mu) = (0,0)$ of the form

$$x \sim -\frac{G_{\mu\mu}}{2G_{\mu x}}\mu + O(\mu^2) \tag{8.36}$$

if $G_{\mu x} \neq 0$. To investigate the other possibility, (8.31), we pose solutions of the form

$$x \sim \sum_{n \geq 1} \beta_n \mu^{\frac{n}{2}} \tag{8.37}$$

for $\mu > 0$ (we will consider the case $\mu < 0$ when we have understood this case). This implies that

$$x^2 \sim \beta_1^2 \mu + 2\beta_1\beta_2\mu^{\frac{3}{2}} + O(\mu^2) \tag{8.38}$$

and

$$x^3 \sim \beta_1^3\mu^{\frac{3}{2}} + 3\beta_1^2\beta_2\mu^2 + O(\mu^{\frac{5}{2}}). \tag{8.39}$$

Substituting into (8.30) and setting $\dot{x} = 0$ we obtain, at order $\mu^{\frac{3}{2}}$,

$$G_{\mu x}\beta_1 + \frac{1}{6}G_{xxx}\beta_1^3 = 0 \tag{8.40}$$

and so, provided $G_{\mu x}G_{xxx} < 0$, we obtain the solution

$$\beta_1 = \pm\sqrt{-\frac{6G_{\mu x}}{G_{xxx}}}. \tag{8.41}$$

Now, at order $\mu^2$,

$$G_{\mu x}\beta_2 + \tfrac{1}{2}G_{\mu\mu} + \tfrac{1}{3!}(3G_{xxx}\beta_1^2\beta_2 + 3G_{xx\mu}\beta_1^2) + \tfrac{1}{4!}G_{xxxx}\beta_1^4 = 0 \tag{8.42}$$

or

$$\beta_2 = \frac{1}{4G_{\mu x}G_{xxx}^2}\left(G_{\mu\mu}G_{xxx}^2 - 6G_{\mu x}G_{xx\mu}G_{xxx} + 3G_{xxxx}G_{\mu x}^2\right). \tag{8.43}$$

We could continue to obtain higher order terms in the asymptotic expansion, but the moral should be clear. Provided $G_{\mu x}G_{xxx} < 0$ a pair of stationary points bifurcates into $\mu > 0$ from $\mu = 0$ in a manner analogous to the curve of stationary points in the saddlenode bifurcation. We leave it as an exercise to show that these stationary points are stable if $G_{xxx} < 0$ and unstable if $G_{xxx} > 0$. If $G_{\mu x}G_{xxx} > 0$ then a similar curve bifurcates into $\mu < 0$ (this can be obtained by trying an asymptotic expansion in $(-\mu)^{\frac{1}{2}}$). We are now in a position to state the main result suggested by this discussion.

(8.4) THEOREM (PITCHFORK BIFURCATION)

*Suppose* $\dot{x} = G(x,\mu)$ *and* $G(0,0) = G_x(0,0) = 0$. *Then if*

$$G_\mu(0,0) = G_{xx}(0,0) = 0, \quad G_{\mu x}(0,0) \neq 0 \text{ and } G_{xxx}(0,0) \neq 0$$

*there exist two curves of stationary points in a neighbourhood of* $(x,\mu) = (0,0)$. *One of these passes through* $(0,0)$ *transverse to the axis* $\mu = 0$ *whilst the other is tangential to* $\mu = 0$ *at* $(0,0)$. *If* $G_{\mu x}G_{xxx} < 0$ *then for each* $\mu$ *with* $|\mu|$ *sufficiently small there is one stationary point near* $x = 0$ *if* $\mu < 0$ *which is stable if* $G_{xxx} < 0$ *and unstable if* $G_{xxx} > 0$, *and if* $\mu > 0$ *there exist three stationary points near* $x = 0$. *Of these the outer pair are stable (resp. unstable) and the inner stationary point is unstable (resp. stable) if* $G_{xxx} < 0$ *(resp.* $G_{xxx} > 0$*).*

*If $G_{\mu x}G_{xxx} < 0$, then there exist three stationary points near $x = 0$ if $\mu < 0$ (the outer pair are stable and the inner one is unstable if $G_{xxx} < 0$) and one stationary point near $x = 0$ if $\mu > 0$ (stable if $G_{xxx} < 0$). Stability properties are reversed if $G_{xxx} > 0$.*

The bifurcation is said to be *supercritical* if the bifurcating pair of stationary points is stable, otherwise the bifurcation is *subcritical.* These various possibilities are sketched in Figure 8.8.

## 8.6 An example

In this section we shall go through the calculation of the centre manifold and determination of the type of bifurcation by projecting the flow (locally) onto the extended centre manifold in full detail. For our example we take the two-dimensional differential equation

$$\dot{x} = (1+\mu)x - 4y + x^2 - 2xy$$
$$\dot{y} = 2x - 4\mu y - y^2 - x^2$$

which has a stationary point at the origin for all real values of $\mu$. The Jacobian matrix evaluated at $(0,0)$ is

$$\begin{pmatrix} (1+\mu) & -4 \\ 2 & -4\mu \end{pmatrix}$$

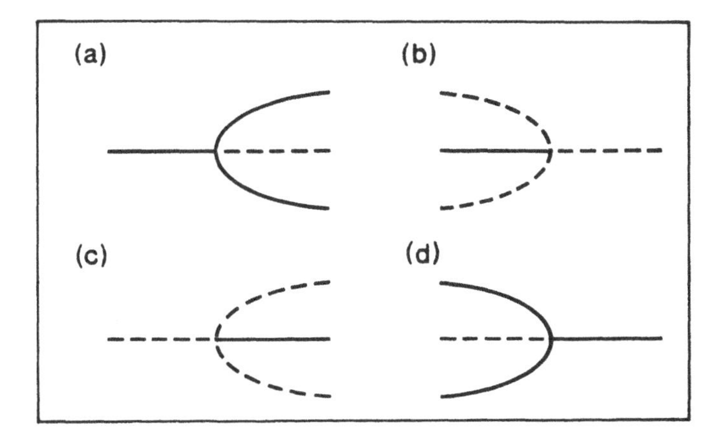


Fig. 8.8 Pitchfork bifurcations: (a) supercritical, $G_{\mu x} > 0$, $G_{xxx} < 0$; (b) subcritical, $G_{\mu x} > 0$, $G_{xxx} > 0$; (c) subcritical, $G_{\mu x} < 0$, $G_{xxx} > 0$; (d) supercritical, $G_{\mu x} < 0$, $G_{xxx} < 0$.

with characteristic equation

$$s^2 - (1 - 3\mu)s - 4\mu(1 + \mu) + 8 = 0.$$

The origin is non-hyperbolic for values of $\mu$ at which $s = 0$ or $s = i\omega$ is a solution of the characteristic equation. Setting $s = 0$ we see that the origin has a simple zero eigenvalue if $-4\mu(1 + \mu) + 8 = 0$, i.e. if $\mu = -2$ or $\mu = 1$. Similarly, setting $s = i\omega$ we find that the origin is non-hyperbolic if $\mu = \frac{1}{3}$ with a pair of purely imaginary eigenvalues. We will ignore this case until the next section and concentrate on the case $\mu = 1$: what sort of bifurcation occurs as $\mu$ passes through one? We begin the calculation by changing coordinates so that the linear part of the flow at the origin when $\mu = 1$ is in canonical form. Setting $\mu = 1$ in the Jacobian matrix we find that the linear part of the equation at the origin is

$$\begin{pmatrix} \dot{x} \\ \dot{y} \end{pmatrix} = \begin{pmatrix} 2 & -4 \\ 2 & -4 \end{pmatrix} \begin{pmatrix} x \\ y \end{pmatrix}$$

and so the linear flow has eigenvalues 0 (as we knew already) and $-2$. The corresponding eigenvectors are $e_0 = (2, 1)^T$ and $e_{-2} = (1, 1)^T$. Now let $P$ be the matrix whose columns are the eigenvectors,

$$P = \begin{pmatrix} 2 & 1 \\ 1 & 1 \end{pmatrix},$$

so that

$$\begin{pmatrix} 2 & -4 \\ 2 & -4 \end{pmatrix} P = P \begin{pmatrix} 0 & 0 \\ 0 & -2 \end{pmatrix}.$$

Hence if we set $\begin{pmatrix} u \\ v \end{pmatrix} = P^{-1} \begin{pmatrix} x \\ y \end{pmatrix}$ we obtain the linear part of the equation for $u$ and $v$ as

$$\begin{pmatrix} \dot{u} \\ \dot{v} \end{pmatrix} = \begin{pmatrix} 0 & 0 \\ 0 & -2 \end{pmatrix} \begin{pmatrix} u \\ v \end{pmatrix}.$$

Furthermore, $P^{-1} = \begin{pmatrix} 1 & -1 \\ -1 & 2 \end{pmatrix}$ and so

$$\begin{pmatrix} u \\ v \end{pmatrix} = P^{-1} \begin{pmatrix} x \\ y \end{pmatrix} = \begin{pmatrix} x - y \\ -x + 2y \end{pmatrix} \text{ and } \begin{pmatrix} x \\ y \end{pmatrix} = P \begin{pmatrix} u \\ v \end{pmatrix} = \begin{pmatrix} 2u + v \\ u + v \end{pmatrix}.$$

Going back to the original equations and rewriting them in terms of the new variables $u$ and $v$ using these equations gives ($\dot{u} = \dot{x} - \dot{y}$, $\dot{v} = -\dot{x} + 2\dot{y}$)

$$\begin{aligned} \dot{u} &= (-1 + \mu)(6u + 5v) - 3u^2 - 4uv - v^2 \\ \dot{v} &= 10(1 - \mu)u + (7 - 9\mu)v - 10u^2 - 10uv - 3v^2. \end{aligned}$$

Since we are interested in the bifurcation at $\mu = 1$ we define a new parameter $\nu$ by $\mu = 1+\nu$ so that the bifurcation occurs when $\nu = 0$ and we are able to treat $|\nu|$ as small in our local investigation. Substituting for $\mu$ and adding the extra equation $\dot{\nu} = 0$ we obtain the extended system

$$\begin{aligned}
\dot{u} &= \nu(6u+5v) - 3u^2 - 4uv - v^2 \\
\dot{v} &= -10\nu u - (2+9\nu)v - 10u^2 - 10uv - 3v^2 \\
\dot{\nu} &= 0.
\end{aligned}$$

Note that since we are now treating $\nu$ in the same way as $u$ and $v$, the first equation has no linear terms, the only linear term in the second equation is $-2v$ and the third equation has no terms at all! The linear centre manifold is thus $E^c(0) = \{(u,v,\nu)|v=0\}$ and the linear stable manifold is $E^s(0) = \{(u,v,\nu)|u=\nu=0\}$. To find an approximation to the nonlinear centre manifold we pose the solution

$$v = h(u,\nu) \quad \text{with} \quad \frac{\partial h}{\partial u}(0,0) = \frac{\partial h}{\partial \nu}(0,0) = 0$$

i.e. $h(u,\nu) = au^2 + bu\nu + c\nu^2 + \dots$. Now, from the $\dot{v}$ equation we have

$$\dot{v} = -10u\nu - (2+9\nu)h(u,\nu) - 10u^2 - 10uh(u,\nu) - 3[h(u,\nu)]^2$$

but since, on the centre manifold, $v = h(u,v)$ we also have that

$$\dot{v} = \dot{u}\frac{\partial h}{\partial u} + \dot{\nu}\frac{\partial h}{\partial \nu} = \frac{\partial h}{\partial u}\left(\nu(6u+5h(u,\nu)) - 3u^2 - 4uh(u,\nu) - [h(u,\nu)]^2\right).$$

Substituting the series expansion of $h$ into these two expressions and equating powers of $u^2$ gives $a = -5$, powers of $u\nu$ give $b = -5$ and powers of $\nu^2$ give $c = 0$. Hence the centre manifold is given by

$$v = -5u(u+\nu) + \text{ cubic terms.}$$

Finally we substitute this expression back into the equation of $\dot{u}$ to obtain the projection of the motion on the centre manifold onto the $u$ axis:

$$\dot{u} = 6\nu u - 3u^2 + \text{ cubic terms}$$

(note that the expression for the centre manifold does not contribute any quadratic terms in this case, so we didn't really need to calculate the centre manifold). Hence on the centre manifold $\dot{u} = G(u,\nu)$ with

$$G(0,0) = G_u(0,0) = G_\nu(0,0) = 0, \; G_{uu} = -6, \; G_{u\nu} = 6 \text{ and } G_{\nu\nu} = 0.$$

This implies that the conditions for a transcritical bifurcation are satisfied: the origin is stable in $\nu < 0$ and there is a separate branch of unstable stationary points, in $\nu > 0$ this other branch becomes stable and the origin becomes unstable. When $\nu = 0$ the equation on the centre manifold is approximately $\dot{u} = -3u^2$ and so the origin is stable if approached from $u > 0$ and unstable if approached from $u < 0$. These results are illustrated schematically in Figure 8.9.

## 8.7 The Implicit Function Theorem

In the previous three sections we have discussed various bifurcations by finding stationary points as asymptotic series in the parameter $\mu$. In order to prove such results rigorously we really need to invoke a result from analysis called the Implicit Function Theorem, which states that solutions to equations exist provided certain conditions hold. We have already used the Implicit Function Theorem (implicitly) when discussing the persistence of hyperbolic stationary points under perturbations of differential equations in Chapter 4. In this section we will give a formal statement of the theorem and indicate how it can be used to put the preceding analysis on a rigorous footing.

(8.5) IMPLICIT FUNCTION THEOREM

*Suppose that* $F : \mathbf{R}^n \times \mathbf{R} \to \mathbf{R}^n$ *is a continuously differentiable function of the variables* $(y_1, \ldots, y_n) \in \mathbf{R}^n$ *and* $z \in \mathbf{R}$, *and that* $F(0,0) = 0$. *If the Jacobian matrix* $DF(0,0)$ *(where* $(DF)_{ij} = \frac{\partial F_i}{\partial y_j}$*) is invertible then*

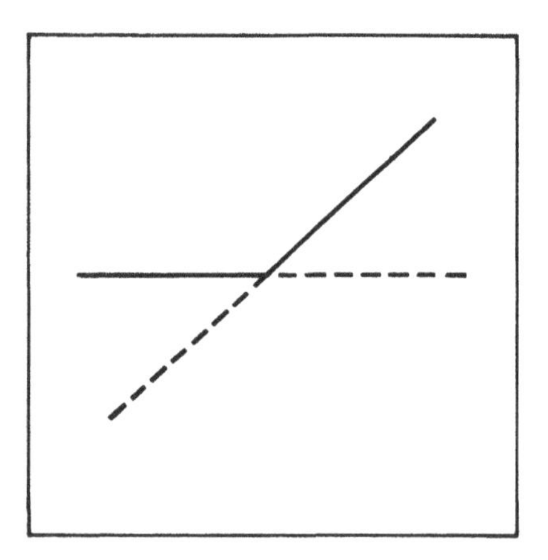

Fig. 8.9

*there exists* $\epsilon > 0$ *and a smooth curve*

$$y_i = Y_i(z), \quad i = 1, \dots, n,$$

*which is the unique solution of* $F(y_1, \dots, y_n, z) = 0$ *in* $|z| < \epsilon$, $|y| < \epsilon$.

All this theorem really says is that if we know a solution to an equation and the Jacobian matrix of the equation is invertible (i.e. has non-zero determinant) at that point, then there are solutions at nearby values of the parameter which are continuously connected to the known solution. A proof of this result can be found in any good book on analysis, but essentially all we are saying is that if $F(0,0) = 0$ then

$$F(y, z) = DF(0,0)y + F_z(0,0)z + \dots$$

and so if $DF(0,0)$ is invertible then solutions to $F(y,z) = 0$ are given approximately (for small $|y|$ and $|z|$) by

$$y \sim [DF(0,0)]^{-1} F_z(0,0)z.$$

This is, of course, precisely the sort of manipulation that we have been doing in the last few sections, all the Implicit Function Theorem says is that this procedure is justified.

To illustrate the use of the Implicit Function Theorem we consider the transcritical bifurcation (the saddlenode bifurcation is trivial because the condition $G_\mu \neq 0$ means that the theorem can be applied directly with $y = \mu$ and $z = x$). So we have

$$\dot{x} = G(x, \mu)$$

with $G(0,0) = G_x(0,0) = G_\mu(0,0) = 0$. Stationary points are given by the equation $G(x, \mu) = 0$ and for small $|x|$ and $|\mu|$

$$2G(x,\mu) = G_{xx}x^2 + 2G_{x\mu}\mu x + G_{\mu\mu}\mu^2 + O((|x| + |\mu|)^2).$$

Since both $G_x$ and $G_\mu$ vanish at $(x, \mu) = (0,0)$ we cannot apply the Implicit Function Theorem to $G(x, \mu)$ to prove the existence of continuous families of solutions. To treat this case, define a new variable $\omega$ by $\omega\mu = x$ and consider

$$g(\omega, \mu) = G(\omega\mu, \mu).$$

Zeros of $g$ correspond to zeros of $G$ and

$$g_\mu(\omega, \mu) = \omega G_x(\omega\mu, \mu) + G_\mu(\omega\mu, \mu),$$
$$g_\omega(\omega, \mu) = \mu G_x(\omega\mu, \mu).$$

Thus both partial derivatives of $g$ are zero if $\mu = 0$. To apply the Implicit Function Theorem define

$$H(\omega,\mu) = \begin{cases} \dfrac{2g(\omega,\mu)}{\mu^2} & \text{if } \mu \neq 0 \\ g_{\mu\mu}(\omega,\mu) & \text{if } \mu = 0 \end{cases}$$

so $H$ is smooth and $H(\omega,0) = 0$ if $g_{\mu\mu}(\omega,0) = 0$. Now,

$$g_{\mu\mu}(\omega,\mu) = \omega^2 G_{xx}(\omega\mu,\mu) + 2\omega G_{\mu x}(\omega\mu,\mu) + G_{\mu\mu}(\omega\mu,\mu)$$

and so $H(\omega,0) = 0$ if $\omega = \omega_\pm$, where

$$\omega_\pm = \frac{-G_{x\mu} \pm \sqrt{G_{x\mu}^2 - G_{xx}G_{\mu\mu}}}{G_{xx}}$$

provided

$$\begin{cases} D = G_{\mu x}^2 - G_{xx}G_{\mu\mu} > 0 \\ G_{xx} \neq 0 \end{cases}.$$

Furthermore,

$$\frac{\partial H}{\partial \omega}(\omega_\pm,0) = 2G_{xx}\omega_\pm + 2G_{x\mu} = \pm 2\sqrt{D}.$$

Thus, provided $D > 0$ we can apply the Implicit Function Theorem (with $y = \omega$ and $z = \mu$) at $(\omega_+,0)$ and $(\omega_-,0)$ to obtain two curves of solutions to $H(\omega,\mu) = 0$ in the form

$$\omega = W_\pm(\mu), \quad W_\pm(0) = \omega_\pm,$$

and hence two branches of stationary points (solutions to $G(x,\mu) = 0$):

$$x = \mu W_\pm(\mu).$$

These two curves intersect at $(x,\mu) = (0,0)$ and the stability results of Section 8.4 are straightforward to verify.

We leave it as an exercise for the interested reader to apply the Implicit Function Theorem to the pitchfork bifurcation.

## 8.8 The Hopf bifurcation

All the bifurcations discussed so far have involved motion on a one-dimensional centre manifold on which stationary points can be created or destroyed as parameters vary. The Hopf bifurcation is several orders of magnitude harder to analyse since it involves a non-hyperbolic stationary point with linearized eigenvalues $\pm i\omega$, and thus a two-dimensional centre manifold, and the bifurcating solutions are periodic rather than stationary. All this implies that the algebra involved becomes significantly harder, although it is not very different in any conceptual sense. In order to make the results comprehensible this section is split into several parts. First we look at a simple example which illustrates the bifurcation and then go on to state the theorem and rehearse the steps of the proof without going into any detail. Then we look at another example suggested by the sketch of the proof and finally go through the gory details. In fact, the essential part of the proof is relatively straightforward as it only needs the type of manipulations that were used in the proof of Poincaré's Linearization Theorem. The major headache is caused by a need to relate the results back to the partial derivatives of the two-dimensional differential equation on the centre manifold. This manipulation, although straightforward, is lengthy and unexciting and is left until the end of the section.

The Hopf Bifurcation Theorem has a slightly curious history. It is named after Hopf, who gave the first proof in $\mathbf{R}^n$ in 1942, but had been proved by Andronov and Leontovich in the late 1930s using techniques due to Poincaré and Bendixson. Indeed, Poincaré makes it clear in *Méthodes nouvelles de la mécanique céleste, Vol. 1* (1892) that he was aware of the result, but finds it too trivial to bother to write down! Some authors refer to the theorem as the Poincaré–Andronov–Hopf bifurcation, or simply the oscillatory bifurcation. The proof presented here is sketched in Wiggins (1991) and owes a great deal to the work of Hassard and Wan (1978), who seem to have been the first authors to use the complex notation adopted below.

*Example 8.7*

Consider the differential equation

$$\dot{x} = \mu x - \omega y - (x^2 + y^2)x$$
$$\dot{y} = \omega x + \mu y - (x^2 + y^2)y.$$

A little linearization about the origin $(x, y) = (0, 0)$ shows that the origin is a stable focus if $\mu < 0$ and an unstable focus if $\mu > 0$ (the eigenvalues of the Jacobian matrix at the origin are $\mu \pm i\omega$). Hence the origin is non-hyperbolic, with linearized eigenvalues $\pm i\omega$, when $\mu = 0$. From the experience of the previous sections we expect some sort of bifurcation to occur when $\mu = 0$. In order to discover what happens it is easiest to change into polar coordinates. This gives

$$\dot{r} = \mu r - r^3, \quad \dot{\theta} = \omega.$$

From the $\dot{\theta}$ equation it is immediately clear that provided $\omega \neq 0$ the only stationary point is at the origin, $r = 0$, so there is no bifurcating stationary solution. However, $\dot{r} = 0$ if $r = 0$ or $\mu = r^2$. Hence if $\mu > 0$ there is a periodic orbit at $r = \sqrt{\mu}$. Plotting $\dot{r}$ against $\mu r - r^3$ we see that this periodic orbit is stable. This gives the picture illustrated in Figure 8.10: a stable periodic orbit with radius $\sqrt{\mu}$ bifurcates from the origin from $\mu = 0$ into $\mu > 0$.

This is an example of a supercritical Hopf bifurcation (supercritical because the bifurcating periodic orbit is stable). The general result is stated below, but first note that the linearized differential equation at the bifurcation point is $\dot{x} = Lx$ where $L = \begin{pmatrix} 0 & -\omega \\ \omega & 0 \end{pmatrix}$ and so $L$ is invertible. Hence, by the Implicit Function Theorem (see Chapter 4 and Section 8.7), there is a stationary point of the flow near the origin for all parameter values close enough to the bifurcation value. This implies that without loss of generality we can consider differential equations for which the origin is always a stationary point for parameter values near zero (which will be taken to be the bifurcation value) since we can always arrange for this to be the case by a simple shift of the origin.

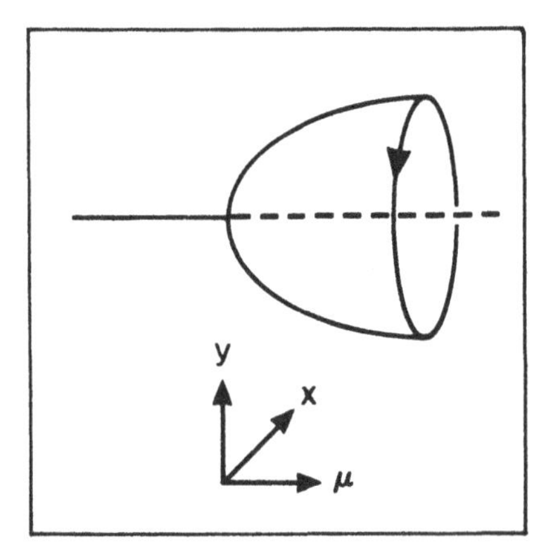


Fig. 8.10 Bifurcation diagram for Example 8.7.

(8.6) THEOREM (HOPF BIFURCATION THEOREM)

*Suppose that*

$$\dot{x} = f(x, y, \mu), \quad \dot{y} = g(x, y, \mu)$$

*with* $f(0,0,\mu) = g(0,0,\mu) = 0$ *and that the Jacobian matrix* $\begin{pmatrix} \frac{\partial f}{\partial x} & \frac{\partial f}{\partial y} \\ \frac{\partial g}{\partial x} & \frac{\partial g}{\partial y} \end{pmatrix}$ *evaluated at the origin when* $\mu = 0$ *is*

$$\begin{pmatrix} 0 & -\omega \\ \omega & 0 \end{pmatrix}$$

*for some* $\omega \neq 0$. *Then if*

$$f_{\mu x} + g_{\mu y} \neq 0$$

*and*

$$a \neq 0$$

*where* $a$ *is a constant defined below, a curve of periodic solutions bifurcates from the origin into* $\mu < 0$ *if* $a(f_{\mu x} + g_{\mu y}) > 0$ *or* $\mu > 0$ *if* $a(f_{\mu x} + g_{\mu y}) < 0$. *The origin is stable for* $\mu > 0$ *(resp.* $\mu < 0$*) and unstable for* $\mu < 0$ *(resp.* $\mu > 0$*) if* $f_{\mu x} + g_{\mu y} < 0$ *(resp.* $> 0$*) whilst the periodic solutions are stable (resp. unstable) if the origin is unstable (resp. stable) on the side of* $\mu = 0$ *for which the periodic solutions exist. The amplitude of the periodic orbits grows like* $|\mu|^{\frac{1}{2}}$ *whilst their periods tend to* $\frac{2\pi}{|\omega|}$ *as* $|\mu|$ *tends to zero. The bifurcation is supercritical if the bifurcating periodic orbits are stable, otherwise it is subcritical.*

The genericity condition, $a \neq 0$, is the usual sort of condition for a bifurcation, involving the partial derivatives of the vector field $(f(x,y,\mu), g(x,y,\mu))$ evaluated at $(0,0,0)$. More explicitly

$$\begin{aligned} a = {} & \tfrac{1}{16}(f_{xxx} + g_{xxy} + f_{xyy} + g_{yyy}) \\ & + \tfrac{1}{16\omega}\left(f_{xy}(f_{xx} + f_{yy}) - g_{xy}(g_{xx} + g_{yy}) - f_{xx}g_{xx} + f_{yy}g_{yy}\right). \end{aligned} \tag{8.44}$$

This horrible expression will be the cause of not a little pain at the end of this section. However, the most important part of the theorem, the existence of bifurcating periodic solutions, is not too hard to prove and the pain referred to above comes from the need to do lots of nasty algebraic manipulation rather than any sophisticated conceptual trickery. As promised we begin with an outline of the strategy of the proof.

*Step 1*

Concentrate only on the linear part of the equation. By assumption, the eigenvalues of the Jacobian matrix at the origin are $\alpha(\mu) \pm i\beta(\mu)$ with $\alpha(0) = 0$ and $\beta(0) = \omega$. Hence, using the linearization techniques of Chapter 3, there is a linear change of coordinates (which may depend upon $\mu$) which allows the equations to be written in the form

$$\dot{x} = \alpha x - \beta y + \text{nonlinear terms}$$
$$\dot{y} = \beta x + \alpha y + \text{nonlinear terms.}$$

Assume that this change of coordinates has been made.

*Step 2*

Rewrite the equations in complex notation by setting $z = x + iy$. This gives an equation of the form

$$\dot{z} = \lambda(\mu)z + \sum_{m\geq 2}\sum_{p=0}^{m} \gamma_{mp}(\mu) z^p z^{*m-p}$$

where $\lambda(\mu) = \alpha(\mu) + i\beta(\mu)$ and $z^*$ is the complex conjugate of $z$. This step is not really necessary, but halves the amount of algebra required in later steps.

*Step 3*

Perform successive near identity changes of coordinate of the form

$$z = w + \sum_{p=0}^{m} a_{mp} w^p w^{*m-p}$$

for $m = 2, 3, 4\ldots$ in such a way that as many of the nonlinear terms in the equation from Step 2 vanish as is possible. This procedure is precisely analogous to the proof of Poincaré's Linearization Theorem. After going through the calculations we are left with the system

$$\dot{z} = \lambda(\mu)z + A(\mu)|z|^2 z + O(|z|^5).$$

*Step 4*

Step three provides us with an autonomous system of differential equations in the plane which can be approached using the techniques of

Chapter 5. Since

$$\dot{z} = \lambda(\mu)z + A(\mu)|z|^2 z + O(|z|^5)$$

setting $z = re^{i\theta}$ gives

$$\dot{z} = (\dot{r} + ir\dot{\theta})e^{i\theta} = (\alpha + i\beta)re^{i\theta} + A(\mu)r^3 e^{i\theta} + O(r^5)$$

and taking real and imaginary parts we obtain

$$\dot{r} = \alpha r + \mathrm{Re}(A)r^3 + O(r^5)$$

$$\dot{\theta} = \beta + \mathrm{Im}(A)r^2 + O(r^4).$$

Provided $\beta(0) = \omega \neq 0$ and $\mathrm{Re}(A) \neq 0$ it is possible to use the Poincaré-Bendixson Theorem from Chapter 5 to show that this system has a periodic orbit in either $\mu < 0$ or $\mu > 0$ depending upon the sign of $\alpha\mathrm{Re}(A)$.

*Step 5*

The last step of the proof is to calculate $\mathrm{Re}A(0)$ explicitly in terms of the partial derivatives of the defining vector field. This can be done by performing the two successive changes of variable in Step 3 explicitly and although trivial this takes a lot of patience and hard work.

Before going through these steps in greater detail we shall consider the example defined in Step 4. The example is almost the same as Example 7, but with more constants floating about to confuse the issue.

*Example 8.8*

From Step 4 we are led to consider

$$\dot{r} = \mu r + ar^3, \quad \dot{\theta} = \omega + b\mu + cr^2$$

for non-zero constants $a$ and $\omega$. For small $r$ and $|\mu|$ the only stationary point is at the origin as $\dot{\theta}$ is not equal to zero. Furthermore the origin is stable if $\mu < 0$ and unstable if $\mu > 0$. From the $\dot{r}$ equation we see that there is a periodic orbit when $r^2 = -\mu/a$, so the periodic orbit bifurcates into $\mu < 0$ if $a > 0$ and into $\mu > 0$ if $a < 0$. The stability of the periodic orbit is easily deduced from the $\dot{r}$ equation: it is unstable if $a > 0$ and it is stable if $a < 0$. Thus there is a subcritical Hopf bifurcation if $a > 0$ (the unstable periodic orbit exists if $\mu < 0$) and a supercritical Hopf bifurcation if $a < 0$ (the stable periodic orbit exists if $\mu > 0$). These possibilities are illustrated in Figure 8.11.

We now come to the proof of the theorem. Steps 1 to 4 are relatively simple. Step 5 is too, but only after going through it several times! It could be left out at a first reading without losing the thread of this chapter.

**Step 1.** Assume that a linear change of coordinates has already been made so that the differential equation has its linear part in normal form at the bifurcation point. Since the origin has been constrained to be a fixed point for all $\mu$ the Taylor expansion of the defining equations can be written as

$$\dot{x} = f(x, y, \mu) = f_{x\mu}\mu x - (\omega - f_{y\mu}\mu)y + \dots \tag{8.45a}$$

$$\dot{y} = g(x, y, \mu) = (\omega + g_{x\mu}\mu)x + g_{y\mu}\mu y + \dots \tag{8.45b}$$

where the dots indicate nonlinear terms in $x$ and $y$ and also terms of order $\mu^2 x$ and $\mu^2 y$, which we will ignore. The characteristic equation of the Jacobian matrix is therefore (to order $\mu$)

$$s^2 - (f_{\mu x} + g_{\mu y})\mu s + \omega^2 + (g_{\mu x} - f_{\mu y})\mu\omega = 0 \tag{8.46}$$

with roots $\alpha(\mu) \pm i\beta(\mu)$ where

$$\alpha(\mu) = \tfrac{1}{2}(f_{\mu x} + g_{\mu y})\mu + O(\mu^2), \tag{8.47}$$

$$\beta(\mu) = \omega\left(1 + \frac{\mu}{2\omega}(g_{\mu x} - f_{\mu y})\right) + O(\mu^2). \tag{8.48}$$

Now, using the techniques of Chapter 3, it is clear that there is a $\mu$ dependent linear change of variables such that the differential equation

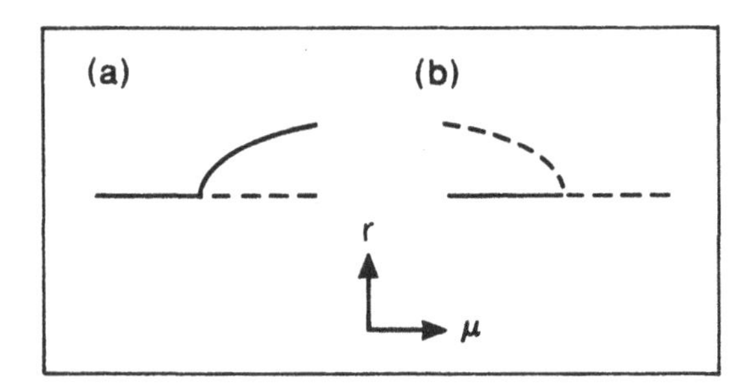


Fig. 8.11 (a) Supercritical Hopf bifurcation; (b) subcritical Hopf bifurcation.

takes the form

$$\dot{x} = \alpha x - \beta y + \ldots$$

$$\dot{y} = -\beta x + \alpha y + \ldots .$$

and that the first genericity condition in the statement of the Hopf Bifurcation Theorem is equivalent to

$$\frac{d}{d\mu}\alpha(\mu)|_{\mu=0} \neq 0. \tag{8.49}$$

This is sometimes called a transversality condition; it implies that the eigenvalues of the linearized flow at the origin cross the imaginary axis when $\mu = 0$ with non-zero velocity. In the steps which follow we will assume that this change of coordinates has been made.

**Step 2.** Now set $z = x + iy$. The differential equation for $z$ is

$$\begin{aligned}\dot{z} = \dot{x} + i\dot{y} &= (\alpha x - \beta y) + i(-\beta x + \alpha y) + \ldots \\ &= \alpha(x + iy) + i\beta(x + iy) + \ldots = \lambda z + \ldots\end{aligned} \tag{8.50}$$

where $\lambda = \alpha + i\beta$. The nonlinear terms can all be written with terms like $z^k z^{*j}$ and so the full equation becomes

$$\dot{z} = \lambda z + \sum_{m\geq 2}\sum_{k=0}^{m} \gamma_{mk}(\mu) z^k z^{*m-k} \tag{8.51}$$

for complex coefficients $\gamma_{mk}$. As an example (which will be needed in Step 5) consider the quadratic terms:

$$z^2 = (x^2 - y^2) + 2ixy, \; zz^* = x^2 + y^2 \text{ and } z^{*2} = (x^2 - y^2) - 2ixy$$

so the quadratic terms in $x$ and $y$ can be rewritten in terms of $z$ by first eliminating the $xy$ terms from the $z^2$ and $z^{*2}$ terms to get

$$x^2 - y^2 = \tfrac{1}{2}(z^2 + z^{*2}), \; xy = \tfrac{1}{4}i(z^{*2} - z^2)$$

and then using the $zz^*$ term with the first of these equations to get

$$x^2 = \tfrac{1}{4}(z^2 + 2zz^* + z^{*2}), \; y^2 = -\tfrac{1}{4}(z^2 - 2zz^* + z^{*2}).$$

Alternatively, simply note that $x = \frac{1}{2}(z + z^*)$ and $y = -\frac{i}{2}(z - z^*)$ and so

$$x^j y^k = \frac{(-i)^k}{2^{j+k}}(z + z^*)^j (z - z^*)^k.$$

**Step 3.** In this step we want to use successive near identity transformations to eliminate as many of the nonlinear terms of the differential equation as possible. Recall that the (complex) differential equation, (8.51), is

$$\dot{z} = \lambda(\mu)z + \sum_{m\geq 2}\sum_{k=0}^{m} \gamma_{mk}(\mu)z^k z^{*m-k}.$$

Consider first the quadratic terms. Define a new complex variable $w$ by

$$w = z + \sum_{p=0}^{2} a_{2k}z^k z^{*2-k}, \tag{8.52}$$

which can be inverted locally to give

$$z = w - \sum_{p=0}^{2} a_{2k}w^k w^{*2-k} + O(|w|^3). \tag{8.53}$$

We aim to choose the coefficients $a_{2k}$ in such a way that the equation for $\dot{w}$ contains as few quadratic terms as possible. Now

$$\dot{w} = \dot{z} + \sum_{p=0}^{2} a_{2k}\left(k\dot{z}z^{k-1}z^{*2-k} + (2-k)\dot{z}^* z^k z^{*1-k}\right) \tag{8.54}$$

and we know the equation for $\dot{z}$, (8.51), and the relationship between $z$ and $w$, (8.53). The only unknown quantity in (8.54) is thus $\dot{z}^*$, which can easily be deduced from (8.51) by complex conjugation giving

$$\dot{z}^* = \lambda^* z^* + \sum_{m\geq 2}\sum_{k=0}^{m} a^*_{mk}z^{*k}z^{m-k}. \tag{8.55}$$

Substituting into (8.54) and retaining only terms of order two or less gives

$$\begin{aligned}\dot{w} = \lambda\left(w - \sum_{k=0}^{2} a_{2k}w^k w^{*2-k}\right) + \sum_{k=0}^{2} \gamma_{2k}w^k w^{*2-k} \\ + \sum_{k=0}^{2} a_{2k}(k\lambda w^k w^{*2-k} + (2-k)\lambda^* w^k w^{*2-k}) + O(|w|^3).\end{aligned} \tag{8.56}$$

Rearranging the summed terms to get a single sum gives

$$\begin{aligned}\dot{w} = \lambda w + \sum_{k=0}^{2}(\gamma_{2k} + a_{2k}[(k-1)\lambda + (2-k)\lambda^*])w^k w^{*2-k} \\ + O(|w|^3)\end{aligned} \tag{8.57}$$

So, provided $(k-1)\lambda + (2-k)\lambda^* \neq 0$ in a neighbourhood of $\mu = 0$, all the second order terms can be eliminated by setting

$$a_{2k} = \frac{-\gamma_{2k}}{(k-1)\lambda + (2-k)\lambda^*}. \tag{8.58}$$

Now, when $\mu = 0$, $\lambda = i\omega$ and so $(k-1)\lambda + (2-k)\lambda^* = (2k-3)i\omega$, which is bounded away from zero for $k = 0, 1, 2$ and so the $a_{2k}$ are well defined in a neighbourhood of $\mu = 0$. Hence, after this near identity change of coordinates (and replacing $w$ by $z$ again), we are left with the differential equation

$$\dot{z} = \lambda z + \sum_{m\geq 3}\sum_{k=0}^{m} \gamma'_{mk} z^k z^{*m-k} \tag{8.59}$$

where the complex coefficients $\gamma'_{mk}$ are obtained by considering the higher order effects of the nonlinear coordinate transformation defined above. Having dealt with the quadratic terms we could now go on to think about the cubic terms, then quartic terms and so on. In Step 5, the modification of $\gamma_{3k}$ due to the quadratic change of variable is given explicitly, but for the moment we need only go through this manipulation once more for general $m^{th}$ order terms. Suppose that after successive near identity transformations of the form $w = z + \sum_{k=0}^{m} a_{mk} z^k z^{*m-k}$ have been used for $m = 2, \ldots, n-1$ the differential equation takes the form

$$\dot{z} = \lambda z + N_{n-1}(z, z^*) + \sum_{m\geq n}\sum_{k=0}^{m} \gamma_{mk} z^k z^{m-k} \tag{8.60}$$

where

$$N_{n-1}(z, z^*) = \sum_{m=3}^{n-1}\sum_{k=0}^{m} b_{mk} z^k z^{*m-k} \tag{8.61}$$

contains those terms of order less than $n$ which cannot be eliminated by near identity transformations of order less than $n$. Now try the next near identity transformation

$$w = z + \sum_{k=0}^{n} a_{nk} z^k z^{*n-k}. \tag{8.62}$$

As before

$$z = w - \sum_{k=0}^{n} a_{nk} w^k w^{*n-k} + O(|w|^{n+1}), \tag{8.63}$$

$$\dot{w} = \dot{z} + \sum_{k=0}^{n} a_{nk}(k\dot{z}z^{k-1}z^{*n-k} + (n-k)\dot{z}^* z^{k-1} z^{*n-k-1}) \tag{8.64}$$

and

$$\dot{z}^* = \lambda^* z^* + O(|z|^2), \quad N_{n-1}(z, z^*) = N_{n-1}(w, w^*) + O(|w|^{n+1}). \tag{8.65}$$

Putting these expressions together we find

$$\begin{aligned}\dot{w} = \lambda \left( w - \sum_{k=0}^{n} a_{nk} w^k w^{*n-k} \right) + N_{n-1}(w, w^*) + \sum_{k=0}^{n} \gamma_{nk} w^k w^{*n-k} \\ + \sum_{k=0}^{n} a_{nk}(k\lambda w^k w^{*n-k} + (n-k)\lambda^* w^k w^{*n-k}) + O(|w|^{n+1}),\end{aligned} \tag{8.66}$$

and collecting up the terms into a single sum gives

$$\begin{aligned}\dot{w} &= \lambda w + N_{n-1}(w, w^*) \\ &\quad + \sum_{k=0}^{n} (a_{nk}[(k-1)\lambda + (n-k)\lambda^*] + \gamma_{nk})\, w^k w^{*n-k} \\ &\quad + O(|w|^{n+1}).\end{aligned} \tag{8.67}$$

Hence, provided $(k-1)\lambda + (n-k)\lambda^* \neq 0$ in a neighbourhood of $\mu = 0$, we can define a change of coordinates by

$$a_{nk} = \frac{-\gamma_{nk}}{(k-1)\lambda + (n-k)\lambda^*} \tag{8.68}$$

in (8.62) to obtain a new differential equation which has no terms of order $z^k z^{*n-k}$. At $\mu = 0$, $\lambda = i\omega$ and the condition becomes

$$(2k - (n+1))\omega \neq 0. \tag{8.69}$$

If $n$ is even this poses no problems; however, if $n = 2p+1$ is odd, it will be impossible to define the required coefficient when $k = p+1$. Hence we can, in principle, remove all terms except those of order $z^{p+1}z^{*p}$ by successive changes of coordinates and we are left an equation of the form

$$\dot{z} = \lambda z + \sum_{q=1}^{p} A_q |z|^{2q} z + O(|z|^{2p+2}). \tag{8.70}$$

In particular, going only as far as the quartic terms,

$$\dot{z} = \lambda(\mu) z + A(\mu)|z|^2 z + O(|z|^5) \tag{8.71}$$

as required.

**Step 4.** The final step in proving the main substance of the theorem, the existence of a bifurcating periodic solution, is to use the Poincaré-Bendixson Theorem on the system found in the previous step. Note that from Step 1 we have that $\lambda(\mu) = \alpha(\mu)+i\beta(\mu)$ where $\alpha(\mu) = \alpha_1\mu+O(\mu^2)$ (with $\alpha_1 \neq 0$ by assumption) and $\beta(\mu) = \omega+\beta_1\mu+O(\mu^2)$ (with $\omega \neq 0$ by assumption). Also $A(\mu) = A(0) + O(\mu)$ where $\text{Re}A(0) = a$, the horrible expression in the statement of the Hopf Bifurcation Theorem, and so $\text{Re}A(0) \neq 0$. In polar coordinates, $z = re^{i\theta}$, the equation derived in Step 3 becomes

$$\dot{r} = \alpha(\mu)r + \text{Re}A(\mu)r^3 + O(r^5) \tag{8.72a}$$

$$\dot{\theta} = \beta(\mu) + \text{Im}A(\mu)r^2 + O(r^4). \tag{8.72b}$$

We begin the analysis of these equations by introducing functions $P$ and $Q$ to make the error terms explicit, so

$$\dot{r} = \alpha(\mu)r + \text{Re}A(\mu)r^3 + P(r,\theta,\mu)r^5 \tag{8.73a}$$

$$\dot{\theta} = \beta(\mu) + \text{Im}A(\mu)r^2 + Q(r,\theta,\mu)r^4. \tag{8.73b}$$

We shall assume that for small $r$ and $|\mu|$ both $P$ and $Q$ are continuously differentiable functions, $2\pi$ periodic in $\theta$. We shall restrict attention to the case

$$\alpha_1 > 0, \quad \text{Re}A(0) = a < 0. \tag{8.74}$$

The other cases can be approached in the same way. Since all the functions involved in this differential equation are continuous, they are bounded for sufficiently small $r$ and $|\mu|$. Given $\epsilon > 0$ sufficiently small this enables us to choose $\delta > 0$ such that for $\mu \in (-\delta, \delta)$ and $r \in (0, \delta)$ the following inequalities hold.

i) $$(1-\epsilon)|\omega| < |\beta(\mu) + \text{Im}A(\mu)r^2 + Q(r,\theta,\mu)r^4| < (1+\epsilon)|\omega|,$$

so $\dot{\theta} \neq 0$ and the origin is the only stationary point in $r < \delta$ for all $\mu \in (-\delta, \delta)$;

ii) $$(1-\epsilon)^2\alpha_1 < \frac{\alpha(\mu)}{\mu} < (1+\epsilon)^2\alpha_1;$$

iii) $$(1-\epsilon) < \left(1 + \frac{P(r,\theta,\mu)r^2}{\text{Re}A(\mu)}\right) < (1+\epsilon);$$

iv) $$(1-\epsilon)|a| < |\text{Re}A(\mu)| < (1+\epsilon)|a|;$$

v) $$(1-\epsilon) < \left(1 + \frac{(Q_\theta r^2 + 6Pr^2 + P_r r^3)}{4\mathrm{Re}A(\mu)}\right) < (1+\epsilon).$$

Now, working locally with $-\delta < \mu < \delta$ and $0 \le r < \delta$ we see that for $\mu < 0$

$$\dot{r} < (1-\epsilon)^2 \alpha_1 \mu r - (1-\epsilon)|a|r^3(1-\epsilon)$$

and so

$$\dot{r} < (1-\epsilon)^2 r(\alpha_1\mu - |a|r^2). \tag{8.75}$$

Since $\alpha_1 > 0$ by assumption this implies that $\dot{r} < 0$ for all $r$ in $(0,\delta)$ when $\mu < 0$ and so the origin is asymptotically stable.

The case $\mu > 0$ is more interesting. We want to construct an annular region in the plane on which we can apply the Poincaré-Bendixson Theorem. By (i), the only stationary point in $r < \delta$ is the origin, so provided the constructed region does not contain the origin it will contain no stationary points. Using (ii), (iii) and (iv) we find that for $\mu \in (0,\delta)$ and $r < \delta$

$$\dot{r} < (1+\epsilon)^2 \alpha_1 \mu r - (1-\epsilon)^2 |a|r^3$$

and so $\dot{r} < 0$ provided

$$r^2 > \left(\frac{1+\epsilon}{1-\epsilon}\right)^2 \frac{\alpha_1\mu}{|a|}. \tag{8.76}$$

Hence, restricting attention to values of $\mu$ satisfying

$$0 < \mu < \left(\frac{1-\epsilon}{1+\epsilon}\right)^2 \frac{|a|\delta^2}{\alpha_1} \tag{8.77}$$

we have $\dot{r} < 0$ provided

$$\left(\frac{1+\epsilon}{1-\epsilon}\right)^2 \frac{\alpha_1\mu}{|a|} < r^2 < \delta^2. \tag{8.78}$$

Furthermore

$$\dot{r} > (1-\epsilon)^2 \alpha_1 \mu r - (1+\epsilon)^2 |a|r^3 \tag{8.79}$$

and so $\dot{r} > 0$ for all $r$ with

$$0 \le r^2 < \left(\frac{1-\epsilon}{1+\epsilon}\right)^2 \frac{\alpha_1\mu}{|a|}. \tag{8.80}$$

Now consider the region

$$D = \left\{(r,\theta)|\ \left(\frac{1-\epsilon}{1+\epsilon}\right)^2 \frac{\alpha_1\mu}{|a|} < r^2 < \left(\frac{1+\epsilon}{1-\epsilon}\right)^2 \frac{\alpha_1\mu}{|a|}\right\}.$$

By (i) this contains no stationary points and by the inequalities derived above trajectories enter but do not leave this region for each value of $\mu$ with $0 < \mu < (\frac{1-\epsilon}{1+\epsilon})^2 \frac{|a|\delta^2}{\alpha_1}$. Hence for each such $\mu$ there is at least one periodic orbit in $D$.

To show that for each such value of $\mu$ there is in fact a unique periodic orbit in $D$ we use a simple adaptation of Dulac's criterion based on the divergence theorem. Suppose that there are two distinct periodic orbits in $D$, $\Gamma_1$ and $\Gamma_2$, say. Then since trajectories cannot cross there is a simply connected annular region $A$ with $\Gamma_1$ and $\Gamma_2$ as boundaries. Now suppose that the differential equation is $\dot{w} = f(w)$, where $w \in \mathbf{R}^2$ and $f : \mathbf{R}^2 \to \mathbf{R}^2$. Then if $n$ represents the outward normals on the curves $\Gamma_i$, $i = 1, 2$,

$$\int_{\Gamma_1} (f.n)dl + \int_{\Gamma_2} (f.n)dl = \int\int_A \nabla.f dA$$

by the divergence theorem. But since both $\Gamma_1$ and $\Gamma_2$ are trajectories, $(f.n) = 0$ and so we find

$$\int\int_A \nabla.f dA = 0.$$

Hence if we can show that $\nabla.f > 0$ or $\nabla.f < 0$ in $D$ we have a contradiction (as $A$ lies entirely in $D$) and so the periodic orbit is unique. To evaluate $\nabla.f$ we need to take the divergence of a function in polar coordinates: if $\dot{r} = R(r,\theta)$ and $\dot{\theta} = \Phi(r,\theta)$ then $f = (R, r\Phi)$ in polar coordinates and so

$$\nabla.f = \frac{1}{r}\frac{\partial}{\partial r}(rR) + \frac{\partial \Phi}{\partial \theta}.$$

Evaluating this for the differential equation obtained in Step 3 we find

$$\begin{aligned}\nabla.f &= \frac{1}{r}[2\alpha(\mu)r + 4\mathrm{Re}A(\mu)r^3 + 6Pr^5 + P_r r^6] + [Q_\theta r^4] \\ &= 2\alpha(\mu) + 4\mathrm{Re}A(\mu)r^2\left(1 + \frac{Q_\theta r^2 + 6Pr^2 + P_r r^3}{4\mathrm{Re}A(\mu)}\right). \end{aligned} \quad (8.81)$$

Using (v) this implies that

$$\nabla.f < 2(1+\epsilon)^2\alpha_1\mu - 4(1-\epsilon)|a|\left(\frac{1-\epsilon}{1+\epsilon}\right)^2 \frac{\alpha_1\mu}{|a|}(1-\epsilon)$$

for $r$ in $D$. A little rearrangement of this expression gives

$$\nabla.f < \frac{2\alpha_1\mu}{(1+\epsilon)^2}[(1+\epsilon)^4 - 2(1-\epsilon)^4]. \quad (8.82)$$

A quick glance at the terms in square brackets should convince you that this expression is negative for sufficiently small $\epsilon$ ($\epsilon < \frac{1}{16}$ is small enough) and so $\nabla . f < 0$ in $D$. Hence the periodic orbit is unique. Furthermore, by the result of Section 6.3, the periodic orbit is stable for the choice of the signs of $\alpha_1$ and $\mathrm{Re}A(\mu)$ made here.

These results can be summarized in the following way for $\alpha_1 > 0$ (see Fig. 8.11). If $\mathrm{Re}A(0) < 0$ then a stable periodic orbit of amplitude

$$R = \left( \frac{\alpha_1 \mu}{|\mathrm{Re}A(0)|} \right)^{\frac{1}{2}}$$

bifurcates from the origin into $\mu > 0$ as $\mu$ passes through zero. The origin itself is a stable focus if $\mu < 0$ and an unstable focus if $\mu > 0$. The periodic orbit has period $2\pi/|\omega|$ as $\mu$ tends to zero from above. This is called a *supercritical Hopf bifurcation.*

If $\mathrm{Re}A(0) > 0$ then an unstable periodic orbit of amplitude

$$\left( \frac{\alpha_1 \mu}{\mathrm{Re}A(0)} \right)^{\frac{1}{2}}$$

bifurcates into $\mu < 0$, where the origin is a stable focus, and there are no periodic orbits in a small enough neighbourhood of the origin if $\mu > 0$. This is called a *subcritical Hopf bifurcation.*

The case where $\alpha_1 < 0$ is similar with the sign of $\mu$ changed.

## 8.9 Calculation of the stability coefficient, ReA(0)

**Step 5.** The stability of the bifurcating periodic solution is determined by the signs of $\alpha_1$ and $\mathrm{Re}A(0)$, where $Re\lambda(\mu) = \alpha_1\mu + O(\mu^2)$ and $A(0)$ is the coefficient on the cubic term which cannot be removed by near identity coordinate transformations. Since it comes from the linear part of the equation, $\alpha_1$ is easy to calculate, but $A(0)$ is a different matter. To calculate $A(0)$ it is necessary to do the two near identity transformations of order two and three and then identify the coefficient as a function of the original equations. This is easier to do in principle than in practice and could, perhaps, be left as an exercise. However, just to show that it *is* possible to do this calculation, and as a check for the enthusiastic reader who may have tried it already, we shall go through the motions here.

Recall that after Step 1 the differential equation has been written as

$$\dot{x} = f(x, y), \quad \dot{y} = g(x, y)$$

where

$$f(x,y) = -\omega y + \sum_{m\geq 2}\sum_{k=0}^{m} \frac{1}{k!(m-k)!} f_{x^k y^{m-k}} x^k y^{m-k} \tag{8.83}$$

and

$$g(x,y) = \omega x + \sum_{m\geq 2}\sum_{k=0}^{m} \frac{1}{k!(m-k)!} g_{x^k y^{m-k}} x^k y^{m-k}. \tag{8.84}$$

In Step 2 a complex variable, $z = x + iy$, was introduced so that

$$\dot{z} = i\omega z + \sum_{m\geq 2}\sum_{k=0}^{m} \gamma_{mk} z^k z^{*m-k}. \tag{8.85}$$

One of the manipulations we shall be forced to do is the determination of the complex coefficients $\gamma_{2k}$ and $\gamma_{3k}$ in terms of the partial differentials of $f(x,y)$ and $g(x,y)$, but this little complication will be left for the moment since we are engaged in higher things.

To calculate $A(0)$ we begin by going through the second order near identity transformation which sets all the quadratic terms in the equation to zero, keeping track of the third order corrections which this produces. Thus define

$$w = z + \sum_{k=0}^{2} a_{2k} z^k z^{*2-k} \tag{8.86}$$

so that

$$\dot{w} = \dot{z} + \sum_{k=0}^{2} a_{2k}(k\dot{z}z^{k-1}z^{*2-k} + (2-k)\dot{z}^* z^k z^{*1-k}). \tag{8.87}$$

Inverting the equation which defines $w$ gives

$$z = w - \sum_{k=0}^{2} a_{2k} w^k w^{*2-k} + O(|w|^4) \tag{8.88}$$

so

$$\dot{z} = i\omega \left( w - \sum_{k=0}^{2} a_{2k} w^k w^{*2-k} \right) + \sum_{k=0}^{2} \gamma_{2k} w^k w^{*2-k}$$
$$+ \sum_{k=0}^{3} \gamma_{3k} w^k w^{*3-k} + O(|w|^4), \tag{8.89}$$

$$\dot{z}z^{k-1}z^{*2-k} = \left[ i\omega \left( w - \sum_{p=0}^{2} a_{2p} w^p w^{*2-p} \right) \right.$$

$$+\sum_{p=0}^{2}\gamma_{2p}w^{p}w^{*2-p}\Bigg]\,w^{k-1}w^{*2-k}$$
$$+O(|w|^4) \tag{8.90}$$

and

$$\dot{z}^{*}z^{k}z^{*1-k}=\Bigg[-i\omega\left(w^{*}-\sum_{p=0}^{2}a_{2p}^{*}w^{*p}w^{2-p}\right)$$
$$+\sum_{p=0}^{2}\gamma_{2p}^{*}w^{*p}w^{2-p}\Bigg]\,w^{k-1}w^{*2-k}$$
$$+O(|w|^4). \tag{8.91}$$

Substituting these expressions into the equation for $\dot{w}$ gives

$$\dot{w}=i\omega\left(w-\sum_{k=0}^{2}a_{2k}w^{k}w^{*2-k}\right)+\sum_{k=0}^{2}\gamma_{2k}w^{k}w^{*2-k}+\sum_{k=0}^{3}\gamma_{3k}w^{k}w^{*3-k}$$
$$+\sum_{k=0}^{2}a_{2k}(k(i\omega)+(2-k)(-i\omega))w^{k}w^{*2-k}$$
$$+\sum_{k=0}^{2}a_{2k}kw^{k-1}w^{*2-k}\left(\sum_{p=0}^{2}\gamma_{2p}w^{p}w^{*2-p}\right)$$
$$+\sum_{k=0}^{2}a_{2k}(2-k)w^{k-1}w^{*2-k}\left(\sum_{p=0}^{2}\gamma_{2p}^{*}w^{*p}w^{2-p}\right)+O(|w|^4) \tag{8.92}$$

where the last two lines represent the modifications to the equation at third order caused by the second order change of variables. Looking at the second order terms we find (as in Step 3) that we can choose all terms at second order to vanish by setting

$$((k-1)-(2-k))i\omega a_{2k}+\gamma_{2k}=0 \tag{8.93}$$

or

$$a_{2k}=\frac{-\gamma_{2k}}{(2k-3)i\omega} \tag{8.94}$$

for $k=0,1,2$. And that with this change of variable (8.92) becomes

$$\dot{w}=i\omega w+\sum_{k=0}^{3}\gamma_{3k}w^{k}w^{*3-k}$$
$$+\sum_{k=0}^{2}a_{2k}kw^{k-1}w^{*2-k}\left(\sum_{p=0}^{2}\gamma_{2p}w^{p}w^{*2-p}\right)$$

$$+\sum_{k=0}^{2} a_{2k}(2-k)w^k w^{*1-k}\left(\sum_{p=0}^{2}\gamma_{2p}^* w^{*p} w^{2-p}\right) + O(|w|^4) \quad (8.95)$$

or, defining new complex coefficients, $\gamma'_{3k}$,

$$\dot{w} = i\omega w + \sum_{k=0}^{3}\gamma'_{3k} w^k w^{*3-k} + O(|w|^4). \quad (8.96)$$

This is the end of the nasty calculations: $A(0) = \gamma'_{32}$. If this is not obvious consider the near identity change of coordinates defined by

$$q = w + \sum_{k=0}^{3} a_{3k} w^k w^{*3-k}. \quad (8.97)$$

Going through the usual rigmarole, but not worrying about the effect upon fourth order terms, we find

$$\dot{q} = i\omega q + \sum_{k=0}^{3} a_{3k}((k-1)(i\omega) + (3-k)(-i\omega))q^k q^{*3-k}$$

$$+\sum_{k=0}^{3}\gamma'_{3k} q^k q^{*3-k} + O(|q|^4) \quad (8.98)$$

so we choose

$$a_{3k} = \frac{-\gamma'_{3k}}{(2k-4)i\omega} \quad (8.99)$$

provided $k \neq 2$ to kill off the third order terms leaving

$$\dot{q} = i\omega q + \gamma'_{32} q^2 q^* + O(|q|^4). \quad (8.100)$$

Comparing this equation with the equation obtained at the end of Step 3 we see that

$$A(0) = \gamma'_{32}. \quad (8.101)$$

To determine $A(0)$ we still need to do two things. First, relate $\gamma'_{32}$ to the original coefficients $\gamma_{2k}$ and $\gamma_{3k}$, and then find the relationship between these coefficients and the partial derivatives of the functions $f(x,y)$ and $g(x,y)$ defined at the end of step 1.

The first step is easy. From (8.95) and (8.96) simply identifying those terms which multiply $w^2 w^*$ gives

$$\gamma'_{32} = \gamma_{32} + a_{21}\gamma_{22} + 2a_{22}\gamma_{21} + 2a_{20}\gamma_{20}^* + a_{21}\gamma_{21}^* \quad (8.102)$$

or, writing out (8.94) component by component,

$$a_{20} = \frac{\gamma_{20}}{3i\omega}, \quad a_{21} = \frac{\gamma_{21}}{i\omega}, \quad a_{22} = \frac{-\gamma_{22}}{i\omega}, \tag{8.103}$$

this becomes

$$\gamma_{32}' = \gamma_{32} - \frac{\gamma_{21}\gamma_{22}}{i\omega} + 2\frac{|\gamma_{20}|^2}{3i\omega} + \frac{|\gamma_{21}|^2}{i\omega}. \tag{8.104}$$

That was the hard work. All that needs to be done next is to identify the coefficients $\gamma_{2k}$ and $\gamma_{32}$ in terms of the original real functions $f$ and $g$ where, at $\mu = 0$,

$$\dot{x} = f(x, y), \quad \dot{y} = g(x, y).$$

The complex coefficients are defined by the equation for $\dot{z}$, where $z = x + iy$ and

$$\dot{z} = f(x, y) + ig(x, y), \quad x = \tfrac{1}{2}(z + z^*), \quad y = -\tfrac{1}{2}i(z - z^*).$$

Expanding $f$ and $g$ as Taylor series and equating quadratic terms with the equation for $\dot{z}$ gives

$$\sum_{k=0}^{2} \gamma_{2k} z^k z^{*2-k} = \sum_{k=0}^{2} (b_{2k} + ic_{2k}) x^k y^{2-k}$$

where

$$b_{22} = \tfrac{1}{2} f_{xx}, \quad b_{21} = f_{xy}, \quad b_{20} = \tfrac{1}{2} f_{yy} \tag{8.105}$$

and

$$c_{22} = \tfrac{1}{2} g_{xx}, \quad c_{21} = g_{xy}, \quad c_{20} = \tfrac{1}{2} g_{yy}. \tag{8.106}$$

Using the expressions for $x$ and $y$ in terms of $z$ and $z^*$, equating the coefficients of $z^2$, $zz^*$ and $z^{*2}$ respectively gives

$$\begin{aligned} \gamma_{22} &= \tfrac{1}{4}(b_{22} + ic_{22}) - \tfrac{1}{4}i(b_{21} + ic_{21}) - \tfrac{1}{4}(b_{20} + ic_{20}) \\ &= \tfrac{1}{8}[(f_{xx} + 2g_{xy} - f_{yy}) + i(g_{xx} - 2f_{xy} - g_{yy})], \end{aligned} \tag{8.107}$$

$$\begin{aligned} \gamma_{21} &= \tfrac{1}{2}(b_{22} + ic_{22}) + \tfrac{1}{2}(b_{20} + ic_{20}) \\ &= \tfrac{1}{4}[(f_{xx} + f_{yy}) + i(g_{xx} + g_{yy})] \end{aligned} \tag{8.108}$$

and

$$\begin{aligned} \gamma_{20} &= \tfrac{1}{4}(b_{22} + ic_{22}) + \tfrac{1}{4}i(b_{21} + ic_{21}) - \tfrac{1}{4}(b_{20} + ic_{20}) \\ &= \tfrac{1}{8}[(f_{xx} - 2g_{xy} - f_{yy}) + i(g_{xx} + 2f_{xy} - g_{yy})]. \end{aligned} \tag{8.109}$$

To deal with the cubic terms we simply repeat the same rigmarole, setting

$$\sum_{k=0}^{3} \gamma_{3k} z^k z^{*3-k} = \sum_{k=0}^{3} (b_{3k} + ic_{3k}) x^k y^{3-k}$$

where

$$b_{33} = \tfrac{1}{3!} f_{xxx},\ b_{32} = \tfrac{1}{2} f_{xxy},\ b_{31} = \tfrac{1}{2} f_{xyy},\ b_{30} = \tfrac{1}{3!} f_{yyy} \tag{8.110}$$

and similar expressions hold for the $c_{3k}$ with $f$ replaced by $g$. The only cubic coefficient we want is $\gamma_{32}$ and reading this straight off gives

$$\gamma_{32} = \tfrac{3}{8}(b_{33} + ic_{33}) - \tfrac{1}{8}i(b_{32} + ic_{32}) + \tfrac{1}{8}(b_{31} + ic_{31}) - \tfrac{3}{8}i(b_{30} + ic_{30})$$

$$= \tfrac{1}{16}[(f_{xxx} + g_{xxy} + f_{xyy} + g_{yyy}) + i(g_{xxx} - f_{xxy} + g_{xyy} - f_{yyy})]. \tag{8.111}$$

Finally, going back to the expression for $\gamma'_{32}$ and using Re$A(0) = Re\gamma'_{32}$ we get

$$\text{Re}A(0) = Re\gamma_{32} - \frac{1}{\omega}(Re\gamma_{22} Im\gamma_{21} + Im\gamma_{22} Re\gamma_{21}). \tag{8.112}$$

Now,

$$Re\gamma_{32} = \tfrac{1}{16}(f_{xxx} + g_{xxy} + f_{xyy} + g_{yyy}), \tag{8.113}$$

$$Re\gamma_{22} Im\gamma_{21} = \tfrac{1}{32}(f_{xx} + 2g_{xy} - f_{yy})(g_{xx} + g_{yy}) \tag{8.114}$$

and

$$Im\gamma_{22} Re\gamma_{21} = \tfrac{1}{32}(g_{xx} - 2f_{xy} - g_{yy})(f_{xx} + f_{yy}) \tag{8.115}$$

so

$$\text{Re}A(0) = \tfrac{1}{16}(f_{xxx} + g_{xxy} + f_{xyy} + g_{yyy})$$
$$+ \tfrac{1}{16\omega}\left(f_{xy}(f_{xx} + f_{yy}) - g_{xy}(g_{xx} + g_{yy}) - f_{xx}g_{xx} + f_{yy}g_{yy}\right). \tag{8.116}$$

This completes the final step of the proof of the Hopf Bifurcation Theorem. The enthusiastic reader might want to try the following exercise.

(8.7) EXERCISE

*Calculate* Im$A(0)$ *and use this to find the order* $\mu$ *corrections to the period of the bifurcating orbit.*

## 8.10 A canard

In French, the word 'canard' has two meanings: first, the English 'duck', but it can also mean 'hoax'. Canard is the name given to a phenomenon associated with the creation of periodic orbits in certain relaxation oscillators (cf. Section 7.13) either because the phase diagrams can be made to look like a duck, or because the phenomenon was widely misunderstood for a long time. Consider the forced van der Pol oscillator

$$\ddot{x} + \epsilon(x^2 - 1)\dot{x} + x = 1 + \mu \tag{8.117}$$

for $\epsilon \gg 1$ and $\mu$ a real parameter. If $\mu = -1$ this system reduces to the relaxation oscillator of Section 7.4, which has a stable periodic orbit with approximate amplitude $A = 2$; but how is this orbit created or destroyed as $\mu$ varies? Writing (8.117) as

$$\dot{x} = y - \epsilon(\tfrac{1}{3}x^3 - x), \quad \dot{y} = -x + 1 + \mu \tag{8.118}$$

we see that the system has a unique stationary point at

$$x = 1 + \mu, \quad y = \epsilon(1 + \mu)(\tfrac{1}{3}(1 + \mu)^2 - 1). \tag{8.119}$$

The stability of this stationary point is determined by the eigenvalues of the Jacobian matrix,

$$\begin{pmatrix} -\epsilon(x^2 - 1) & 1 \\ -1 & 0 \end{pmatrix}$$

which has characteristic equation

$$s^2 + \epsilon\mu(2 + \mu)s + 1 = 0 \tag{8.120}$$

at the stationary point. Thus the eigenvalues of the Jacobian matrix have strictly negative real parts if $\mu > 0$ or $\mu < -2$ (and so the stationary point is stable) and strictly positive real parts if $-2 < \mu < 0$ (for which values of $\mu$ the stationary point is unstable). If $\mu = -2$ or $\mu = 0$ the characteristic equation is $s^2 + 1 = 0$, which has solutions $s = \pm i$. Hence we expect to find a Hopf bifurcation from $\mu = 0$ and from $\mu = -2$. To investigate this possibility we begin by transforming the stationary point to the origin so that we can apply the results of the previous sections. Setting

$$u = x - (1 + \mu) \quad \text{and} \quad v = y - \epsilon(1 + \mu)(\tfrac{1}{3}(1 + \mu)^2 - 1) \tag{8.121}$$

we obtain

$$\dot{u} = v - \epsilon u\left(\mu(\mu + 2) + (1 + \mu)u + \tfrac{1}{3}u^2\right) \tag{8.122a}$$

$$\dot{v} = -u. \tag{8.122b}$$

which, by design, has a stationary point at the origin. We shall concentrate on the case $\mu = 0$; the case $\mu = -2$ is similar and is left as an exercise.

If $\mu = 0$ the Jacobian matrix at the origin is

$$\begin{pmatrix} 0 & 1 \\ -1 & 0 \end{pmatrix},$$

which is in the desired form if we take $\omega = -1$. Furthermore, writing $\dot{u} = f(u, v, \mu)$ and $\dot{v} = g(u, v, \mu)$ we obtain

$$f_{u\mu} + g_{v\mu} = -2\epsilon < 0 \tag{8.123}$$

for $(u.v.\mu) = (0, 0, 0)$, and so $\alpha_1 < 0$. Looking at the higher derivatives

$$f_{uuu} = -2\epsilon, \quad f_{uu} = -2\epsilon \tag{8.124}$$

and all the other second and third derivatives of $f$ and $g$ vanish at the origin if $\mu = 0$. Hence

$$a = -\tfrac{1}{8}\epsilon < 0. \tag{8.125}$$

Looking back at the results for the Hopf bifurcation we see that this implies that $\alpha_1 = -\epsilon$ and $|\text{Re}A(0)| = \frac{1}{8}\epsilon$. Hence a stable periodic orbit exists for $\mu < 0$ with an amplitude which scales (for small $|\mu|$) like $8|\mu|^{\frac{1}{2}}$.

So, where is the canard? Figure 8.12a shows a sketch of the amplitude of the periodic orbit against $\mu$. Note that the amplitude of the orbit

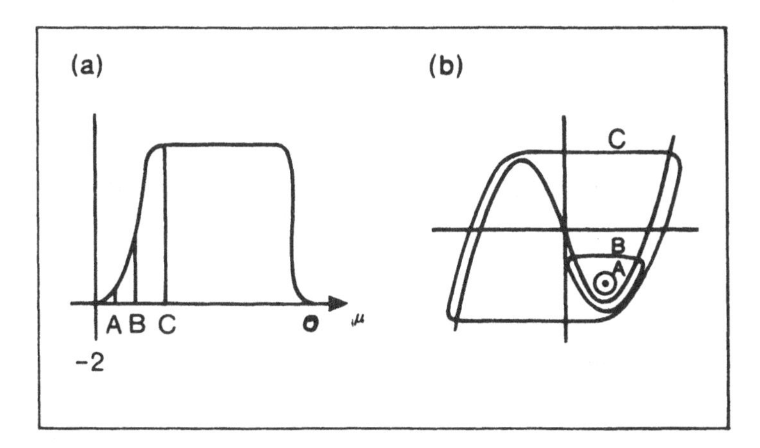


Fig. 8.12 (a) Amplitude vs. parameter for the canard; (b) the relaxation curve showing the position of the periodic orbit at three parameter values.

jumps to 2 very abruptly (in fact, over a range of parameters which scales like $\exp(-\epsilon)$ with $\epsilon \gg 1$).This abrupt change can be understood by thinking about the relaxation curve $y = \frac{1}{3}x^3 - x$ as in Section 7.13. Initially, the periodic orbit lies in the crook of the curve at $x = 1$. As $\mu$ decreases the amplitude of the orbit increases with the usual scaling until it hits the relaxation curve (see Fig. 8.12b). Thereafter it drops off the minimum and so we have the relaxation oscillation of Section 7.13. This change is continuous but very fast, and it was not fully understood until the early 1980s! A full description is beyond the scope of this book.

## Exercises 8

1. Find the stationary points and determine their stability for the following examples. In each case sketch a bifurcation diagram and find the bifurcation values of the parameter.

i) $\dot{x} = x(\mu - x^2)(\mu - 2x^2)$;
ii) $\dot{x} = \mu - x^2 + 4x^4$;
iii) $\dot{x} = x(x^2 - \mu)(x^2 + \mu^2 - 1)$;
iv) $\dot{x} = \mu - |x|$.

2. For each of the following examples find an approximation to the centre manifold at the origin and hence determine the local behaviour of solutions. (You should develop the centre manifold to whatever order is required to determine the nature of the flow near the origin.)

i) $\dot{x} = -2x + y - x^2, \quad \dot{y} = x(y - x)$.
ii) $\dot{x} = y - 3x^2 + xy, \quad \dot{y} = -3y + y^2 + x^2$.

3. Find the value of $\mu$ for which there is a bifurcation at the origin for the system

$$\dot{x} = y - x - x^2, \quad \dot{y} = \mu x - y - y^2.$$

Find the evolution equation on the extended centre manifold correct to third order and hence deduce that the bifurcation is a transcritical bifurcation.

4. Describe the bifurcations in the systems

$$\dot{x} = x(\mu - x - x^2)$$

and

$$\dot{x} = x(\mu - x^2).$$

Discuss the effect of adding a small constant term to the right hand side of each of these equations, i.e. for the second example consider the equation $\dot{x} = x(\mu - x^2) + \epsilon$ for some small, fixed $\epsilon$.

5. Consider the equations

$$\begin{aligned}\dot{x} &= 2 - x + y \\ \dot{y} &= (2a - y^2)(4a^2 - 4 + 2x + y^2).\end{aligned}$$

Plot the loci of the stationary points in the $(a, x)$ plane and hence deduce that there are bifurcations when $a = -\frac{1}{2}$, $a = 0$ and $a = \frac{1}{2}$. Investigate the bifurcation at $a = 0$ in detail: find the extended centre manifold and determine the type of the bifurcation. Assuming that there are no other bifurcations, sketch a consistent bifurcation diagram in the $(a, x)$ plane, indicating where the different branches of stationary points are sources, sinks and saddles.

6. Show that the system

$$\ddot{x} - (\mu - x^2)\dot{x} + x = 0$$

has a Hopf bifurcation at $(0, 0, 0)$. Indicate briefly, on the basis of energy considerations, whether you would expect the bifurcation to be subcritical or supercritical. Confirm your prediction by calculating the stability coefficient.

Given that the equation

$$\ddot{x} - \mu(1 - x^2)\dot{x} + x = 0$$

has a periodic orbit of finite amplitude, estimate the amplitude and period of this orbit. What periodic orbits does the equation have at $\mu = 0$?

7. Discuss the nature of the stationary point at the origin for

$$\begin{aligned}\dot{x} &= \mu x + \omega(x + y) - (2x^2 + 2xy + y^2)x \\ \dot{y} &= \mu y - \omega(2x + y) - (2x^2 + 2xy + y^2)y\end{aligned}$$

as the parameter $\mu$ is varied. Find a transformation which reduces the problem to normal form and hence show that there is a supercritical bifurcation at $\mu = 0$.

8. Sketch the different phase portaraits which can arise near the origin in the system

$$\dot{x} = \mu x + y - xf(r), \quad \dot{y} = -x + \mu y - yf(r),$$

where $r^2 = x^2 + y^2$ and $f$ is a continuous function such that $f(0) = 0$ and $f(t) > 0$ for all $t > 0$.

9. Find a transformation of the form

$$\begin{aligned} x &= \xi + \alpha_1\xi^2 + \beta_1\xi\eta + \gamma_1\eta^2 \\ y &= \eta + \alpha_2\xi^2 + \beta_2\xi\eta + \gamma_2\eta^2 \end{aligned}$$

that reduces the equation

$$\begin{aligned} \dot{x} &= y + a_1x^2 + b_1xy + c_1y^2 \\ \dot{y} &= a_2x^2 + b_2xy + c_2y^2 \end{aligned}$$

to the form

$$\begin{aligned} \dot{\xi} &= \eta + O(3) \\ \dot{\eta} &= A\xi^2 + B\xi\eta + O(3), \end{aligned}$$

where $O(3)$ denotes terms of third order and higher. A typical small perturbation of this equation is

$$\dot{\xi} = \eta, \quad \dot{\eta} = \lambda\xi + \mu\eta + A\xi^2 + B\xi\eta,$$

where $\lambda$ and $\mu$ are small parameters. Discuss the local bifurcations in the $(\lambda, \mu)$ plane for the cases $(A, B) = (1, 1)$ and $(A, B) = (-1, 1)$. By considering the bifurcations on a closed loop about the origin in parameter space, show that there must be other bifurcations.

10. Use the Implicit Function Theorem to obtain conditions for a pitchfork bifurcation to occur.

# 9

# *Bifurcation theory II: periodic orbits and maps*

Since the flow near a periodic orbit can be described by a return map (which is as smooth as the original flow) the bifurcations of periodic orbits of differential equations and fixed points or periodic orbits of maps can be treated as one and the same topic. As we saw in Chapter 6, the linearized map near a fixed point of a nonlinear map is a good model of the behaviour near the fixed point provided the fixed point is hyperbolic, i.e. no eigenvalues of the linear map lie on the unit circle. The eigenvectors associated with eigenvalues inside the unit circle correspond to stable directions and those associated with eigenvalues outside the unit circle correspond to unstable directions. Furthermore, since hyperbolic fixed points of maps are persistent under small perturbations of the map (a result analogous to the persistence of hyperbolic stationary points in flows; cf. Chapter 4) there can be no local bifurcations near a hyperbolic fixed point so, as in the previous chapter, we are forced to consider non-hyperbolic fixed points. This is made easier by the centre manifold for maps, which plays the same role in this chapter as the Centre Manifold Theorem for flows did in the previous chapter.

(9.1) THEOREM (CENTRE MANIFOLD THEOREM FOR MAPS)

*Consider the map $x_{n+1} = f(x_n)$ where $x \in \mathbf{R}^n$ and $f : \mathbf{R}^n \to \mathbf{R}^n$ is a smooth invertible map (i.e. a diffeomorphism) with a fixed point at $x = 0$. Divide the eigenvalues, $\lambda$ of $Df(0)$, into three sets in the following way: $\lambda \in \sigma_u$ if $|\lambda| > 1$, $\lambda \in \sigma_s$ if $|\lambda| < 1$ and $\lambda \in \sigma_c$ if $|\lambda| = 1$. Let $E^s$, $E^u$ and $E^c$ be the corresponding linear subspaces spanned by the generalized eigenvectors. Then there exist smooth unstable, stable and centre manifolds tangential to the corresponding linear subspaces at $x = 0$. Each manifold is invariant, but the centre manifold is not necessarily unique.*

This theorem (which we will not prove) enables us to work on the centre manifold of the extended system

$$x_{n+1} = f(x_n, \mu_n), \quad \mu_{n+1} = \mu_n \tag{9.1}$$

near the bifurcation value of the parameter (or parameters) $\mu$. The bifurcation value is easy to determine since it is the parameter value at which the map has a fixed point for with an eigenvalue of the Jacobian matrix on the unit circle.

In this chapter we shall look at the simple bifurcations of maps, beginning with the case for which there is a single simple eigenvalue of $+1$, then looking at the case of a single simple eigenvalue of $-1$. The former case will give bifurcations analagous to those already discussed for flows. The latter case gives a completely new bifurcation: period-doubling. Finally we consider the Hopf bifurcation for maps, where a closed invariant curve bifurcates from the fixed point (which gives a torus in phase space if the fixed point corresponds to a periodic orbit of some flow). The motion on this invariant curve can be either periodic or dense on the curve.

## 9.1 A simple eigenvalue of +1

Suppose that if $\mu = 0$ a map $w_{n+1} = f(w_n, \mu)$ has a non-hyperbolic fixed point (which we can take to be at the origin) and that the linearized map, $Df$, has a simple eigenvalue of $+1$ and no other eigenvalues on the unit circle. Then there is a one-dimensional centre manifold on which the motion is of the form

$$x_{n+1} = g(x_n, \mu), \quad x \in \mathbf{R} \tag{9.2}$$

where, assuming that a coordinate change has been made so that the fixed point is at $x = 0$ and the bifurcation value is $\mu = 0$,

$$g(0,0) = 0 \quad \text{and} \quad \frac{\partial g}{\partial x}(0,0) = 1. \tag{9.3}$$

Thus we can expand $g$ as a Taylor series about $(x, \mu) = (0, 0)$ to get

$$g(x,\mu) = x + g_\mu \mu + \tfrac{1}{2}(g_{xx}x^2 + 2g_{x\mu}\mu x + g_{\mu\mu}\mu^2) + \ldots \tag{9.4}$$

or

$$g(x,\mu) = A(\mu) + (1 + B(\mu))x + C(\mu)x^2 + O(x^3) \tag{9.5}$$

where

$$A(\mu) = g_\mu \mu + O(\mu^2), \; B(\mu) = g_{x\mu}\mu + O(\mu^2)$$

$$\text{and } C(\mu) = \tfrac{1}{2} g_{xx} + O(\mu).$$

Fixed points of $g$ satisfy the equation $x = g(x,\mu)$ and so they are solutions to

$$A(\mu) + B(\mu)x + C(\mu)x^2 + O(x^3) = 0. \tag{9.6}$$

We can now consider (9.6) in precisely the same way as we did for the continuous time cases in the previous chapters. Comparing the expression for fixed points of the map on the centre manifold with the expression obtained at the beginning of Chapter 8 for stationary points of flows we see that they are exactly the same. This means that rather than go through the same manipulations as before we can simply translate the results obtained in the previous chapter (Sections 8.3, 8.4 and 8.5) to the present case. The only difference is that the stability criteria need to be recalculated, since fixed points are stable if the eigenvalues of the linear map lie inside the unit circle (as opposed to in the left-half plane for stationary points).

(9.2) THEOREM (SADDLENODE BIFURCATION)

*Suppose that for $\mu = 0$ the origin, $x = 0$, is a non-hyperbolic fixed point of the (one-dimensional) map $x_{n+1} = g(x_n, \mu)$ and that $g_x(0,0) = 1$. Then provided*

$$g_\mu(0,0) \neq 0 \text{ and } g_{xx}(0,0) \neq 0$$

*there is a continuous curve of fixed points in a neighbourhood of $(x,\mu) = (0,0)$ which is tangent to $\mu = 0$ at $(0,0)$. If $g_\mu g_{xx} < 0$ (resp. $> 0$) then there are no fixed points near $x = 0$ if $\mu < 0$ (resp. $\mu > 0$) whilst for each $\mu > 0$ (resp. $\mu < 0$) sufficiently small there are two hyperbolic fixed points near $x = 0$. The upper fixed point is stable and the lower is unstable if $g_{xx} < 0$ and the stabilities are reversed if $g_{xx} > 0$.*

This is almost exactly the same situation as in Theorem 8.2 (see Fig. 9.1). To complete the proof we need only verify the stability conditions.

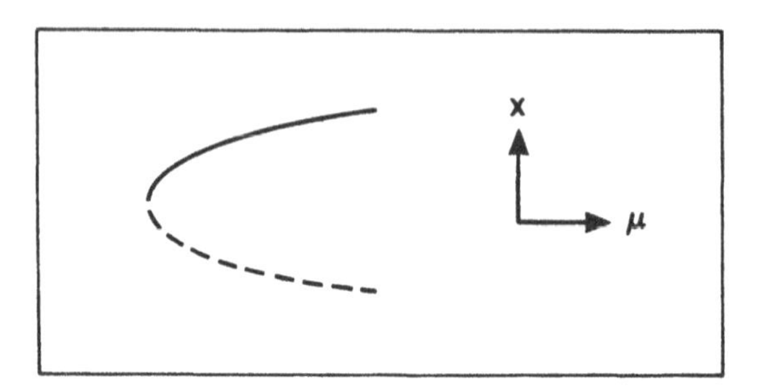


Fig. 9.1 A saddlenode bifurcation.

From the form of the map we find that the Jacobian matrix (or linear map) is

$$\frac{\partial g}{\partial x}(x,\mu) = 1 + g_{x\mu}\mu + g_{xx}x + \dots \tag{9.7}$$

and from Section 8.3 we know that the fixed points, $x_\pm$ are

$$x_\pm \sim \pm\sqrt{\frac{-2g_\mu}{g_{xx}}}\mu^{\frac{1}{2}} \tag{9.8}$$

for $g_\mu g_{xx} < 0$ and $\mu > 0$, and a similar expression in $\mu < 0$ if $g_\mu g_{xx} > 0$. In either case the dominant terms in the Jacobian are $1 + g_{xx}x$ and so (for small $|\mu|$) $x_+$ is stable if $g_{xx} < 0$ and unstable if $g_{xx} > 0$. The stability properties of $x_-$ are the opposite to those of $x_+$.

Now, as in the previous chapter, we can consider what happens if one or other of the genericity conditions for the saddlenode bifurcation does not hold.

(9.3) THEOREM (TRANSCRITICAL BIFURCATION)

*Suppose that if $\mu = 0$ the origin, $x = 0$, is a non-hyperbolic fixed point of the (one-dimensional) map $x_{n+1} = g(x_n, \mu)$ and that $g_x(0,0) = 1$. Then if*

$$g_\mu = 0, \quad g_{xx} \neq 0 \quad \text{and} \quad g_{\mu x}^2 - g_{xx}g_{\mu\mu} \neq 0$$

*there are two curves of fixed points in a neighbourhood of $(0,0)$ which intersect transversely at $(0,0)$. If $g_{xx} < 0$ then the upper fixed point is stable and the lower fixed point is unstable, whilst if $g_{xx} > 0$ stability properties are reversed.*

Once again, we need only verify the stability properties. If $\Delta^2 = g_{\mu x}^2 - g_{xx}g_{\mu\mu} > 0$ (and $\Delta > 0$) then from Section 8.4 we see that the pair of fixed points are

$$x_{\pm} \approx -\frac{g_{\mu x} \pm \Delta}{g_{xx}}\mu \tag{9.9}$$

and the dominant terms in the Jacobian matrix are therefore $1 + g_{\mu x}\mu + g_{xx}x_{\pm}$. Evaluating this for $x_+$ gives $1 - \Delta\mu$, which is less than one if $\mu > 0$ and greater than one if $\mu < 0$. Hence $x_+$ is stable if $\mu > 0$ (where it is the upper fixed point if $g_{xx} < 0$ and the lower fixed point if $g_{xx} > 0$) and unstable if $\mu < 0$ (where it is the lower fixed point if $g_{xx} < 0$ and the upper fixed point if $g_{xx} > 0$).

Finally we come to the pitchfork bifurcation.

(9.4) THEOREM (PITCHFORK BIFURCATION)

*Suppose that if $\mu = 0$ the origin, $x = 0$, is a non-hyperbolic fixed point of the (one-dimensional) map $x_{n+1} = g(x_n, \mu)$ and that $g_x(0,0) = 1$. Then if*

$$g_\mu = g_{xx} = 0, \quad g_{\mu x} \neq 0 \quad \text{and} \quad g_{xxx} \neq 0$$

*there is a branch of fixed points which passes through $(0,0)$ transverse to $\mu = 0$; these fixed points are stable in $\mu < 0$ and unstable in $\mu > 0$ if $g_{\mu x} > 0$ with stabilities reversed if $g_{\mu x} < 0$. A second branch of fixed points bifurcates from $(0,0)$ tangential to $\mu = 0$ into $\mu > 0$ if $g_{x\mu}g_{xxx} < 0$ or into $\mu < 0$ if $g_{x\mu}g_{xxx} > 0$ (so for each value of $\mu > 0$ sufficiently small there are three fixed points near $x = 0$ if $g_{x\mu}g_{xxx} < 0$). This second branch of fixed points is stable if it exists in $\mu < 0$ and $g_{x\mu} < 0$ or in $\mu > 0$ with $g_{x\mu} > 0$ (a supercritical bifurcation). Otherwise it is unstable (a subcritical bifurcation).*

It is easy to construct simple examples of maps with these types of bifurcation from the examples in the continuous time cases of the previous chapter. We shall go through these examples quickly; further examples can be found in the exercises at the end of this chapter.

*Example 9.1*

Consider the map

$$x_{n+1} = x_n + \mu - x_n^2.$$

Fixed points of the map are solutions to $\mu - x^2 = 0$, i.e. for $\mu < 0$ there are no fixed points and for $\mu > 0$ there are two fixed points: $x_\pm = \pm\sqrt{\mu}$. The Jacobian matrix is simply $1 - 2x$ and so at $\mu = 0$, $x = 0$ is a fixed point with Jacobian equal to 1. Hence it is non-hyperbolic, and there is a bifurcation at $\mu = 0$. Writing $g(x,\mu)$ for the righ thand side of the equation defining the map we see that $g(0,0) = 0$, $g_x(0,0) = 1$ (the general conditions for a non-hyperbolic fixed point) and furthermore $g_\mu = 1$ and $g_{xx} = -2$. Hence the bifurcation is a saddlenode, creating a pair of fixed points in $\mu > 0$, $x_+$, which is stable, and $x_-$, which is unstable. Note that there is a further bifurcation if $x_+ = 1$ (i.e. if $\mu = 1$) since the Jacobian matrix is $-1$ at this point. The bifurcation associated with an eigenvalue $-1$ is described in the next section.

*Example 9.2*

Consider

$$x_{n+1} = x_n(1 + \mu - x_n).$$

Fixed points are solutions of $x(\mu - x) = 0$ so provided $\mu \neq 0$ there are two fixed points, one at $x = 0$ and the other at $x = \mu$. The Jacobian of the map is $1 + \mu - 2x$ and so $x = 0$ is stable provided $-2 < \mu < 0$ and $x = \mu$ is stable provided $0 < \mu < 2$. Writing $g(x,\mu)$ for the right hand side of the defining equation as before we see that $x = 0$ is a non-hyperbolic fixed point of the map if $\mu = 0$ and $g_\mu = 0$, $g_{xx} = -2$, $g_{x\mu} = 1$ and $g_{\mu\mu} = 0$. Hence there is a transcritical bifurcation if $\mu = 0$. When $\mu = \pm 2$ there is a further bifurcation; once again the Jacobian is $-1$ and we defer discussion of this case to the next section.

*Example 9.3*

Consider

$$x_{n+1} = x_n(1 + \mu - x_n^2).$$

Fixed points are solutions of $x(\mu - x^2) = 0$ so $x = 0$ is always a fixed point and if $\mu > 0$ there are two other fixed points, $x_\pm = \pm\sqrt{\mu}$. It is not hard to see that this is a supercritical pitchfork bifurcation.

These three examples are sometimes referred to as *normal forms* for the three bifurcations: they are the simplest models which undergo the bifurcation at $\mu = 0$. The evolution of these maps as $\mu$ passes through zero and the corresponding phase portraits for flows (if the problem comes from the return map near a periodic orbit) is shown in Figures 9.1–9.2.

## 9.2 Period-doubling bifurcations

Suppose now that the map has (for $\mu = 0$) a fixed point for which the linear map has a simple eigenvalue of $-1$ and no other eigenvalues on the unit circle. If the map were truly linear this would lead to the equation $x_{n+1} = -x_n$ for which *all* points (except for $x = 0$) are periodic of period two since $x_{n+2} = x_n$. This suggests that we might expect to find some sort of oscillation of period two associated with this type of bifurcation. The map on the centre manifold is of the form $x_{n+1} = g(x_n, \mu)$ where $g(0,0) = 0$ and $g_x(0,0) = -1$. Expanding $g$ locally about $(x,\mu) = (0,0)$ we find

$$x_{n+1} = A(\mu) + (-1 + B(\mu))x + C(\mu)x^2 + D(\mu)x^3 + O(x^4) \qquad (9.10)$$

with

$$A(\mu) = g_\mu \mu + \tfrac{1}{2} g_{\mu\mu}\mu^2 + O(\mu^3), \; B(\mu) = g_{\mu x}\mu + O(\mu^2),$$

$$C(\mu) = \tfrac{1}{2} g_{xx} + O(\mu) \text{ and } D(\mu) = \tfrac{1}{6} g_{xxx} + O(\mu).$$

Fixed points are therefore solutions of

$$A(\mu) + (-2 + B(\mu))x + C(\mu)x^2 + D(\mu)x^3 + O(x^4) = 0. \qquad (9.11)$$

For $(x,\mu)$ near $(0,0)$ this has a single fixed point at $x^*$ where

$$x^* = \tfrac{1}{2} A(\mu) + O(\mu^2) = \tfrac{1}{2} g_\mu \mu + O(\mu^2). \qquad (9.12)$$

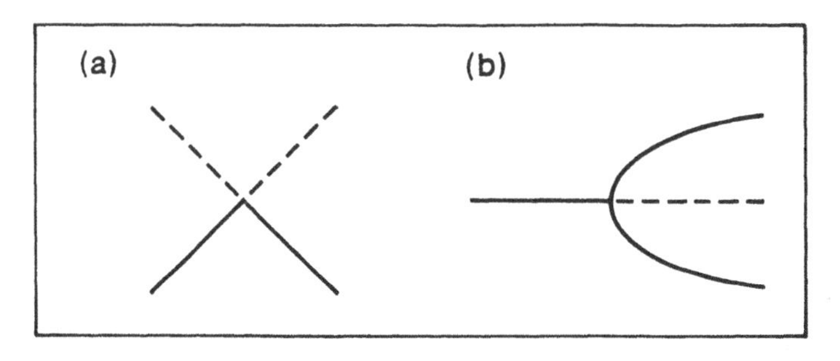


Fig. 9.2 (a) A transcritical bifurcation; (b) a pitchfork bifurcation.

To find the stability of this solution we look at the Jacobian matrix of the map at the fixed point, the Jacobian is, for this one-dimensional map, simply the derivative of the map with respect to $x$:

$$Dg(x) = -1 + B(\mu) + 2C(\mu)x + 3D(\mu)x^2 + O(x^3) \tag{9.13}$$

and on substituting the approximate value of the fixed point, $x^* = \frac{1}{2}g_\mu\mu + O(\mu^2)$,

$$\begin{aligned} Dg(x^*) &= -1 + B(\mu) + C(\mu)g_\mu\mu + O(\mu^2) \\ &= -1 + (g_{\mu x} + \tfrac{1}{2}g_\mu g_{xx})\mu + O(\mu^2). \end{aligned} \tag{9.14}$$

Hence if $g_{\mu x} + \frac{1}{2}g_\mu g_{xx} > 0$ this fixed point is stable if $\mu > 0$ and unstable if $\mu < 0$, whilst stability properties are reversed if this expression takes the opposite sign. Our first condition for a non-degenerate bifurcation is the transversality condition

$$g_{\mu x} + \tfrac{1}{2}g_\mu g_{xx} \neq 0, \tag{9.15}$$

which is equivalent to saying that the eigenvalue of the linear map at the fixed point passes through the value $-1$ with non-zero velocity as $\mu$ varies.

To investigate the possibility of bifurcating solutions associated with this loss of stability we begin by making a change of variables to set $x^*$ at the origin. Setting $y = x - x^*$ and noting that the effect of this transformation is to set the 'constant', or $\mu$-dependent, term to zero ($y = 0$ is a fixed point of the transformed map), a small calculation gives

$$y_{n+1} = (-1 + B'(\mu))y_n + C'(\mu)y_n^2 + D'(\mu)y_n^3 + O(y_n^4) \tag{9.16}$$

where

$$B'(\mu) = B(\mu) + A(\mu)C(\mu) + O(\mu^2), \ C'(\mu) = C(\mu) + O(\mu)$$

and

$$D'(\mu) = D(\mu) + O(\mu).$$

(9.5) EXERCISE

*Using (9.12) prove this statement and show that*

$$B'(\mu) = (g_{\mu x} + \tfrac{1}{2}g_\mu g_{xx})\mu + O(\mu^2), \ C'(0) = \tfrac{1}{2}g_{xx} \text{ and } D'(0) = \tfrac{1}{6}g_{xxx}.$$

We already know that the only fixed point of the map in a neighbourhood of $(y, \mu) = (0, 0)$ is $y = 0$, corresponding to the fixed point $x = x^*$, but in the preamble to this section we suggested that there might be periodic points of period two. These will be fixed points of the second iterate of the map, so consider $y_{n+2}$:

$$y_{n+2} = (-1 + B'(\mu))y_{n+1} + C'(\mu)y_{n+1}^2 + D'(\mu)y_{n+1}^3 + O(y_{n+1}^4) \quad (9.17)$$

and so, substituting for $y_{n+1}$ and rearranging terms

$$y_{n+2} = (1 + a(\mu))y_n + b(\mu)y_n^2 + c(\mu)y_n^3 + O(y_n^4) \quad (9.18)$$

where

$$a(\mu) = -2B'(\mu) + O(\mu^2),\ b(\mu) = B'C' + O(\mu^2)$$

and

$$c(\mu) = -2(C'^2 + D') + O(\mu). \quad (9.19)$$

Fixed points of the second iterate of the map are therefore $y = 0$, which we knew about already (it is a fixed point of the map and hence it is also a fixed point of the second iterate) and solutions of

$$a(\mu) + b(\mu)y + c(\mu)y^2 + O(y^3) = 0. \quad (9.20)$$

The quadratic part of this equation has solutions provided

$$b^2 - 4ac = -16B'(C'^2 + D') + O(\mu^2) > 0. \quad (9.21)$$

But $B'(\mu) = (g_{\mu x} + \frac{1}{2}g_\mu g_{xx})\mu + O(\mu^2)$ and the term multiplying $\mu$ is precisely the term which we have already assumed to be non-zero in the transversality condition above and $(C'^2 + D') = \frac{1}{4}g_{xx}^2 + \frac{1}{6}g_{xxx}$. Hence provided

$$\tfrac{1}{2}g_{xx}^2 + \tfrac{1}{3}g_{xxx} \neq 0 \quad (9.22)$$

the quadratic has a pair of solutions in either $\mu > 0$ or $\mu < 0$ depending upon the sign of the two non-vanishing expressions. This then is the genericity condition associated with the period-doubling bifurcation; since both these solutions are local and are fixed points of the second iterate of the map but not of the map itself they must form an orbit of period two. The stability of this bifurcating orbit can be read off from the equation for $y_{n+2}$: they are unstable if the fixed point $y = 0$ is stable and stable if the fixed point at $y = 0$ is unstable. We leave this as an exercise. One further remark: calculation of the locus of the points of period two shows that the curve is tangential to the $y$-axis at $(y, \mu) = (0, 0)$.

(9.6) THEOREM (PERIOD-DOUBLING BIFURCATION)

*Suppose that if $\mu = 0$ the map $w_{n+1} = f(w_n, \mu)$, $w \in \mathbf{R}^n$, has a fixed point at $w = 0$ and that the linear map at $(w, \mu) = (0, 0)$ has a simple eigenvalue of $-1$ and no other eigenvalues lie on the unit circle. Then the equation on the centre manifold is $x_{n+1} = g(x_n, \mu)$, $x \in \mathbf{R}$, with $g(0,0) = 0$ and $\frac{\partial g}{\partial x}(0,0) = -1$. If*

$$u = 2g_{\mu x} + g_\mu g_{xx} \neq 0$$

*and*

$$v = \tfrac{1}{2}g_{xx}^2 + \tfrac{1}{3}g_{xxx} \neq 0$$

*(where the partial derivatives are evaluated at $(0,0)$) a curve of periodic points of period two bifurcate from $(0,0)$ into $\mu > 0$ if $uv < 0$ or $\mu < 0$ if $uv > 0$. The fixed point from which these solutions bifurcate is stable in $\mu > 0$ and unstable in $\mu < 0$ if $u > 0$, with the signs of $\mu$ reversed if $u < 0$. The bifurcating cycle of period two is stable if it coexists with an unstable fixed point and vice versa. The bifurcation is said to be supercritical if the bifurcating solution of period two is stable and subcritical if the bifurcating solution is unstable.*

*Example 9.4*

Consider the map

$$x_{n+1} = g(x_n, \mu) = -(1+\mu)x_n - x_n^3.$$

When $\mu = 0$ the origin $x = 0$ is a non-hyperbolic fixed point and the linear map at the origin has an eigenvalue of $-1$. Furthermore, with the notation of Theorem 9.6, $u = -2$ and $v = -2$. Hence we expect to see an orbit of period two which bifurcates into $\mu < 0$ from the origin, which is a stable fixed point in $\mu < 0$ since $u < 0$. It is easy to confirm these facts by direct calculation for this example. First consider the origin. Looking at the linear part of the map we see that the fixed point at the origin is stable if $\mu < 0$ and unstable if $\mu > 0$ (for small $|\mu|$). Furthermore (again for small $|\mu|$) the origin is the only fixed point of the map. Now consider the second iterate of the map,

$$x_{n+2} = -(1+\mu)[-(1+\mu)x_n - x_n^3] - [-(1+\mu)x_n - x_n^3]^3.$$

Expanding out this expression, keeping only terms of order three and less, we find

$$x_{n+2} \approx (1+\mu)^2 x_n + (1+\mu)x_n^3[1+(1+\mu)^2]$$

or

$$x_{n+2} \approx (1+2\mu)x_n + 2x_n^3.$$

This equation has fixed points at $x = 0$ and $x^2 = -\mu$, so there is a bifurcating branch of solutions from $x = 0$ into $\mu < 0$. The stability of these solutions (as fixed points of the second iterate of the map) is determined by the Jacobian $(1+2\mu) + 6x^2 = 1 - 4\mu$, so since $\mu < 0$ the orbit of period two is unstable. This confirms that there is a subcritical period-doubling bifurcation if $\mu = 0$.

Figure 9.3 shows the behaviour of this map and the corresponding bifurcation for a periodic orbit of a flow. Note that this bifurcation can only occur in flows in $\mathbf{R}^n$, $n \geq 3$, since trajectories cannot cross and, as we saw in Chapter 6, the eigenvalues of a fixed point of a return map for a two-dimensional flow are both positive.

## 9.3 The Hopf bifurcation for maps

A Hopf bifurcation occurs in maps if a simple pair of complex conjugate eigenvalues of the linear map cross the unit circle. There are various possible cases depending on whether these eigenvalues cross at roots of unity or not. If they do not (the non-resonant case) it is possible to choose a change of coordinates so that the map takes a particularly simple form analogous to the continuous time case described in the previous

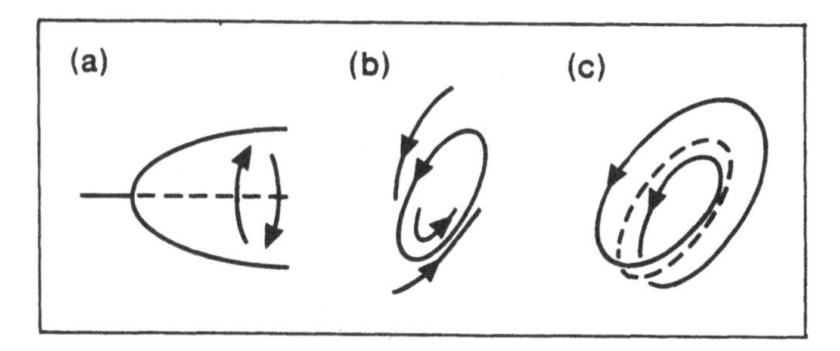


Fig. 9.3 (a) $(x, \mu)$ bifurcation diagram; (b) the flow for $\mu < 0$; (c) the flow for $\mu > 0$.

chapter. However, if the eigenvalues do cross the unit circle at a root of unity then it is impossible to eliminate certain terms in the power series expansion of the map. This leads to complications known as resonance, which have their roots in the fact that the required coordinate changes have small divisor problems similar to those encountered in Poincaré's Linearization Theorem. These resonant modifications do not manifest themselves in any serious way provided the eigenvalues on the unit circle, $\lambda_0$ and $\lambda_0^*$ do not satisfy $\lambda_0^n = 1$ for $n = 1, 2, 3, 4$ (the strong resonances). In the next section we shall describe the differences between weak resonance ($\lambda_0^n = 1$ for some $n > 4$) and the non-resonant case ($\lambda_0^n \neq 1$ for all $n \geq 1$). The strong resonances are still not completely understood; Arnold (1983) has a good discussion of some of the cases.

Suppose that for $\mu = 0$ the map $x_{n+1} = f(x_n, \mu)$ has a fixed point at the origin and that the linear map evaluated at $(x, \mu) = (0, 0)$ has a pair of complex eigenvalues $\lambda_0$ and $\lambda_0^*$ on the unit circle with $\lambda_0^n \neq 1$ for $n = 1, 2, 3, 4$ and that all the other eigenvalues lie off the unit circle. Then the equation on the centre manifold is $y_{n+1} = g(y_n, \mu)$, $y \in \mathbf{R}^2$, with $g(0,0) = 0$, and $Dg(0,0)$ has eigenvalues $\lambda_0$ and $\lambda_0^*$. By assumption, $\lambda_0 = \cos\theta_0 + i\sin\theta_0$, $\theta_0 \neq \frac{2\pi p}{q}$ for $q = 1, 2, 3, 4$. We shall assume the standard transversality condition: that the eigenvalues pass through the unit circle with non-zero velocity. This implies that the linear part of the map in a neighbourhood of $\mu = 0$ can be written as

$$M(\mu) \begin{pmatrix} \cos\theta(\mu) & -\sin\theta(\mu) \\ \sin\theta(\mu) & \cos\theta(\mu) \end{pmatrix} \tag{9.23}$$

where $M(\mu) = 1 + \alpha\mu + O(\mu^2)$ and $\theta(\mu) = \theta_0 + \theta_1\mu + O(\mu^2)$. In other words, the eigenvalues of the linear map are $M(\mu)\exp(\pm 2\pi i\theta(\mu))$ and the transversality condition is

$$\frac{d|\lambda|}{d\mu}(0) = \frac{dM}{d\mu}(0) = \alpha \neq 0. \tag{9.24}$$

Now, as in the Hopf bifurcation for continuous time, we choose to work in the complex plane, identifying $\mathbf{R}^2$ with $\mathbf{C}$ in the standard way and treating $z$ and $z^*$ as independent variables. This leads us to consider the maps

$$z_{n+1} = G(z_n, z_n^*, \mu) = \lambda(\mu)z_n + \sum_{m\geq 2}\sum_{r=0}^{m} \gamma_{mr}(\mu) z_n^r z_n^{*m-r}, \tag{9.25}$$

where

$$\lambda(\mu) = M(\mu)exp\left(i\theta(\mu)\right). \tag{9.26}$$

This type of equation should look familiar and indeed we now simply repeat the arguments of the section on the Hopf bifurcation for differential equations in order to obtain the following result.

(9.7) THEOREM (HOPF BIFURCATION THEOREM FOR MAPS)

*Suppose that the map*

$$(x_{n+1}, y_{n+1}) = (F(x_n, y_n, \mu), G(x_n, y_n, \mu))$$

*satisfies* $F(0,0,\mu) = G(0,0,\mu) = 0$ *on some neighbourhood of* $\mu = 0$ *and that if* $\mu = 0$ *the Jacobian matrix of the map at* $(x,y) = (0,0)$ *is*

$$\begin{pmatrix} \cos\theta_0 & -\sin\theta_0 \\ \sin\theta_0 & \cos\theta_0 \end{pmatrix}$$

*where* $exp(ni\theta_0) \neq 1$ *for* $n = 1,2,3,4$. *If*

$$\cos\theta_0(F_{\mu x} + G_{\mu y}) + \sin\theta_0(G_{\mu x} - F_{\mu y}) \neq 0$$

*and*

$$\mathrm{Re}A(0) \neq 0$$

*where partial derivatives are evaluated at* $(x,y,\mu) = (0,0,0)$ *and* $A(\mu)$ *is a complex function of* $\mu$ *defined below, then an invariant circle bifurcates into either* $\mu > 0$ *or into* $\mu < 0$, *depending upon the signs of the non-zero expressions above. This invariant circle is attracting if it bifurcates into the region of* $\mu$ *for which the origin is unstable (a supercritical bifurcation) and repelling if it bifurcates into the region for which the origin is stable (a subcritical bifurcation).*

*Proof*: As with the continuous time case the proof splits neatly into a number of steps:
Step 1: Check that the transversality condition holds (this will give the first expression).

Step 2: By assumption, the origin is a fixed point of the map for all $\mu$ near $\mu = 0$, so we can make a linear coordinate change to bring the linearized map at the origin into the form described above the statement of the theorem. Then, working in the complex plane, we do a series of near identity coordinate changes so that the map becomes

$$z_{n+1} = \lambda(\mu)z_n + A(\mu)|z_n|^2 z + O(|z_n|^4).$$

Step 3. Show that an invariant circle bifurcates from the origin when

$\mu = 0$ for this system providing $\text{Re}A(0) \neq 0$.
Step 4. Derive an expression for $\text{Re}A(0)$ in terms of the partial derivatives of $F$ and $G$.

**Step 1.** Note that by assumption $F_x = G_y = \cos\theta_0$ and $G_x = -F_y = \sin\theta_0$. Expanding $F$ and $G$ as Taylor series, remembering that the origin is always a fixed point for the values of $\mu$ considered, we find that the Jacobian matrix of the map at $(x, y, \mu) = (0, 0, 0)$ is

$$\begin{pmatrix} F_x + F_{\mu x}\mu & F_y + F_{\mu y}\mu \\ G_x + G_{\mu x}\mu & G_y + G_{\mu y}\mu \end{pmatrix}$$

up to terms of order $|\mu|$. Now, $|\lambda(\mu)|^2$ is simply the determinant of this matrix (since it is the product of the eigenvalues), so

$$|\lambda(\mu)|^2 = (F_x + F_{\mu x}\mu)(G_y + G_{\mu y}\mu) - (F_y + F_{\mu y}\mu)(G_x + G_{\mu x}\mu)$$

$$= F_xG_y - F_yG_x + \mu(F_xG_{\mu y} + G_yF_{\mu x} - F_yG_{\mu x} - G_xF_{\mu y}) + O(\mu^2).$$

From this it follows that $\frac{d}{d\mu}|\lambda| \neq 0$ if $(x, y, \mu) = (0, 0, 0)$ provided

$$\cos\theta_0(G_{\mu y} + F_{\mu x}) + \sin\theta_0(G_{\mu x} - F_{\mu y}) \neq 0 \tag{9.27}$$

(remember that $F_x = \cos\theta_0$ and so on).

**Step 2.** We now assume that a linear change of variables has been made such that the linear part of the map takes the form

$$M(\mu)\begin{pmatrix} \cos\theta(\mu) & -\sin\theta(\mu) \\ \sin\theta(\mu) & \cos\theta(\mu) \end{pmatrix}$$

as defined above the statement of the theorem. Note that by assumption the map is already in this form if $\mu = 0$ so there is no change of variables in this case. Now write $z = x + iy$ so that the map becomes

$$z_{n+1} = \lambda(\mu)z_n + \sum_{m\geq 2}\sum_{r=0}^{m} \gamma_{mr}(\mu)z_n^r z_n^{*m-r} \tag{9.28}$$

where the coefficients $\gamma_{mr}$ are functions of $\mu$ whose coefficients can be written in terms of the partial derivatives of the map with respect to $x$, $y$ and $\mu$ as in the continuous time case. We now try near identity changes of coordinates to get rid of as many terms in the expansion as possible. Since we have done this type of successive elimination of terms twice before we jump straight to the general case. Suppose that after a sequence of near identity changes of variable of order two, three and so on we have the system

$$z_{n+1} = \lambda(\mu)z_n + N_{s-1}(z_n, z_n^*, \mu) + \sum_{m\geq s}\sum_{k=0}^{m} \gamma'_{mk}(\mu) z_n^k z_n^{*m-k} \tag{9.29}$$

where the terms in $N_{s-1}(z, z^*, \mu)$ contain those terms of order $r$ ($2 \leq r < s$) in $z$ and $z^*$ which have not been removed by near identity transformations of order less than $s$ and the coefficients $\gamma'_{mk}$ are the modified coefficients resulting from the near identity tranformations already done. Now try the next near identity transformation,

$$w = z + \sum_{k=0}^{s} a_{sk} z^k z^{*s-k} \tag{9.30}$$

with inverse

$$z = w - \sum_{k=0}^{s} a_{sk} w^k w^{*s-k} + O(|w|^{s+1}). \tag{9.31}$$

From the definition of $w$

$$w_{n+1} = z_{n+1} + \sum_{k=0}^{s} a_{sk} z_{n+1}^k z_{n+1}^{*s-k} \tag{9.32}$$

and since $z_{n+1} = \lambda(\mu)z_n + O(|z_n|^2)$ and $z_{n+1}^* = \lambda(\mu)^* z_n^* + O(|z_n|^2)$,

$$\begin{aligned} w_{n+1} &= \lambda(\mu)z_n + N_{s-1}(z_n, z_n^*, \mu) + \sum_{m\geq s}\sum_{k=0}^{m} \gamma'_{mk}(\mu) z_n^k z_n^{*m-k} \\ &\quad + \sum_{k=0}^{s} a_{sk}\lambda(\mu)^k \lambda(\mu)^{*s-k} z_n^k z_n^{*s-k} + O(|z_n|^{s+1}). \end{aligned} \tag{9.33}$$

Now, $\lambda z_n = \lambda w_n - \sum_{k=0}^{s} \lambda a_{sk} w_n^k w_n^{*s-k} + O(|w_n|^{s+1})$ so

$$\begin{aligned} w_{n+1} &= \lambda(\mu)w_n + N_{s-1}(w_n, w_n^*, \mu) \\ &\quad + \sum_{k=0}^{s} \left(\gamma'_{sk}(\mu) + [\lambda(\mu)^k\lambda(\mu)^{*s-k} - \lambda(\mu)]a_{sk}\right) w_n^k w_n^{*s-k} \\ &\quad + O(|w_n|^{s+1}). \end{aligned} \tag{9.34}$$

Hence if we can choose

$$a_{sk} = \frac{\gamma'_{sk}(\mu)}{\lambda(\mu) - \lambda(\mu)^k \lambda(\mu)^{*s-k}} \tag{9.35}$$

the terms of order $w^k w^{*s-k}$ can be eliminated. This can be done, at least formally, provided $\lambda \neq \lambda^k \lambda^{*s-k}$ in some neighbourhood of $\mu = 0$.

Evaluating these terms at $\mu = 0$, using the fact that $\lambda_0 = \lambda_0^{*-1}$, this condition becomes

$$\lambda_0^{s+1-2k} \neq 1. \tag{9.36}$$

If $\lambda^q \neq 1$ for all $q \geq 1$ then the only problems arise if the exponent is zero. This happens if $s+1 = 2k$, so $s$ must be odd, $s = 2p+1$, $p \geq 1$, with $k = p+1$ and $s - k = p$. Thus, in general, we cannot get rid of terms of order $|z|^{2p}z$ but all other terms can be set to zero by successive near identity changes of coordinates. Thus if $\lambda_0^q \neq 1$ for all $q \geq 1$ a series of near identity changes of coordinates can be made to bring the map into the form

$$z_{n+1} = \lambda(\mu)z_n + \sum_{p\geq 1} \alpha_p(\mu)|z_n|^{2p}z_n. \tag{9.37}$$

On the other hand, if $\lambda_0^q = 1$ for some $q \geq 1$ (the resonant case) we are unable to get rid of the $|z|^{2p}z$ terms as above, but we also have problems if $s+1-2k = q$. The first such term occurs for $k = 0$, giving a term with $s = q-1$, $z^{*q-1}$ which cannot be eliminated. The second such term ($k = 1$) has $s = q+1$. Hence, after a series of near identity changes of coordinate the map can be written in the form

$$z_{n+1} = \lambda(\mu)z_n + \sum_{p\geq 1} \alpha_p(\mu)|z_n|^{2p}z_n + \beta(\mu)z_n^{*q-1} + O(|z_n|^{q+1}). \tag{9.38}$$

None the less, provided $q \geq 5$ (the weak resonances) this takes the form

$$z_{n+1} = \lambda(\mu)z_n + \alpha_1(\mu)|z_n|^2 z + O(|z_n|^4) \tag{9.39}$$

as required.

**Step 3.** Assuming that $\lambda_0^q \neq 1$, $q = 1, 2, 3, 4$ the map can be written as

$$z_{n+1} = \lambda(\mu)z_n\left(1 + \frac{\alpha_1(\mu)}{\lambda(\mu)}|z_n|^2\right) + O(|z_n|^4) \tag{9.40}$$

so

$$|z_{n+1}| = |\lambda(\mu)||z_n|\left|\left(1 + \frac{\alpha_1(\mu)}{\lambda(\mu)}|z_n|^2\right)\right| + O(|z|^4). \tag{9.41}$$

Now

$$\begin{aligned}\left|\left(1 + \frac{\alpha_1(\mu)}{\lambda(\mu)}|z_n|^2\right)\right|^2 &= \left(1 + \frac{\alpha_1(\mu)}{\lambda(\mu)}|z_n|^2\right)\left(1 + \frac{\alpha_1^*(\mu)}{\lambda^*(\mu)}|z_n|^2\right) \\ &= 1 + 2\mathrm{Re}\left(\frac{\alpha_1(\mu)}{\lambda(\mu)}\right)|z_n|^2 + O(|z_n|^4)\end{aligned}$$

and so

$$\left|\left(1+\frac{\alpha_1(\mu)}{\lambda(\mu)}|z_n|^2\right)\right| = 1+\mathrm{Re}\left(\frac{\alpha_1(\mu)}{\lambda(\mu)}\right)|z_n|^2 + O(|z_n|^4)$$

for small $|z_n|$. Setting $z_n = r_n exp(i\varphi_n)$ this implies that

$$r_{n+1} = M(\mu)r_n\left[1+\mathrm{Re}\left(\frac{\alpha_1(\mu)}{\lambda(\mu)}\right)r_n^2\right] + O(r_n^4) \tag{9.42}$$

where $M(\mu)$ is the modulus of $\lambda$ defined earlier. This difference equation for the modulus of $z$ is independent of the argument of $z$ to lowest order, so a fixed point for the modulus will correspond to an invariant circle for the two-dimensional map. Fixed points are given by

$$r = M(\mu)r\left[1+\mathrm{Re}\left(\frac{\alpha_1(\mu)}{\lambda(\mu)}\right)r^2\right] + O(r^4) \tag{9.43}$$

so we have the fixed point at $r = 0$, which is the fixed point at the origin for the two-dimensional map and is stable if $|M(\mu)| < 1$ and unstable if $|M(\mu)| > 1$. Since $M(\mu) = 1 + \alpha\mu + O(\mu^2)$, where $\alpha$ is the non-zero constant given by the transversality condition, the origin is stable in $\mu < 0$ and unstable in $\mu > 0$ if $\alpha > 0$ and the stability properties are reversed if $\alpha < 0$. A non-zero fixed point of the modulus of $z$ satisfies

$$\mathrm{Re}A(\mu)r^2 = 1 - M(\mu) = -\alpha\mu + O(\mu^2) \tag{9.44}$$

where

$$A(\mu) = \alpha_1(\mu)/\lambda(\mu). \tag{9.45}$$

Hence there is a non-trivial fixed point for the modulus of $z$ if

$$(\alpha\mu)/\mathrm{Re}A(0) > 0 \tag{9.46}$$

corresponding to an invariant circle for the two-dimensional map on the centre manifold.

(9.8) EXERCISE

*Determine the stability of this fixed point.*

**Step 4.** All that remains to be done now is to determine the expression $\mathrm{Re}A(0)$ in terms of the partial derivatives of $F$ and $G$. As before (in the continuous time case) this is a painful and unexciting process. We have left it as a series of exercises which should enable the reader to check at the end of each calculation whether s/he is on the right track.

First note that $\mathrm{Re}A(0) = \alpha_1(0)/\lambda_0 = \alpha_1(0)\lambda_0^*$. Hence

$$\mathrm{Re}A(0) = \cos\theta_0.\mathrm{Re}(\alpha_1(0)) + \sin\theta_0.\mathrm{Im}(\alpha_1(0)). \tag{9.47}$$

We need to calculate $\gamma_{32}'(0)$ obtained after performing the first near identity coordinate change.

(9.9) EXERCISE

*Show that*

$$\gamma_{32}'(0) = \gamma_{32} + \frac{|\gamma_{21}|^2}{1-\lambda_0^*} + \frac{2|\gamma_{20}|^2}{\lambda_0^2-\lambda_0^*} + \gamma_{21}\gamma_{22}\frac{2\lambda_0 - 1}{\lambda_0(1-\lambda_0)}.$$

(9.10) EXERCISE

*Now,* $\alpha_1(0) = \gamma_{32}'(0)$. *So to calculate* $\alpha_1(0)$ *we need the original coefficients* $\gamma_{mk}$ *in terms of the derivatives of* $F$ *and* $G$. *Show that*

$$\begin{aligned}
\gamma_{22} &= \tfrac{1}{8}[(F_{xx} + 2G_{xy} - F_{yy}) + i(G_{xx} - 2F_{xy} - G_{yy})],\\
\gamma_{21} &= \tfrac{1}{4}[(F_{xx} + F_{yy}) + i(G_{xx} + G_{yy})],\\
\gamma_{20} &= \tfrac{1}{8}[(F_{xx} - 2G_{xy} - F_{yy}) + i(G_{xx} + 2F_{xy} - G_{yy})]
\end{aligned}$$

*and*

$$\gamma_{32} = \frac{1}{16}[(3F_{xxx}+F_{xyy}+G_{xxy}+3G_{yyy})+i(3G_{xxx}+G_{xyy}-F_{xxy}-3F_{yyy})].$$

(9.11) EXERCISE

*Finally, substitute these equations into the expression for* Re $A(0)$ *to obtain a really nasty expression which must not vanish if the above analysis is to work.*

If the Hopf bifurcation occurs in a map associated with the return map near a periodic orbit of an autonomous flow then the bifurcation is called a secondary Hopf bifurcation. In this case the invariant curve corresponds to an invariant torus for the flow as shown in Figure 9.4. This completes the description of the Hopf bifurcation for maps. We have shown that there is an invariant circle with radius proportional to $\sqrt{|\mu|}$ which bifurcates from the origin into $\mu > 0$ or $\mu < 0$ depending upon the signs of two non-zero constants in the non-resonant and weakly resonant cases. The dynamics on this invariant circle have not been discussed. This is what we attempt to understand next.

## 9.4 Arnol'd (resonant) tongues

In the discussion of the Hopf bifurcation above there are two essential ingredients: a pair of complex conjugate eigenvalues of the linear map pass through the unit circle with non-zero velocity at a particular angle, $\theta_0$. In some ways it is more natural to see this as a two parameter problem: one parameter controlling the radial component of the eigenvalue through the unit circle and the other controlling the argument of the eigenvalue when it intersects this circle. This leads us to consider maps with linear part $\lambda(\mu,\nu)z$ in the complex plane close to a weakly resonant Hopf bifurcation $\theta_0 = 2\pi p/q$, $q \geq 5$. For convenience we choose the parametrization so that

$$\lambda(\mu,\nu) = (1+\mu)\exp\left(i\left[\frac{2\pi p}{q}+\nu\right]\right) \tag{9.48}$$

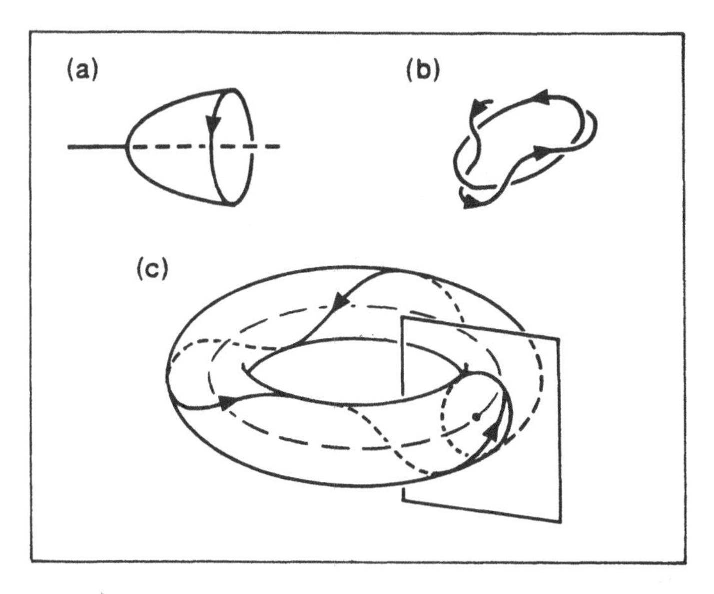


Fig. 9.4 Hopf bifurcation: (a) $(x, y, \mu)$ bifurcation diagram; (b) flow for $\mu < 0$; (c) flow for $\mu > 0$.

so $|\lambda(\mu,\nu)| = 1+\mu$ and $\text{Arg}\lambda(\mu,\nu) = \frac{2\pi p}{q} + \nu$. By precisely the same argument as in Step 3 of the proof of the Hopf Bifurcation Theorem there is a change of variable such that the map takes the form

$$\begin{aligned} z_{n+1} = \lambda(\mu,\nu)z_n + \sum_{p\geq 1} \alpha_p(\mu,\nu)|z_n|^{2p}z_n \\ + \beta(\mu,\nu)z_n^{*q-1} + O(|z_n|^{q+1}). \end{aligned} \tag{9.49}$$

As before we want to write this equation in terms of the evolution of the modulus of $z_n$ and also find the equation for the argument of $z$. This means that we have to be a little more careful than before. Setting $z = r\exp(i\varphi)$ we have

$$\begin{aligned} r_{n+1}e^{i\varphi_{n+1}} = \\ \lambda r_n e^{i\varphi_n}\left(1 + \sum_{k\geq 1}\frac{\alpha_k}{\lambda}r_n^{2k} + \frac{\beta}{\lambda}r_n^{q-2}e^{-iq\varphi_n} + O(r_n^q)\right) \end{aligned} \tag{9.50}$$

where $\lambda = (1+\mu)\exp(i(\frac{2\pi p}{q}+\nu))$ and $\alpha_k$ and $\beta$ are functions of $\mu$ and $\nu$. Taking the modulus of both sides,

$$r_{n+1} = (1+\mu)r_n\left|1 + \sum_{k\geq 1}\frac{\alpha_k}{\lambda}r_n^{2k} + \frac{\beta}{\lambda}r_n^{q-2}e^{-iq\varphi_n} + O(r_n^q)\right|. \tag{9.51}$$

Considering the modulus in (9.51) for small $r$ we find that

$$\begin{aligned} \left|1 + \sum_{k\geq 1}\frac{\alpha_k}{\lambda}r_n^{2k} + \frac{\beta}{\lambda}r_n^{q-2}e^{-iq\varphi_n} + O(r_n^q)\right| = \\ 1 + \sum_{k\geq 1} a_k r_n^{2k} + b r_n^{q-2} + O(r_n^q) \end{aligned} \tag{9.52}$$

where the coefficients $(a_k)$ are functions of the $\alpha_j$ and $\lambda$ and so functions of $\mu$ and $\nu$ alone, whilst $b = b(\mu,\nu,\varphi)$ and has the form

$$A(\mu,\nu)\cos(q\varphi) + B(\mu,\nu)\sin(q\varphi). \tag{9.53}$$

(9.12) EXERCISE

*Prove this last remark.*

Now, the non-trivial fixed point which gives the radius of the bifurcating circle is a solution of

$$1 = (1+\mu)\left(1 + \sum_{k\geq 1} a_k r^{2k} + br^{q-2} + O(r^q)\right) \tag{9.54}$$

and so there is a slight dependence on $\varphi$ through the $br^{q-2}$ term. This will be crucially important in the ensuing discussion. Now let $r_0(\mu)$ be the solution of

$$1 = (1+\mu)\left(1 + \sum_{k=1}^{[\frac{q-2}{2}]} a_k r_0^{2k}\right) \tag{9.55}$$

where $[\frac{q-2}{2}]$ is the largest natural number less than $\frac{q-2}{2}$. Assuming $a_1(0,0) < 0$ the (approximate) invariant circle exists for $\mu > 0$ with $r_0^2 = \mu/(-a_1) + O(\mu^2)$. On the invariant circle

$$e^{i\varphi_{n+1}} = \lambda e^{i\varphi_n}\left(1 + \sum_{k=1}^{[\frac{q-2}{2}]} a_k r_0^{2k} + br_0^{q-2} + O(r_0^{q-1})\right). \tag{9.56}$$

Hence

$$\begin{aligned}\varphi_{n+1} &= \varphi_n + \mathrm{Arg}(\lambda)\\ &\quad + \mathrm{Arg}\left(1 + \sum_{k=1}^{[\frac{q-2}{2}]} a_k r_0^{2k} + br_0^{q-2} + O(r_0^{q-1})\right).\end{aligned} \tag{9.57}$$

After some further manipulation of the third term, this expression can be written as

$$\varphi_{n+1} = \varphi_n + \frac{2\pi p}{q} + \nu + \Omega(\mu,\nu) + \mu^{\frac{q-2}{2}}\tilde{b}(\mu,\nu,\varphi_n) + O(\mu^{\frac{q-1}{2}}) \tag{9.58}$$

where $\tilde{b}$ is just a rescaled version of $b$ in (9.53). So, using the obvious notation and ignoring higher order terms

$$\begin{aligned}\varphi_{n+1} &= \varphi_n + \frac{2\pi p}{q} + \nu + \Omega(\mu,\nu)\\ &\quad + \mu^{\frac{q-2}{2}}(\tilde{A}(\mu,\nu)\cos q\varphi_n + \tilde{B}(\mu,\nu)\sin q\varphi_n).\end{aligned} \tag{9.59}$$

We can now look for fixed points of this map with period $q$, corresponding to periodic orbits of period $q$ which lie on the invariant circle. To lowest order

$$\varphi_{n+q} = \varphi_n + q\nu + q\Omega(\mu,\nu)$$

$$+ q\mu^{\frac{q-2}{2}}(\tilde{A}(\mu,\nu)\cos q\varphi_n + \tilde{B}(\mu,\nu)\sin q\varphi_n) \quad (9.60)$$

and hence fixed points satisfy

$$\nu + \Omega(\mu,\nu) + \mu^{\frac{q-2}{2}}(\tilde{A}(\mu,\nu)\cos q\varphi_n + \tilde{B}(\mu,\nu)\sin q\varphi_n) = 0. \quad (9.61)$$

For sufficiently small $|\nu + \Omega|$ this equation clearly has two families of solutions which are small perturbations of zeros of

$$\tilde{A}\cos(q\varphi) + \tilde{B}\sin(q\varphi) = 0. \quad (9.62)$$

One of these families of solutions is stable and the other is unstable (see Fig. 9.5a). These annihilate each other on curves of the form

$$|\nu + \Omega| = (\tilde{A}^2 + \tilde{B}^2)^{\frac{1}{2}}\mu^{\frac{q-2}{2}}. \quad (9.63)$$

This situation is shown if Figure 9.5b: the circle represents the unit circle in the complex plane, from each point $\exp(2\pi ip/q)$ there is an Arnol'd tongue centred upon the curve $\nu = \Omega(\mu,\nu)$ and of width $|\mu|^{\frac{q-2}{2}}$ in which the bifurcating invariant circle is resonant: there are two periodic orbits of period $q$ on the circle, one stable and the other unstable. Typically a one-parameter family of diffeomorphisms for which a pair of eigenvalues of the linear map pass through the unit circle will also pass through an infinite sequence of such Arnol'd tongues. Thus there will be a sequence of parameter intervals on which the map is said to be phase-locked: there is an invariant curve, but points on the curve are attracted to a stable periodic orbit of period q. The study of circle maps (such as the map in $\varphi$) is a rich source of interesting dynamic phenomena. Some properties of simple circle maps are described in the exercises. A more complete description of these results, together with the best smoothness

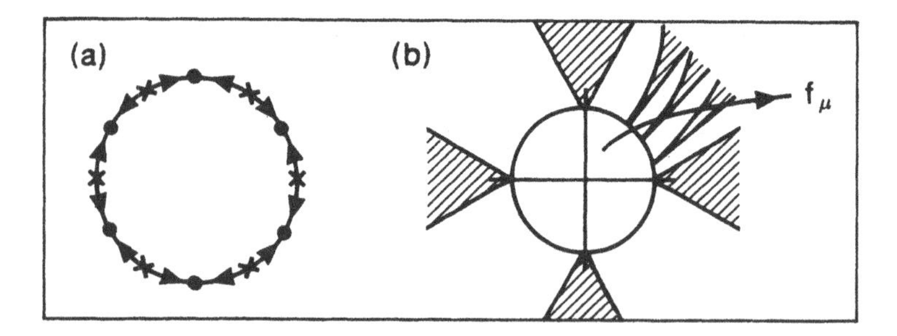


Fig. 9.5 (a) A resonant circle ($q = 6$); (b) Arnol'd tongues and a typical parameter path, $f_\mu$.

hypotheses, can be found in Iooss (1978). Arrowsmith and Place (1990) also contains a fuller discussion of the derivation of (9.59).

## Exercises 9

1. In each of the following maps, $x_{n+1} = f(x_n, \mu)$, determine the type of the bifurcation which occurs at the given value of $\mu$.

i) $f(x,\mu) = \mu - x^2$, $\mu = -\frac{1}{4}$;
ii) $f(x,\mu) = \mu - x^2$, $\mu = \frac{3}{4}$;
iii) $f(x,\mu) = \mu \sin x$, $\mu = 1$;
iv) $f(x,\mu) = \mu \sin x$, $\mu = -1$;
v) $f(x,\mu) = \mu \tan^{-1} x$, $\mu = -1$;
vi) $f(x,\mu) = \mu \tan x$, $\mu = 1$.

2. Suppose that $f(x,\mu) = -f(-x,\mu)$. Show that the map $x_{n+1} = f(x_n, \mu)$ has a fixed point at the origin for all $\mu$. If $f(y,\mu) = -y$, $y \neq 0$, show that $y$ is a point of period two. Hence describe the bifurcation at $\mu = 1$ for $f(x,\mu) = -\mu x + x^3$ and for $f(x,\mu) = -\mu x - x^3$.

3. Compare the locus of the points of period two found in the examples of the previous exercise with the approximation $\mu = K(0)x^2$ derived at the end of Section 9.6.

4. The Hénon map is a map on $\mathbf{R}^2$ defined by

$$x_{n+1} = y_n, \quad y_{n+1} = -bx_n + \mu - y_n^2,$$

where $b$ and $\mu$ are real parameters. Find the fixed points of this map as a function of the parameters. Find the locus of saddlenode bifurcations and the locus of period-doubling bifurcations of the fixed points in the $(b, \mu)$ plane.

5. Suppose that $f$ maps the interval $[a, b]$ into itself and that $f$ is continuous and strictly increasing. Show that there is at least one fixed point of $f$ in $[a, b]$. Show further that the orbit of every point in $[a, b]$ tends to a fixed point of $f$.

6. Suppose that $f$ maps $[a, b]$ into itself and that $f$ is continuous and strictly decreasing. Show that $f$ has at least one fixed point in $[a, b]$. Show further that the orbit of every point in $[a, b]$ tends either to a fixed point of $f$ or to an orbit of period two. Give an example of a strictly

decreasing, continuous map of the interval which has no points of period two. Give an example which has two points of period two.

7. Construct an example of a map $g(x,\mu)$ for which $g(0,0) = 0$, $g_x(0,0) = -1$ and $2g_{\mu x} + g_\mu g_{xx} = 0$. Investigate the number of fixed points and points of period two near $(x,\mu) = (0,0)$. How does the behaviour of your example differ from the standard picture of period-doubling described in Theorem 9.6? *[Hint: You may find it easiest to work with a symmetric example and use the results of Exercise 2.]*

8. Let $F : \mathbf{R} \to \mathbf{R}$ be the map $F(x) = x + \alpha$ and $\pi : \mathbf{R} \to S^1$ be the projection $\pi(x) = \exp(2\pi i x)$. Show that the map $f : S^1 \to S^1$ given by

$$\pi(F(x)) = f(\pi(x))$$

is well-defined. Prove that

$$\lim_{n\to\infty} \frac{F^n(x) - x}{n} = \alpha.$$

9. Let $f : S^1 \to S^1$ be an orientation preserving diffeomorphism of the circle and let $\pi$ denote the projection operator defined in Exercise 8. A lift, $F : \mathbf{R} \to \mathbf{R}$, of $f$ is a map which satisfies $\pi F = f\pi$. Show

i) two lifts of the same map $f$ differ by an integer;
ii) $F'(x) > 0$ and $F(x+1) = F(x) + 1$;
iii) $F^n(x+1) - (x+1) = F^n(x) - x$;
iv) if $|x - y| < 1$ then $|F^n(x) - F^n(y)| < 1$.

10. Let $f : S^1 \to S^1$ be an orientation preserving diffeomorphism of the circle and let $F$ be a lift of $f$. Use (iv) of Exercise 9 to show that

$$r(F) = \lim_{n\to\infty} \frac{F^n(x) - x}{n}$$

is independent of $x$. Let $F_1$ and $F_2$ be two lifts of $f$ with $F_1(x) = F_2(x) + k$ for some $k \in \mathbf{Z}$. Show that $F_1^n(x) = F_2^n(x) + nk$ and hence that $r(F_1) = r(F_2) + k$.

11. Let $F$ be a lift of an orientation preserving diffeomorphism of the circle, $f$. If $f$ has a periodic point, $\theta$, with $f^p(\theta) = \theta$ show that $F^p(x) = x + k$, $k \in \mathbf{Z}$, for some $x \in \mathbf{R}$. Hence show that

$$\lim_{n\to\infty} \frac{F^{pn}(x) - x}{pn} = \frac{k}{p}.$$

Given $j = pn + s$, $1 \leq s < p$, show that $|F^j(x) - F^{pn}(x)|$ is bounded and hence deduce that

$$\lim_{j\to\infty} \frac{F^j(x) - x}{j} = \frac{k}{p}.$$

12. Suppose that $f$ is an orientation preserving diffeomorphism of the circle with lift $F$. If $f$ has no periodic points use Exercise 9 (iii) to show that for all $n > 0$ there exists $C_n$ such that

$$C_n < F^n(x) - x < C_n + 1,$$

and so

$$C_n < F^{jn}(x) - F^{(j-1)n}(x) < C_n + 1.$$

By adding these inequalities for $j = 1, 2, \ldots, m$ show that

$$\frac{C_n}{n} < \frac{F^{mn}(x) - x}{mn} < \frac{C_n + 1}{n}$$

and hence that

$$\left| \frac{(F^{mn}(x) - x)}{mn} - \frac{(F^n(x) - x)}{n} \right| < \frac{1}{n}.$$

Interchanging the roles of $m$ and $n$ find a second inequality and deduce that $(F^n(x) - x)/n$ tends to a limit.

13. The rotation number of an orientation preserving diffeomorphism of the circle $f$ is the fractional part of $r(F)$ for any lift, $F$, of $f$. Use Exercises 11 and 12 to show that the rotation number is well-defined.

14. Let $f$ be the circle map $f(\theta) = \theta + \alpha \pmod 1$, $\theta \in [0, 1)$. For $\alpha \in [0, 1)$ prove

i) the rotation number of $f$ is $\alpha$;
ii) if $\alpha$ is rational then every point $\theta \in [0, 1)$ lies on a periodic orbit;
iii) if $\alpha$ is irrational then the orbit of every point $\theta \in [0, 1)$ is dense.

# 10

## *Bifurcational miscellany*

As the title suggests, this chapter is devoted to a number of examples and different approaches to bifurcations. Except for the first section, there is nothing new in any theoretical sense and this chapter could be ignored completely. None the less, a number of curious phenomena are described which are worthy of mention.

### 10.1 Unfolding degenerate singularities

In previous chapters we have found general conditions for given bifurcations to occur. All these bifurcations are associated with non-hyperbolic stationary points for flows or non-hyperbolic fixed points for return maps. Consider one-dimensional flows for the moment, then if there is a non-hyperbolic stationary point the flow can be written (locally) as

$$\dot{x} = x^n + O(|x|^{n+1}) \tag{10.1}$$

for some $n \geq 2$. We can think of this as being a degenerate situation because a 'typical' point would have a local expansion of the form

$$\dot{x} = \sum_{k=0}^{n-1} \mu_k x^k + x^n + O(|x|^{n+1}). \tag{10.2}$$

The vanishing of $\mu_0$ implies that the origin is stationary and the vanishing of $\mu_1$ implies that this stationary point is non-hyperbolic. This latter equation is called an unfolding of the degenerate singularity $x^n$, and one could hope that by understanding the behaviour of systems in a neighbourhood of $\mu_k = 0$, $k = 0, 1, \ldots, n-1$, we would understand all possible local behaviour for systems near the degenerate system. Thus we are asking a different sort of question than in the previous chapters: there we were interested in the changes that would occur in typical examples, whilst here we want a complete description of the behaviour of systems in a neighbourhood of the degenerate system. A second difference lies in the fact that in previous chapters we have been working with a single real parameter, whereas here we appear to need $n$ independent parameters. The most simple unfolding is for $n = 2$, in which case we have

$$\dot{x} = \mu_0 + \mu_1 x + x^2 + O(|x|^3). \tag{10.3}$$

Thus there are two stationary points provided $\mu_1^2 > 4\mu_0$, and if $\mu_1^2 = 4\mu_0$ there is a saddlenode bifurcation which creates or destroys the two stationary points. It should be obvious that there is really only one exciting thing that can happen, since although we have two parameters, the change of coordinates $y = x - \frac{\mu_1}{2}$ (i.e. completing the square) gives

$$\dot{y} = \nu + y^2 \tag{10.4}$$

where $\nu = \mu_0 - \frac{\mu_1^2}{4}$ and so we effectively only had one parameter: the sign of $\nu$ determines the number of stationary points. Thus $\nu + y^2$ is a minimal unfolding of $y^2$, since only one parameter is really needed to capture all possible behaviours in a neighbourhood of the degenerate system.

New phenomena begin to be observed when unfolding the cubic singularity

$$\dot{x} = -x^3. \tag{10.5}$$

Note that the sign of the coefficient of $x^3$ here is arbitrary since we can change the sign by reversing the direction of time. The general unfolding of this singularity is

$$\dot{x} = \mu_0 + \mu_1 x + \mu_2 x^2 - x^3 \tag{10.6}$$

but it is always possible to get rid of either the $x$ or $x^2$ term by completing a square, so a minimal unfolding is

$$\dot{y} = \nu_0 + \nu_1 y - y^3. \tag{10.7}$$

(10.1) EXERCISE

*Find the coordinate change that brings the $\dot{x}$ equation into the form given by the minimal unfolding.*

Stationary points of the minimal unfolding are solutions of the algebraic equation

$$\nu_0 + \nu_1 y - y^3 = 0 \tag{10.8}$$

and so there are either one, two or three real solutions. A cubic can only have two real solutions if one of these solutions is a double root, i.e. if

$$\nu_0 + \nu_1 y - y^3 = 0 \text{ and } \nu_1 - 3y^2 = 0.$$

Thus we find two real solutions if $\nu_1 > 0$ and, substituting $y^2 = \nu_1/3$ into the cubic equation,

$$\nu_1^3 = \frac{27}{4}\nu_0^2. \tag{10.9}$$

Hence there are three solutions in the cusp region of Figure 10.1. Outside this region there is only one real solution.

(10.2) EXERCISE

*Sketch the bifurcation diagram for (10.7) for $\nu_1$ fixed and positive with $\nu_0$ varying.*

The reader who has come across ideas in catastrophe theory will recognise the two examples here (unfolding the singularities $x^2$ and $x^3$) as the most simple catastrophes: the fold and the cusp.

## 10.2 Imperfection theory

In some experiments there are assumptions about symmetry properties of the setup. If these are broken the bifurcations observed can be rad-

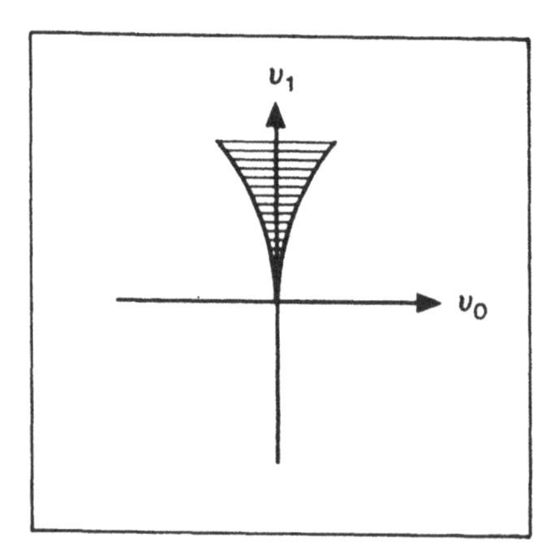


Fig. 10.1

ically different. For example, consider a perfect rod being compressed vertically (see Fig. 10.2a). At some critical value of the compression force the rod will buckle, but there is no way in which we can tell whether it will buckle to the left or to the right. This is an example of a pitchfork bifurcation, with symmetry between branches. Suppose that the rod has some small imperfection (as in Fig. 10.2b), then we would expect the buckling to occur more smoothly, and always towards the same side. A simple model of the perfect bifurcation might be

$$\dot{x} = \mu x - x^3, \tag{10.10}$$

which is invariant under the transformations $x \to -x$, whilst for the second (imperfect) situation we might simply add a bias, $\nu$, so

$$\dot{x} = \nu + \mu x - x^3. \tag{10.11}$$

This is, of course, precisely the minimal unfolding of the singularity $-x^3$ considered in the previous section. However, here we want to consider $\nu$ as a small, fixed parameter, $\nu > 0$ for definiteness, and $\mu$ small. Stationary points (in the $(\mu, x)$ plane) lie on the curve

$$\mu = x^2 - \frac{\nu}{x} \tag{10.12}$$

as shown in Figure 10.3a, which also shows the stability properties of solutions (which we leave as an exercise). Notice that the bifurcation fork of the pitchfork has been broken into two distinct curves, and that if the parameter ($\mu$) is increased slowly, we are unlikely to notice the lower branch of solutions, seeing only a gradual change of the displacement as the compression force changes. Indeed, without the above analysis, relying only on experimental data, we would not even realize that there is a bifurcation.

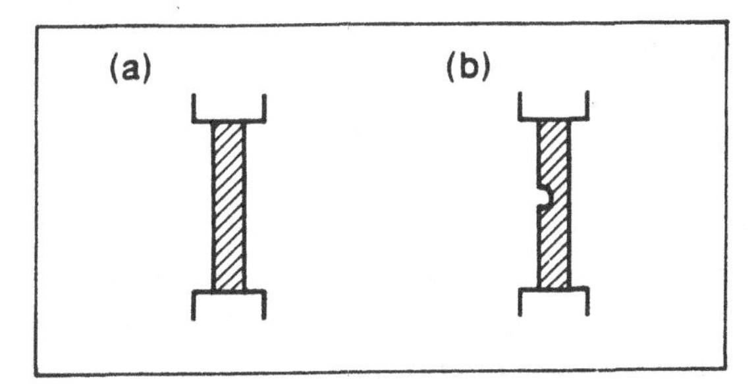


Fig. 10.2 (a) A perfect rod; (b) an imperfect rod.

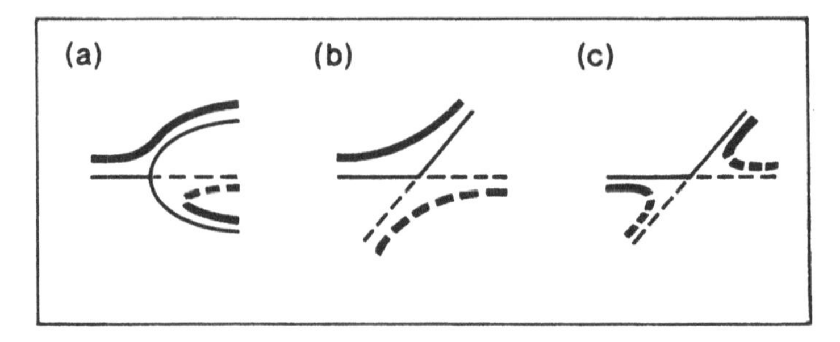


Fig. 10.3 (a) Pitchfork bifurcation with symmetry-breaking; (b) transcritical bifurcation with symmetry-breaking ($\nu > 0$); (c) transcritical bifurcation with symmetry-breaking ($\nu < 0$).

A similar bit of analysis can be done for the transcritical bifurcation, $\dot{x} = \mu x - x^2$ shown in Figure 10.3b. If a small, constant perturbation $\nu$ is applied which breaks the constraint that $x = 0$ is a solution we obtain the system

$$\dot{x} = \nu + \mu x - x^2, \tag{10.13}$$

which has stationary points on curves in the $(\mu, x)$ plane given by

$$\mu = x - \frac{\nu}{x}. \tag{10.14}$$

Once again, determining this locus is a simple exercise (see Fig. 10.3b,c) and the simple crossing of branches associated with the transcritical bifurcation is broken into two disjoint curves. If $\nu > 0$ there are no bifurcations, whilst if $\nu < 0$ there is a pair of saddlenode bifurcations. In a sense, both of these examples show that the saddlenode bifurcation (sometimes called a fold bifurcation by analogy with catastrophe theory) is the standard bifurcation: under perturbation both the transcritical and pitchfork bifurcations break into systems which have only saddlenode bifurcations.

## 10.3 Isolas

In Chapter 8 the quantity $F_{x\mu}^2 - F_{xx}F_{\mu\mu}$ arises in the condition for transcritical bifurcations; this expression must be positive in order for the (lowest order) approximation

$$\dot{x} = \tfrac{1}{2}(F_{xx}x^2 + 2F_{x\mu}\mu x + F_{\mu\mu}\mu^2) \tag{10.15}$$

to have stationary points (real solutions to the right hand side of this equation equalling zero). In the previous section we saw the effect of small constant perturbations to this situation, but one can equally well ask what happens if $F_{x\mu}^2 - F_{xx}F_{\mu\mu}$ is negative. If there is no external perturbation (or imperfection) then the only local stationary point exists for $\mu = 0$ (at the origin) and disappears if $\mu \neq 0$. However, if we now add a small perturbation, $\nu$, as in the previous section, we are led to consider the equation

$$\dot{x} = \nu + \tfrac{1}{2}(F_{xx}x^2 + 2F_{x\mu}\mu x + F_{\mu\mu}\mu^2), \tag{10.16}$$

which has stationary points in the $(\mu, x)$ plane if

$$\tfrac{1}{2}(F_{xx}x^2 + 2F_{x\mu}\mu x + F_{\mu\mu}\mu^2) = -\nu. \tag{10.17}$$

This equation defines a curve of ellipses in the $(\mu, x)$ plane provided $\nu$ takes the right sign (see below), corresponding to a pair of stationary points which are created and then destroyed in saddlenode bifurcations. These are often hard to find in experiments since they may exist only in small regions of parameter space. We can find the locus of the two curves of saddlenode bifurcations by observing, as in Section 10.1, that these occur if the equation for the stationary points has a double root, i.e. if both (10.17) is satisfied and

$$F_{xx}x + F_{x\mu}\mu = 0. \tag{10.18}$$

Putting these two equations together we find that the locus of the saddlenode bifurcations is

$$(F_{\mu x}^2 - F_{xx}F_{\mu\mu})\mu^2 = F_{xx}\nu \tag{10.19}$$

and since, by assumption throughout this section, $F_{\mu x}^2 - F_{xx}F_{\mu\mu} < 0$ this gives a parabola in $\nu > 0$ if $F_{xx} < 0$ and $\nu < 0$ if $F_{xx} > 0$.

## 10.4 Periodic orbits in Lotka-Volterra models

In Chapter 5 two-dimensional population models were described and it was shown that isolated periodic orbits could not exist in these models; if the system admits periodic orbits then these orbits are non-isolated (or Hamiltonian). For many years it was believed that in three species models it should be possible to find isolated periodic orbits, but there are few general techniques for proving the existence of periodic orbits in dimension greater than two and a proof of the existence of such orbits was not forthcoming until 1979. Coste, Peyraud and Coullet, of the University of Nice in France, realized that the Hopf Bifurcation Theorem does prove the existence of isolated periodic orbits, so the problem reduces to showing that three species Lotka-Volterra models can undergo standard Hopf bifurcations. The proof is really an extended exercise in linear algebra, but it illustrates the application of the Hopf Bifurcation Theorem and the Centre Manifold Theorem, so we will go through the argument in some detail. Clearly we need a stationary point in the positive quadrant which will undergo a Hopf bifurcation. With some loss of generality (which is made up for by the simplification in the manipulations involved) the Lotka-Volterra models can be written as

$$\dot{X}_i = X_i \sum_{j=1}^{3} A_{ij}(1 - X_j) \tag{10.20}$$

for $i = 1,2,3$, where $A$ is a $3 * 3$ matrix and the coordinates $X_i$ denote the populations of the three species. This has been written in such a way as to make the fact that there is a stationary point at $(X_1, X_2, X_3) = (1,1,1)$ obvious. The birth rate of the species $X_i$ is therefore $\sum_{k=1}^{3} A_{ik}$ and so the only constraint on the coefficients of the matrix $A$ is that this sum should be positive for each $i = 1,2,3$, which implies that the origin is an unstable node. Since we shall be concerned with the stability and bifurcation of the stationary point at $(1,1,1)$ it is sensible to transform the coordinate system so that this point is at the origin. Set $Y_i = X_i - 1$, $i = 1,2,3$, then the equations become

$$Y_i = -(1 + Y_i) \sum_{j=1}^{3} A_{ij} Y_j \tag{10.21}$$

and so the Jacobian matrix at the stationary point $(1,1,1)$ is simply $-A$. To prove the existence of isolated periodic orbits in the model we now have to choose $A$ in such a way that this stationary point has a Hopf bifurcation. We shall choose

$$A = \begin{pmatrix} 3-\mu & 3 & 2 \\ 2 & 1-\mu & 0 \\ 0 & 2 & 1-\mu \end{pmatrix}, \tag{10.22}$$

which has eigenvalues $\lambda_{\pm} = -\mu \pm i$ and $s = 5 - \mu$. Thus $-A$, which is the Jacobian matrix for $(1,1,1)$ has eigenvalues $-\lambda_{\pm}$ and $-s$. The stationary point has a pair of purely imaginary eigenvalues if $\mu = 0$ and

$$\frac{dRe(\lambda_{\pm})}{d\mu}(0) = 1 \tag{10.23}$$

so the transversality condition for the Hopf bifurcation at $\mu = 0$ is certainly satisfied. Note that $(1,1,1)$ is stable (all eigenvalues have negative real part) for $\mu < 0$ and is unstable if $\mu > 0$ (at least for sufficiently small values of $\mu$). It only remains for us to check the genericity condition ($\text{Re}A(0) \neq 0$, where $\text{Re}A(0)$ is the nasty expression involving second and third derivatives of the equation on the centre manifold in the statement of the Hopf Bifurcation Theorem). To do this we need to do three things: first, make a linear transformation so that the Jacobian matrix comes into the Jordan normal form at $\mu = 0$, then find an approximation to the centre manifold, and finally find the defining equations on the centre manifold for $\mu = 0$ to calculate $\text{Re}A(0)$. Since $\text{Re}A(0)$ depends only upon the equations with $\mu = 0$ we can set $\mu = 0$ from the outset and consider

$$Y_i = (1 + Y_i) \sum_{j=1}^{3} B_{ij} Y_j \tag{10.24}$$

where $B = -A$ at $\mu = 0$, i.e.

$$B = \begin{pmatrix} -3 & -3 & -2 \\ -2 & -1 & 0 \\ 0 & -2 & -1 \end{pmatrix}. \tag{10.25}$$

The eigenvalues of $B$ are $\pm i$ and $-5$ so we are looking for a change of coordinate to bring the linear part of the differential equation into the normal form

$$N = \begin{pmatrix} 0 & -1 & 0 \\ 1 & 0 & 0 \\ 0 & 0 & -5 \end{pmatrix}. \tag{10.26}$$

This is done by using a matrix built from the eigenvectors of $B$, and it is a simple exercise to show that the (complex) eigenvector associated with the eigenvalue $i$, $e_i$, and the (real) eigenvector associated with the eigenvalue $-5$, $e_{-5}$, can be chosen such that

$$e_i = \begin{pmatrix} 1 \\ -(1-i) \\ -2i \end{pmatrix} \quad \text{and} \quad e_{-5} = \begin{pmatrix} 4 \\ 2 \\ 1 \end{pmatrix}.$$

Now, write $e_i = e_R + ie_I$ where $e_R$ and $e_I$ are real vectors; then the matrix whose columns are $e_I$, $e_R$ and $e_{-5}$ should give the required coordinate change. So, let

$$D = \begin{pmatrix} 0 & 1 & 4 \\ 1 & -1 & 2 \\ -2 & 0 & 1 \end{pmatrix} \quad \text{with} \quad 13D^{-1} = \begin{pmatrix} 1 & 1 & -6 \\ 5 & -8 & -4 \\ 2 & 2 & 1 \end{pmatrix},$$

then we leave it as an exercise for the reader to check that $BD = DN$ or $D^{-1}BD = N$. So, if we define new variable $y_i$ with $Y = Dy$ the differential equation becomes

$$\dot{y}_i = \sum_{k=1}^{3} D_{ik}^{-1}\dot{Y}_k = \sum_{k=1}^{3} D_{ik}^{-1}\left(\sum_{j=1}^{3} B_{kj}Y_j + \sum_{j=1}^{3} B_{kj}Y_jY_k\right). \tag{10.27}$$

Now replacing $Y$ by $Dy$ and using $D^{-1}BD = N$ this gives

$$\dot{y}_i = \sum_{k=1}^{3} N_{ik}y_k + \sum_{k=1}^{3} D_{ik}^{-1}\left(\sum_{j=1}^{3} D_{kj}y_j\right)\left(\sum_{j=1}^{3}\sum_{r=1}^{3} B_{kj}D_{jr}y_r\right)$$

or

$$\dot{y}_i = \sum_{k=1}^{3} N_{ik}y_k + \sum_{k=1}^{3} D_{ik}^{-1}\left(\sum_{j=1}^{3} D_{kj}y_j\right)\left(\sum_{j=1}^{3}\sum_{r=1}^{3} D_{kj}N_{jr}y_r\right). \tag{10.28}$$

From the definition of $N$ we have that $Ny = (-y_2, y_1, -5y_3)^T$ and so the linear part of the equation is in the required form:

$$\dot{y}_1 = -y_2 + \ldots, \quad \dot{y}_2 = y_1 + \ldots, \quad \dot{y}_3 = -5y_3 + \ldots .$$

The only tricky bit is working out the nonlinear terms. These are a product of two sums:

$$\sum_{k=1}^{3} D_{ik}^{-1}P_kQ_k \tag{10.29}$$

where

$$P_k = \sum_{j=1}^{3} D_{kj}y_j \quad \text{and} \quad Q_k = \sum_{j=1}^{3}\sum_{r=1}^{3} D_{kj}N_{jr}y_r. \tag{10.30}$$

So

$$P_k = D_{k1}y_1 + D_{k2}y_2 + D_{k3}y_3 \tag{10.31}$$

and

$$Q_k = -D_{k1}y_2 + D_{k2}y_1 - 5D_{k3}y_3. \tag{10.32}$$

To find the nonlinear terms we must calculate the products $P_kQ_k$, $k = 1, 2, 3$. Using the definition of the matrix $D$ above these give

$$\begin{aligned} P_1Q_1 &= (y_2 + 4y_3)(y_1 - 20y_3) \\ &= y_1y_2 + 4y_1y_3 - 20y_2y_3 - 80y_3^2, \end{aligned} \tag{10.33}$$

$$\begin{aligned} P_2Q_2 &= (y_1 - y_2 + 2y_3)(-y_2 - y_1 - 10y_3) \\ &= -y_1^2 + y_2^2 - 20y_3^2 + 8y_2y_3 - 12y_1y_3, \end{aligned} \tag{10.34}$$

and

$$\begin{aligned} P_3Q_3 &= (-2y_1 + y_3)(2y_2 - 5y_3) \\ &= -5y_3^2 + 2y_2y_3 + 10y_1y_3 - 4y_1y_2. \end{aligned} \tag{10.35}$$

With this information we can find the equations for $\dot{y}_i$, $i = 1, 2, 3$ from

$$\dot{y}_i = \sum_{k=1}^{3}(N_{ik}y_k + D_{ik}^{-1}P_kQ_k). \tag{10.36}$$

After some unexciting additions and subtractions, which we leave as an exercise, we obtain

$$\dot{y}_1 = -y_2 + \frac{1}{13}(-y_1^2 + y_2^2 - 130y_3^2 - 23y_1y_2 + 52y_1y_3) \tag{10.37a}$$

$$\dot{y}_2 = y_1 + \frac{1}{13}(8y_1^2 - 8y_2^2 - 220y_3^2 + 21y_1y_2 + 76y_1y_3 - 172y_2y_3) \tag{10.37b}$$

$$\dot{y}_3 = -5y_3 + \frac{1}{13}(-2y_1^2 + 2y_2^2 - 205y_3^2 - 2y_1y_2 - 6y_1y_3 - 22y_2y_3). \tag{10.37c}$$

Here endeth the first step of the procedure. These three equations give the differential equation in the new coordinates for which the linear part of the equation is in normal form. The next step is to find the centre manifold, which is of the form $y_3 = h(y_1, y_2)$ where $h$ is tangential to $y_3 = 0$ at $(y_1, y_2) = (0, 0)$. Hence we try an approximation to the centre manifold of the form

$$y_3 = h(y_1, y_2) = ay_1^2 + by_1y_2 + cy_2^2 + h.o.t. \tag{10.38}$$

and use the $\dot{y}_3$ equation to give (if $y_3 = h(y_1, y_2)$)

$$\dot{y}_3 = -5(ay_1^2 + by_1y_2 + cy_2^2) + \frac{1}{13}(-2y_1^2 + 2y_2^2 - 2y_1y_2) + h.o.t. \tag{10.39}$$

and also $\dot{y}_3 = \dot{y}_1\frac{\partial h}{\partial y_1} + \dot{y}_2\frac{\partial h}{\partial y_2}$ so

$$\dot{y}_3 = -y_2(2ay_1 + by_2) + y_1(by_1 + 2cy_2) + h.o.t. \tag{10.40}$$

Identifying the coefficients of $y_1^2$, $y_1y_2$ and $y_2^2$ in (10.39) and (10.40) we find that

$$-5a - \frac{2}{13} = b \tag{10.41}$$

$$-5b - \frac{2}{13} = -2a + 2c \tag{10.42}$$

and

$$-5c + \frac{2}{13} = -b \tag{10.43}$$

from which we get, with no real effort,

$$c = -a = \frac{8}{(13)(29)} \quad \text{and} \quad b = -\frac{18}{(13)(29)}. \tag{10.44}$$

Hence the centre manifold, to second order terms, is

$$h(y_1, y_2) = \frac{2}{(13)(29)}(-4y_1^2 - 9y_1y_2 + 4y_2^2) + h.o.t. \tag{10.45}$$

Now all we have to do is substitute this expression for $y_3$ on the centre manifold into the equations for $\dot{y}_1$ and $\dot{y}_2$ to obtain the equation on the centre manifold correct to third order. This gives

$$\begin{aligned} \dot{y}_1 = & -y_2 + \frac{1}{13}(-y_1^2 - 23y_1y_2 + y_2^2) \\ & + \frac{8}{(13)(29)}(-4y_1^3 - 9y_1^2y_2 + 4y_1y_2^2) \\ & + \text{ higher order terms} \end{aligned} \tag{10.46a}$$

$$\begin{aligned} \dot{y}_2 = y_1 & + \frac{1}{13}(8y_1^2 + 21y_1y_2 - 8y_2^2) \\ & + \frac{2}{(13)^2(29)}(-304y_1^3 + 4y_1^2y_2 + 860y_1y_2^2 - 688y_2^3) \\ & + \text{ higher order terms.} \end{aligned} \tag{10.46b}$$

Also, from Chapter 8 we know that if $\dot{y}_1 = f(y_1, y_2)$ and $\dot{y}_2 = g(y_1, y_2)$ then

$$\text{Re}A(0) = \frac{1}{16}(f_{xxx} + f_{xyy} + g_{xxy} + g_{yyy})$$

$$+\frac{1}{16\omega}[f_{xy}(f_{xx} + f_{yy}) - g_{xy}(g_{xx} + g_{yy}) - f_{xx}g_{xx} + f_{yy}g_{yy}]$$

where $i\omega$ is the purely imaginary eigenvalue of the linear problem, so $\omega = 1$ here. Evaluating this quantity, noting that $f_{xxx} = -f_{xyy}$ and the whole of the term involving only second derivatives vanishes, we obtain

$$\mathrm{Re}A(0) = \frac{1}{16}\cdot\frac{2}{(13)^2(29)}(-688+4) = -\frac{1368}{(13)^2(29)}. \qquad (10.47)$$

Hence $\mathrm{Re}A(0) < 0$ and the eigenvalue (as a function of $\mu$) $\mu \pm i$, giving the normal form for the $r$ polar coordinate

$$\dot{r} = \mu r - \frac{1368}{(13)^2(29)}r^3 + O(r^4). \qquad (10.48)$$

Hence the bifurcation is supercritical and there is a stable (isolated) periodic orbit if $\mu > 0$ for each sufficiently small $|\mu|$.

(10.3) EXERCISE

*Coste, Peyraud and Coullet consider more general forms of the matrix $A$, taking*

$$A = \begin{pmatrix} 2t-\mu & 2t & t+\frac{1}{2} \\ 2 & 1-\mu & 0 \\ 0 & 2 & 1-\mu \end{pmatrix}$$

*where $t$ is a real parameter. This matrix has eigenvalues $-\mu \pm i$ and $2(t+1)-\mu$. The worked example above is for the special case $t = 3/2$. The masochistic reader might want to classify the Hopf bifurcations from the stationary point $(1,1,1)$ as a function of $t$. When is it subcritical and when is it supercritical? For which values of $t$ is* $\mathrm{Re}A(0) = 0$*?*

## 10.5 Subharmonic resonance revisited

In Section 7.11 we used perturbation theory to analyse the behaviour of a forced oscillator and found that there are parameter regimes where the period of the solution is smaller than the natural period of the unforced oscillator (subharmonic resonance). The non-resonance condition at $O(\epsilon)$, in terms of a complex function $A(T)$, is

$$2A_T = (1 - \tfrac{1}{2}a^2 + i\delta)A - |A|^2A + \tfrac{1}{2}aA^{*2} \qquad (10.49)$$

where we shall take $a$ and $\delta$ to be positive real parameters. Setting $A = re^{i\theta}$ gives two real equations

$$r_T = \tfrac{1}{4}r(2 - a^2 - ar\cos 3\theta - 2r^2) \qquad (10.50a)$$

$$\theta_T = \tfrac{1}{2}(\delta + \tfrac{1}{2}ar\sin 3\theta). \qquad (10.50b)$$

These equations were analysed graphically, noting that since the dependence on $\theta$ is through $\cos 3\theta$ and $\sin 3\theta$ we need only consider values of $\theta$ between 0 and $\frac{2\pi}{3}$, since the dynamics in the other two-thirds of the plane can be obtained from this by rotating through an angle $\frac{2\pi}{3}$ and $\frac{4\pi}{3}$. When describing the dynamics of this system as a function of the two (positive) parameters we made vague noises about what can happen depending on whether $\delta$ is large or small. Using some bifurcation theory we ought to be able to make this more explicit.

Looking back at Section 7.12 it should be clear that there are two problems that we need to address: where did the periodic orbit which exists in $a^2 < 2$ come from (the answer must lie in the Hopf Bifurcation Theorem) and when do the non-trivial stationary points exist?

In Cartesian coordinates, $A = x + iy$, the non-resonance condition (10.49) becomes

$$x_T = \tfrac{1}{2}\left((1 - \tfrac{1}{2}a^2)x - \delta y - (x^2 + y^2)x + \tfrac{1}{2}a(x^2 - y^2)\right) \tag{10.51a}$$

$$y_T = \tfrac{1}{2}\left(\delta x + (1 - \tfrac{1}{2}a^2)y - (x^2 + y^2)y - axy\right) \tag{10.51b}$$

and so the origin is always a stationary point of the system and the linearized flow about the origin has eigenvalues

$$\tfrac{1}{2}(1 - \tfrac{1}{2}a^2 \pm i\delta). \tag{10.52}$$

Thus if $a^2 = 2$ the real part of the eigenvalues vanishes and the origin is non-hyperbolic with linear eigenvalues $\pm i\frac{\delta}{2}$. This suggests that there is a Hopf bifurcation at $a^2 = 2$. To verify this, and to discover whether the bifurcation is subcritical or supercritical, we need to check the two conditions for the Hopf bifurcation. First note that the origin is stable if $a^2 > 2$ and unstable if $a^2 < 2$. Now, the first condition for the Hopf Bifurcation Theorem is that

$$\frac{d\mathrm{Re}\lambda(a)}{da} \neq 0 \tag{10.53}$$

at the bifurcation value, $a = \sqrt{2}$. Since $\mathrm{Re}\lambda = \frac{1}{2}(1 - \frac{a^2}{2})$, the derivative is $-\frac{a}{2}$, which is clearly non-zero. The second condition is the nasty one. When $a = \sqrt{2}$ the equation is

$$x_T = \tfrac{1}{2}\left(-\delta y - (x^2 + y^2)x + \tfrac{1}{\sqrt{2}}(x^2 - y^2)\right) = f(x, y) \tag{10.54a}$$

$$y_T = \tfrac{1}{2}\left(\delta x - (x^2 + y^2)y - \sqrt{2}xy\right) = g(x, y) \tag{10.54b}$$

and so the linear part is already in the correct form to apply the Hopf Bifurcation Theorem. The second condition is that

$$\frac{1}{16}(f_{xxx}+g_{xxy}+f_{xyy}+g_{yyy})$$
$$+\frac{1}{16\omega}(-f_{xx}g_{xx}+f_{yy}g_{yy}+f_{xy}(f_{xx}+f_{yy})-g_{xy}(g_{xx}+g_{yy}))$$

is non-zero at the bifurcation point $(x,y)=(0,0)$. Here, $\omega$ is the coefficient multiplying $v$ in the linear part of $g(x,y)$, i.e. $\omega=\frac{\delta}{2}$. Calculating this quantity gives

$$\frac{1}{32}(-6-2-2-6)=-\frac{1}{2}<0. \tag{10.55}$$

Hence the bifurcation is supercritical and the periodic orbits bifurcate into the region of parameter space in which the origin is unstable, i.e. $a^2<2$.

We can now move on to determining the parameter values for which the equation has non-trivial stationary points ($r\neq 0$). Looking back to the equation in polar coordinates, (10.50), we see that $\theta_T=0$ if

$$\sin 3\theta=-\frac{2\delta}{ar} \tag{10.56}$$

and since $\sin 3\theta>-1$ this implies that any stationary point must satisfy

$$r>\frac{2\delta}{a}$$

and that

$$\cos^2 3\theta=1-\frac{4\delta^2}{a^2r^2}. \tag{10.57}$$

Now, $r_T=0$ if

$$ar\cos 3\theta=2-a^2-2r^2 \tag{10.58}$$

so, squaring (10.58) and using (10.57),

$$(2-a^2)^2-4(2-a^2)r^2+4r^4=a^2r^2-4\delta^2$$

or

$$4r^4-(4(2-a^2)+a^2)r^2+(2-a^2)^2+4\delta^2=0. \tag{10.59}$$

This is a quadratic equation for $r^2$ which we can solve to obtain

$$r^2=\frac{8-3a^2\pm\sqrt{a^2(16-7a^2)-64\delta^2}}{8}. \tag{10.60}$$

This equation has two real roots provided

$$a^2(16-7a^2)>64\delta^2 \tag{10.61}$$

(see Fig. 10.4). We now have to check that both of these roots are positive and that $r^2 > \frac{4\delta^2}{a^2}$. In the region for which roots are real, $8 - 3a^2 > 0$ and so at least one of the real roots is positive. The other root has $r^2 > \frac{4\delta^2}{a^2}$ provided

$$8 - 3a^2 - \sqrt{a^2(16 - 7a^2) - 64\delta^2} > \frac{32\delta^2}{a^2}$$

or, setting $b = \frac{\delta}{a}$,

$$8 - 3a^2 - 32b^2 > \sqrt{a^2(16 - 7a^2 - 64b^2)}. \qquad (10.62)$$

The right hand side of this equation is always positive, so provided the left hand side of the expression is also positive, i.e.

$$8 - 3a^2 - 32b^2 > 0, \qquad (10.63)$$

we can square both sides to obtain, after rearranging terms a little, that there are two non-trivial solutions for $r$ if

$$(8 - 3a^2 - 32b^2)^2 - a^2(16 - 7a^2 - 64b^2) > 0. \qquad (10.64)$$

Now, $2(8 - 3a^2 - 32b^2) > 16 - 7a^2 - 64b^2$, so $8 - 3a^2 - 32b^2$ is positive in the region of interest, $16 - 7a^2 - 64b^2 > 0$. This means that the only extra condition is the squared equation above. Simplifying this condition gives

$$4 - 4a^2 + a^4 - 32b^2 + 16a^2b^2 + 64b^4 > 0,$$

which we recognise immediately as

$$(a^2 + 8b^2 - 2)^2 > 0. \qquad (10.65)$$

Hence this condition is automatically satisfied and so there are two non-trivial values of $r$ for stationary solutions provided

$$16 - 7a^2 - 64b^2 > 0 \qquad (10.66)$$

(the condition for two real roots to exist).

This condition together with the line $a^2 = 2$ of the Hopf bifurcation divides parameter space into four open regions (Fig. 10.4).

(i) $a^2 < 2$, $a^2(16 - 7a^2) < 64\delta^2$.
In this region there is the stable periodic orbit created in the Hopf bifurcation and a stationary point at the origin (which is an unstable focus). This corresponds to beating in the original forced oscillator.
(ii) $a^2 > 2$, $a^2(16 - 7a^2) < 64\delta^2$.

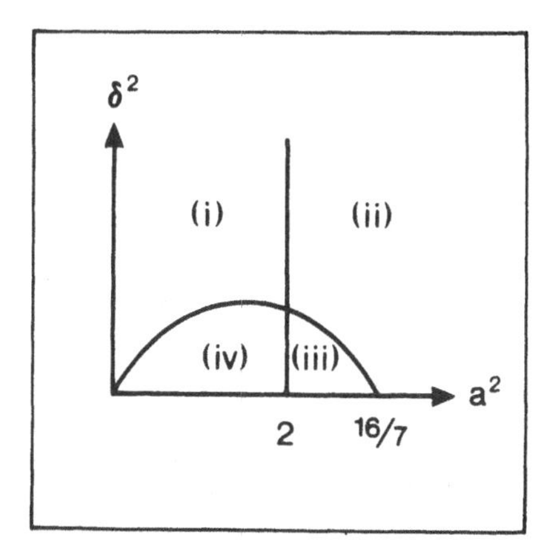


Fig. 10.4 The different regions of parameter space.

The origin is an asymptotically stable focus (there are no other stationary points and no periodic orbits). This corresponds to quenching in the original equations.
(iii) $a^2 > 2$, $a^2(16 - 7a^2) > 64\delta^2$.
The origin is a stable focus and there are six other stationary points, created in three pairs (due to the dependence on $3\theta$) if $a^2(16-7a^2) = 64\delta^2$. Each pair consists of one stable stationary point and a saddle. This corresponds to hard excitation in the original problem.
(iv) $a^2 < 2$, $a^2(16 - 7a^2) > 64\delta^2$.
This is the really interesting region: the origin is an unstable focus and the phase portrait must also contain the three pairs of non-trivial stationary points and the periodic orbit.

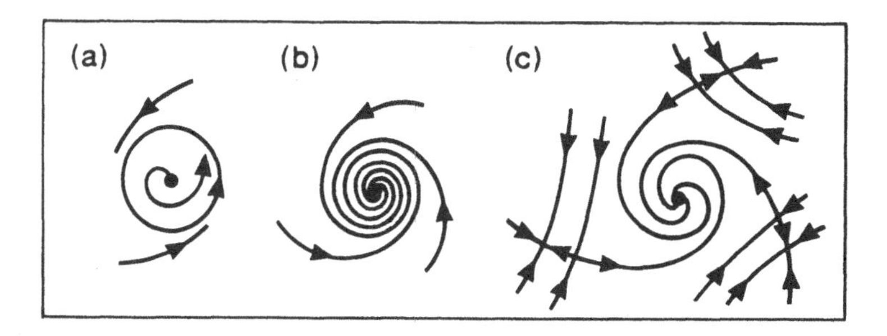


Fig. 10.5 (a) Region (i); (b) region (ii); (c) region (iii).

Fig. 10.6 Region (iv).

Figure 10.5 shows the phase portraits of the system for parameters in regions (i), (ii) and (iii). Figure 10.6 shows the phase portrait for systems in region (iv).

# 11

# *Chaos*

Solutions of simple nonlinear systems can behave in extremely complicated ways. This observation, and the subsequent mathematical treatment of 'chaos', is one of the most exciting recent developments of mathematics. Loosely speaking, a chaotic solution is aperiodic but bounded and nearby solutions separate rapidly in time. This latter property, called sensitive dependence upon initial conditions, can be thought of as a loss of memory of the system of the past history of any solution. It implies that long term predictions of the system are almost impossible despite the deterministic nature of the equations. Historically the possibility of aperiodic solutions with complicated geometric structure was known to both Poincaré and Birkhoff in the late nineteenth and early twentieth centuries, but it was not until computer simulation of differential equations became feasible that the subject really took off. This is probably because it is extremely difficult to get any intuitive feel for how a system behaves simply by looking at the equations. The computer allows one to see the type of result one might try to prove and motivates the development of conjectures and theorems.

Two important papers appeared in the 1960s, one on the applied side of the subject and one on the pure. In 1963, Lorenz published a paper called *Deterministic Non-periodic Flows* in which he described the numerical results he had obtained by integrating a simple third order system of ordinary differential equations on a computer (this was not the first such paper, but it has become the most influential). These equations were derived from a simple model of the weather and Lorenz was trying to show that the solutions of differential equations could be (in practice) unpredictable despite being deterministic. The equations he considered are

$$\dot{x} = 10(y - x) \tag{11.1a}$$

$$\dot{y} = -y + 28x - xz \tag{11.1b}$$

$$\dot{z} = -\tfrac{8}{3}z + xy. \tag{11.1c}$$

Numerical solutions of these equations are shown in Figure 11.1, where it is clear that although the solutions have some geometric structure, they do not appear to settle down to any simple periodic orbit. We shall return to a more detailed discussion of systems like the Lorenz equations in the next chapter.

The second paper, *Differentiable Dynamical Systems*, was published in 1967 by Smale. It is a review article about a particular class of maps for which it is possible to prove rigorous mathematical results about the existence of chaotic solutions. These two papers show up some of the problems encountered when talking about chaos. For the applied mathematician or physicist a system is not chaotic unless it has what is called a strange attractor, so the complicated behaviour must be visible. A pure mathematician may be more interested in the topological structure of solutions and so may call a system chaotic even if the chaotic set is not attracting; it is enough to know that it is there. Hence the class of systems studied by Smale and co-workers is not especially interesting to the physicist since the chaotic behaviour is not attracting, whilst the examples from physics such as the Lorenz equations are so complicated that it is very difficult to prove any rigorous results about the system. This difference in philosophy is reflected in the choice of definitions used to characterize chaos. In order to prove results we shall choose a topological approach, since it is remarkably difficult to prove theorems about strange attractors in all but the most artificial of ex-

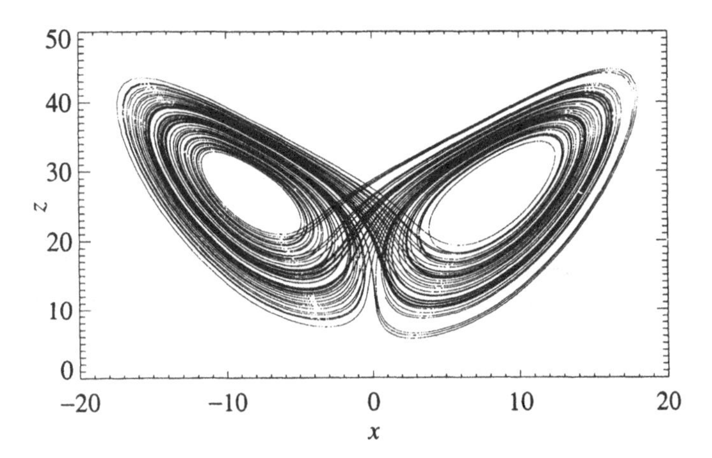


Fig. 11.1 The Lorenz attractor.

amples. None the less, we shall remain aware of the dichotomy between the study of physical examples and the rigours of theory and wherever possible give results about attractors.

This chapter begins with some basic definitions and examples, then goes on to develop the theory of some particularly simple maps which have complicated (or chaotic) behaviour.

## 11.1 Characterizing chaos

The first step towards developing any theory is to try to establish the most important features of the problem. As remarked above this is not completely clear for chaotic systems and depends crucially on whether one takes a topological point of view or chooses to ignore anything which is not an attractor (see, for example, Exercise 1 at the end of this chapter). An example may help to crystallize ideas.

*Example 11.1*

Consider the simple map $T : [0,1) \to [0,1)$ defined by

$$T(x) = 2x \text{ (mod 1)}.$$

Seen as a map from the interval to itself this map has a discontinuity at $x = \frac{1}{2}$ which can cause problems (see Fig. 11.2). However, this example

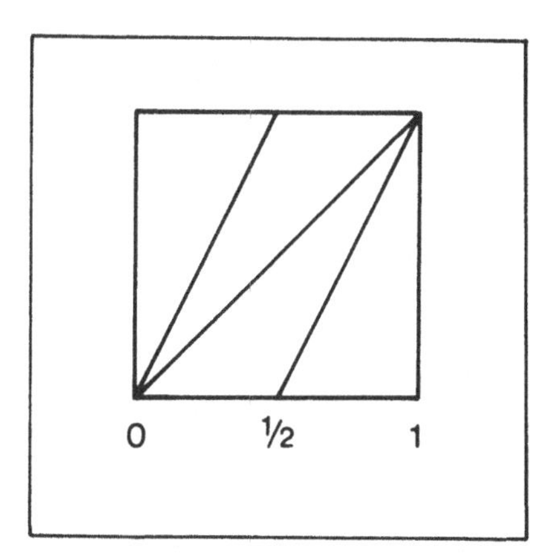


Fig. 11.2 The map $T(x) = 2x(\text{mod } 1)$.

can be thought of as being derived from the complex map

$$z_{n+1} = z_n^2$$

by noting that if $|z_n| = 1$ then $|z_{n+1}| = 1$, so setting $z_n = e^{2\pi i x_n}$ we find that $x_{n+1} = 2x_n$ and since $x$ is an angle it can be thought of as taking values between 0 and 1. Hence although the map is a discontinuous map of the interval it represents a continuous map of the circle. Any point $x \in [0,1]$ can be written as a binary expansion in the form

$$x = \sum_{i=1}^{\infty} \frac{a_i}{2^i}$$

where $a_i \in \{0,1\}$, and so

$$2x = a_1 + \sum_{i=1}^{\infty} \frac{a_{i+1}}{2^i}.$$

Hence, if we identify 1 with 0,

$$T(x) = \sum_{i=1}^{\infty} \frac{a_{i+1}}{2^i}.$$

The effect of $T$ is now clear: if $x$ has a binary expansion $[a_1, a_2, a_3, ...]$ then $T(x)$ has binary expansion $[a_2, a_3, ...]$, $T^2(x)$ has binary expansion $[a_3, ...]$ and so on. It is now possible to deduce that

i) $T$ has $2^n$ periodic points of period $n$ (but not necessarily least period $n$), since there are $2^n$ different sequences of 0s and 1s of period $n$;
ii) if $x \neq y$ then $|T^n(x) - T^n(y)|$ grows initially like $2^n|x-y|$, since the slope of the map is 2 everywhere;
iii) given any neighbourhood $J$ of $x$, there exists a subinterval $K$ of $J$ and $n > 0$ such that $T^n(J) = (0,1)$ and $T^n|J$ is a homeomorphism, this is proved below;
iv) periodic points are dense in [0,1], this follows from (iii): given $x$ and a neighbourhood $J$, define $K = [a,b]$ and $n$ as in (iii), then $T^n(a_+) = 0 \leq a$ and $T^n(b_-) = 1 \geq b$, now since $T^n$ is continuous on $K$ this implies that $T^n$ has a fixed point in $K$ (by the Intermediate Value Theorem) and hence $T$ has a periodic point in $K$. Alternatively, simply note that the periodic points are points with periodic binary expansions. Such points are dense in $[0,1]$;
v) $T$ has a dense orbit; this is by construction: let $A_{mk}$ be a list of the possible sequences of 0s and 1s of length $m$, so $1 \leq k \leq 2^m$,

and let $x$ be the point with binary expansion

$$[A_{11}, A_{12}, A_{21}, A_{22}, A_{23}, A_{24}, A_{31}, A_{32}, ...]$$

then $T^n(x)$ passes arbitrarily close to every point in (0,1).

The only property that we have not proved is (iii). The proof uses techniques which will be useful in a more general context, so we will dignify the statement by calling it a lemma. Note that there is an ambiguity in the binary expansion of some points, for example $[1,0,0,0,\ldots] = [0,1,1,1,\ldots]$. This reflects the fact that $T$ is really defined on the circle on which the points 0 and 1 are identified. We will not worry about this here.

(11.1) LEMMA

*Consider the map* $T : [0,1) \to [0,1)$ *defined by* $T(x) = 2x \ (mod\ 1)$, *with the convention that* $T(\frac{1}{2}) = 0$. *Then for all* $x \in (0,1)$ *and all* $\epsilon > 0$ *there exists an open neighbourhood,* $J$, *of* $x$ *of length less than* $\epsilon$ *such that* $T^n(J) = (0,1)$ *and* $T^n|J$ *is a homeomorphism.*

*Proof*: Suppose that $x$ has binary expansion $[a_1, a_2, \ldots]$ and that $x$ is not a preimage of $\frac{1}{2}$ (so the binary expansion does not end with an infinite string of 0s or 1s). All points close to $x$ have binary expansions which agree with the binary expansion of $x$ up to some stage, $N$. By choosing $N$ sufficiently large we can arrange for all points with binary expansions $[a_1, \ldots, a_N, b_1, b_2, \ldots]$ to lie inside an $\epsilon$ neighbourhood of $x$. Since $x$ is not a preimage of $\frac{1}{2}$ we can choose $M > N$ such that $a_{M+1} = 0$. Now let $y$ be the point with binary expansion

$$[a_1, \ldots, a_M, 0, 0, 0, 0, 0, \ldots].$$

It should be obvious that $y < x$ since

$$x - y = \sum_{M+1}^{\infty} \frac{a_i}{2^i} > 0$$

and, by assumption, $a_i \neq 0$ for some $i > M$.

Similarly, let $z$ be the point with binary expansion

$$[a_1, \ldots, a_M, 1, 0, 0, 0, 0, \ldots]$$

and note that since $a_{M+1} = 0$, $z > x$. (You should prove this: show in general that if

$$x = [a_1, \ldots, a_k, 0, a_{k+1}, \ldots] \text{ and } z = [a_1, \ldots, a_k, 1, b_{k+1}, \ldots],$$

i.e. if the binary expansion of $x$ and $z$ agree in the first $k$ terms and then $x$ has a 0 whilst $z$ has a 1, then $x < z$ regardless of the subsequent expansion.)

We now claim that $T^m(y,z) = (0,\frac{1}{2})$ and that $T^m|(y,z)$ is a homeomorphism, which proves the lemma since $T(0,\frac{1}{2}) = (0,1)$ and $T$ is a homeomorphism on $(0,\frac{1}{2})$. The claim is proved in two parts.

i) First we show that if the binary expansions of two points agree for the first $k$ terms then $T^k$ is a homeomorphism when restricted to the interval between these points.

ii) The definition of the action of $T$ on the binary expansion of points then allows us to conclude that $T^M(y) = 0$ and $T^M(z) = \frac{1}{2}$. This, together with the fact that $T^M$ is a homeomorphism on $(y,z)$, completes the proof.

The proof of (i) is the most difficult; (ii) is obvious. Suppose that the binary expansions of $y$ and $z$ agree for the first $k$ terms. This implies that the binary expansions of all points between them also agree in the first $k$ terms. (If this is not obvious to you, prove it.) If the first term of the binary expansion of a point is 0 then that point lies in $[0,\frac{1}{2}]$, whilst if it starts with 1 then the point lies in $[\frac{1}{2},1]$. Consider $T(y,z)$. Since all points between $y$ and $z$ start with the same symbol, $T|(y,z)$ is a homeomorphism and (if $k>1$) all points in the image of $(y,z)$ start with the same symbol and $T(y,z) = (T(y),T(z))$ since $T$ is monotonic and increasing on $(y,z)$. Repeating this argument $k$ times we find that $T^k|(y,z)$ is a homeomorphism onto its image, which is $(T^k(y),T^k(z))$.

Now, for the choice $k = M$ and $y$ and $z$ defined in the beginning of the proof, we see that $T^M|(y,z)$ is a homeomorphism onto its image, which is $(0,\frac{1}{2})$ by construction. Hence $T^{M+1}(y,z) = (0,1)$.

This example has picked out various properties that one might wish to use as definitions of chaos. It is now only a question of choosing some minimal list to define what we will mean by chaotic. The definition of chaos below is based upon a generalization of the previous example, which is called a horseshoe.

(11.2) DEFINITION

*Let $f$ be a continuous map of the interval, $I$. The map $f$ has a horseshoe iff there exists $J$ contained in $I$ and disjoint open subintervals $K_1$ and $K_2$ of $J$ such that $f(K_i) = J$ for $i = 1,2$.*

Before describing the consequences of a map having a horseshoe we give one further definition.

(11.3) DEFINITION

*A continuous map $f : I \to I$ is chaotic iff $f^n$ has a horseshoe for some $n \geq 1$.*

Although this definition of chaos may seem to be a definition by example we will see later that it is, in fact, equivalent to definitions which give the impression of being considerably more general. In some sense, for maps of the interval (and diffeomorphisms of the plane) the horseshoe is not simply an example of chaotic behaviour, it is *the* example of chaotic behaviour. The consequences of a map having a horseshoe are almost precisely the same as the results obtained for Example 11.1.

(11.4) LEMMA

*Suppose that the continuous map $f : I \to I$ has a horseshoe. Then*

i) *$f^n$ has at least $2^n$ fixed points;*
ii) *$f$ has periodic points of every period;*
iii) *$f$ has an uncountable number of aperiodic orbits.*

The proof of this lemma is the same as the proof of the equivalent statements for Example 11.1. However, $f$ may not be monotonic on $K_1$ and $K_2$ and may have stable periodic orbits which complicate some of the more exciting parts of the proof. We leave the proof as an exercise.

As remarked above there are many other definitions of chaos. For completeness we shall give three other characterizations of chaos and give an idea of how they relate to the definition given above.

(11.5) DEFINITION

*Let $f$ be a continuous map of the interval or a smooth map of the plane. A set $S$ is $(n, \epsilon)$-separated iff for all $x$ and $y$ in $S$, $x \neq y$, there exists $k$, $0 \leq k \leq n$, such that $|f^k(x) - f^k(y)| > \epsilon$.*

Thus the cardinality of $(n, \epsilon)$-separated sets gives an idea of the number of distinguishable orbits of $f$ at resolution $\epsilon$. Now let $C(f, \epsilon, n)$

denote the maximal cardinality of $(n, \epsilon)$-separated sets for $f$. If the orbit structure of a map is complicated then we expect $C(f, \epsilon, n)$ to grow, so define the growth rate of $C(f, \epsilon, n)$ to be

$$h(f, \epsilon) = \limsup_{n \to \infty} \frac{1}{n} \log C(f, \epsilon, n). \tag{11.2}$$

This is positive if $C(f, \epsilon, n)$ grows exponentially with $n$.

(11.6) DEFINITION

*Let $f$ be a map of the interval. Then the topological entropy of $f$, $h(f)$, is defined by*

$$h(f) = \lim_{\epsilon \to 0} h(f, \epsilon).$$

Thus the topological entropy gives the asymptotic growth rate of the cardinality of $(n, \epsilon)$-separated sets for $f$. A common mathematical definition of chaos is simply that a map is chaotic iff $h(f)$ is positive. The topological entropy of maps can be very difficult to calculate, but as the next result shows, it is equivalent to the definition of chaos above.

(11.7) THEOREM

*Let $f : I \to I$ be a continuous map of the interval. Then $h(f) > 0$ iff $f^n$ has a horseshoe for some $n > 0$.*

The proof of this theorem goes far beyond the scope of this book, but the statement shows that the property of having a horseshoe is a great deal deeper than one might imagine.

The definitions given so far are topological; they do not distinguish between observable or attracting chaotic behaviour and cases where the chaotic set is unstable. For some purposes attractors are the only objects of interest and it is more useful to use a characterization of chaos which takes this into account.

(11.8) DEFINITION

*Let $f : I \to I$ be a continuous map of the interval. For all $x \in I$ define the Liapounov exponent of $x$, $L(f, x)$ to be*

$$L(f, x) = \limsup_{n \to \infty} \frac{1}{n} \log |Df^n(x)|.$$

Note that by the chain rule

$$\log |Df^n(x)| = \sum_{k=0}^{n-1} \log |Df(f^k(x))|.$$

If $L(f,x)$ is positive, then the average local expansion of points near $x$ is exponential: $|f^n(x) - f^n(y)| \sim e^{nL(f,x)}$ for $y$ near $x$. A map is then called (Liapounov exponent) chaotic if $L(f,x)$ is positive for almost all $x$ (with respect to Lebesgue measure).

Finally, one last set of definitions based upon topological ideas is given. These form the basis for the definition of chaos used by Devaney (1989).

(11.9) DEFINITION

*Let $f : I \to I$ be a continuous map of the interval and let $\Lambda$ be an $f$-invariant set. Then $f$ is topologically transitive on $\Lambda$ iff for all open sets $U$ and $V$ such that $U \cap \Lambda \neq \emptyset$ and $V \cap \Lambda \neq \emptyset$ there exists $n > 0$ such that $f^n(U) \cap V \neq \emptyset$.*

(11.10) DEFINITION

*Let $f$ and $\Lambda$ be as in Definition (11.9). Then $f$ has sensitive dependence on initial conditions (sdic) on $\Lambda$ iff there exists $\delta > 0$ such that for all $x \in \Lambda$ and $\epsilon > 0$ there exists $y \in \Lambda$, $y \neq x$ and $n \geq 0$ such that $|x-y| < \epsilon$ and $|f^n(x) - f^n(y)| > \delta$.*

A further definition of chaos is to say that $f$ is chaotic if there is an $f$-invariant set $\Lambda$ such that $f$ has sdic on $\Lambda$, $f$ is topologically transitive on $\Lambda$ and periodic orbits are dense in $\Lambda$. For all intents and purposes this definition gives the same results as the definition we are using if the set $\Lambda$ is defined appropriately, but there are examples which are chaotic by this definition but not by the definition involving horeseshoes above. This is because sdic only demands that *some* nearby points move away from the orbit of a given point under iteration and that this divergence need not be exponential, whilst the equivalence between having horseshoes and positive topological entropy shows that for horseshoes one needs exponential local divergence.

## 11.2 Period three implies chaos

The aim of this section is to show how the simple definitions of the previous section can be used to prove quite general results about continuous maps of the interval. In particular we shall show that if a continuous map of the interval has a periodic orbit of period three then it is chaotic. The main mathematical tool of this section is the Intermediate Value Theorem (IVT).

(11.11) THEOREM (INTERMEDIATE VALUE THEOREM)

*Let $f : [a,b] \to \mathbf{R}$ be a continuous map of the finite interval $[a,b]$ with $f(a) = c$ and $f(b) = d$. Then for any $x$ between $c$ and $d$ there exists $y \in [a,b]$ such that $f(y) = x$.*

The proof of this result can be found in any elementary text book of analysis. The IVT can be used to prove the following simple fixed point lemma.

(11.12) LEMMA (FIXED POINT LEMMA)

*Let $f : [a,b] \to \mathbf{R}$ be a continuous map of the finite interval $[a,b]$ with $f(a) \geq a$ and $f(b) \leq b$ (or $f(a) \leq a$ and $f(b) \geq b$). Then there exists $y \in [a,b]$ such that $f(y) = y$.*

*Proof*: Set $g(x) = f(x) - x$. Then $g : [a,b] \to \mathbf{R}$ is continuous and $g(a) \geq 0$ and $g(b) \leq 0$ (or $g(a) \leq 0$ and $g(b) \geq 0$) so by the IVT there exists $y \in [a,b]$ such that $g(y) = 0$. Hence $f(y) = y$.

These two results will enable us to prove the following theorem, first proved (under slightly different definitions) by Li and Yorke (1975).

(11.13) THEOREM

*Suppose $f : [a,b] \to \mathbf{R}$ is continuous and has an orbit of (least) period three. Then $f$ is chaotic.*

*Proof*: Let $x_i$, $i = 1,2,3$, be the three distinct points of period three with $x_1 < x_2 < x_3$. Then either

$$f(x_1) = x_2,\ f(x_2) = x_3 \text{ and } f(x_3) = x_1$$

or

$$f(x_1) = x_3,\ f(x_2) = x_1 \text{ and } f(x_3) = x_2.$$

These two cases are equivalent after reversing the direction of the $x$-axis, so without loss of generality consider the first possibility.

Since $f(x_2) > x_2$ and $f(x_3) < x_3$ the fixed point theorem implies that there exists a point $z \in (x_2, x_3)$ with $f(z) = z$. Also

$$f(x_1) = x_2 < z \quad \text{and} \quad f(x_2) = x_3 > z$$

so by the IVT there exists a point $y \in (x_1, x_2)$ with $f(y) = z$. Thus we have points

$$x_1 < y < x_2 < z < x_3$$

which satisfy

$$f(x_1) = x_2,\ f(y) = z,\ f(x_2) = x_3,\ f(z) = z \text{ and } f(x_3) = x_1$$

purely from the continuity of $f$ and the existence of the periodic orbit of period three. These five relations will be enough to show that $f^2$ has a horseshoe.

Consider $f^2|[y,z]$. Since $f$ is continuous, $f^2$ is continuous and the five relations above show that

$$f^2(y) = z,\ f^2(x_2) = x_1 < y, \text{ and } f^2(z) = z.$$

Hence, by the IVT again, there exists a (smallest) $r \in (y, x_1)$ such that $f^2(r) = y$ and a (largest) $s \in (x_1, z)$ such that $f^2(s) = y$. Let $K_1 = (y, r)$ and $K_2 = (s, z)$. Then $K_1$ and $K_2$ are both in $J = (y, z)$ and $f^2(K_1) = f^2(K_2) = J$. Hence $f^2$ has a horseshoe and so $f$ is chaotic.

(11.14) COROLLARY

*Let $f$ be as in Theorem 11.13. Then $f$ has periodic points of period $2n$ for all $n \in \mathbf{N}$.*

This corollary follows from the fact that $f^2$ has a horseshoe and hence (Lemma 11.4) orbits of all integer period. Hence $f$ has orbits of all period $2n$. In fact, by working a little harder it is possible to prove that if $f$ has an orbit of period three then it has orbits of *all* possible periods.

(11.15) EXERCISE

*Let $f$ be as above. With the labelling of the orbit of period three as in the proof of Theorem 11.13, show that there is an interval $J_1 \subseteq (x_1, x_2)$ with $f(J_1) = (x_2, x_3)$ and an interval $J_2 \subseteq (x_2, x_3)$ with $f(J_2) = (x_1, x_3)$. Use arguments similar to those of Example 11.1 to show that $f$ has orbits of period $n$ for all $n \geq 1$. [The methods used to prove this will be described formally in Section 11.8.]*

This exercise is a particular case of a beautiful theorem. Consider the order on the natural numbers defined by

$$1 \prec 2 \prec 4 \prec 2^3 \prec \ldots \prec 2^n \prec 2^{n+1} \prec \ldots$$

$$\ldots \prec 2^{n+1}.9 \prec 2^{n+1}.7 \prec 2^{n+1}.5 \prec 2^{n+1}.3 \prec \ldots$$

$$\ldots \prec 2^n.9 \prec 2^n.7 \prec 2^n.5 \prec 2^n.3 \prec \ldots \prec 9 \prec 7 \prec 5 \prec 3$$

i.e. powers of 2 ascending followed by $\ldots 2^{n+1}$ times the odd numbers (except 1) descending followed by $2^n$ times the odd numbers (except 1) descending ... followed by the odd numbers descending to three. This order is called the Sharkovskii order.

(11.16) THEOREM (SHARKOVSKII)

*Let $f : [a, b] \to \mathbf{R}$ be a continuous map of the interval. Suppose $f$ has an orbit of period $n$, then $f$ has an orbit of period $k$ for all $k \prec n$ in the Sharkovskii order.*

I still find this theorem remarkable. The proof involves some new techniques which are left to the end of this chapter.

## 11.3 Unimodal maps I: an overview

Unimodal or one-hump maps are probably the best understood maps of the interval which can have chaotic behaviour. They form an important class of examples for which it is known how and why certain phenomena occur and hence are worth studying both in themselves and as a means of understanding more complicated classes of maps. Roughly, a map $f : I \to I$ is unimodal if there exists a point $c \in I$ such that $f$ is monotonically increasing for $x < c$ and decreasing for $x > c$. An example which displays all the possible topological behaviour associated with unimodal maps is the family $g_\mu(x) = \mu - x^2$. In this section we shall consider some of the features of this typical example for different values of $\mu$.

(11.17) EXERCISE

*Show that $g_\mu(x) = \mu - x^2$ is a unimodal map of the interval for $-\frac{1}{4} \leq \mu \leq 2$ and find the maximal interval mapped into itself by $g_\mu$ in this parameter region as a function of $\mu$.*

Computers can be an important means of examining the behaviour of simple maps, although they do have drawbacks. Figure 11.3 shows the picture obtained numerically by choosing many values of the parameter $\mu$ between 0 and 2, and iterating the turning point of $g_\mu$, $x = 0$, for 30 iterations without printing anything (to allow any transient behaviour to die down) and then printing the next 200 values of $x$. This is repeated for different values of $\mu$ to give a picture of the evolution of attracting behaviour with $\mu$. There are a number of features of this diagram, which has become one of the icons of modern dynamical systems, which we would like to understand.

First, note that for $\mu$ less than some value, $\mu_\infty$, the system seems to settle down to a periodic orbit of period $2^n$ for some $n \geq 0$, and that as $\mu$ increases there is a sequence of period-doubling bifurcations which appear to accumulate on $\mu_\infty$. For values of $\mu$ greater than $\mu_\infty$ the attracting behaviour is more complicated, but there are at least two features which are immediately striking. If $\mu = 2$ the system is chaotic: $g_2$ has a horseshoe. As $\mu$ decreases from 2 towards $\mu_\infty$ a band structure emerges; initially, for $\mu$ near 2 iterates of the map lie in a single band, but as $\mu$ is decreased this splits into two bands, then four bands and so on. It is as if there is some sort of noisy periodicity of period $2^n$ with $n$ increasing as $\mu$ tends to $\mu_\infty$ from above. For parameter values above $\mu_\infty$ there are obviously windows in which the attracting behaviour is periodic. Looking at the window of parameter values for which $g_\mu$ has an orbit of period three (and a little further beyond) we see that there appears to be a copy of the original diagram, but on a smaller scale. Each periodic orbit has its own period-doubling cascade and band structure associated with it. Figure 11.3 has therefore suggested three problems which might be worth considering in more detail: period-doubling cascades for small $\mu$, band structure for larger $\mu$ and the existence of parameter regions in which some periodic orbit of the map goes through the same process. In this section we shall sketch reasons for believing that these features are more general than they might appear at first sight. Subsequent sections will go into more detail of the mathematical structure of orbits for unimodal maps.

Given a nonlinear map, the first step is almost always to try to find the fixed points and their bifurcations. This is often the only step which can be done analytically and will at least give some concrete information about the system. For the family of maps $g_\mu(x) = \mu - x^2$, fixed points are solutions of the quadratic equation $x = \mu - x^2$ and so there are two fixed points, $x_\pm$, where

$$x_\pm = \tfrac{1}{2}(-1 \pm \sqrt{1+4\mu}),$$

provided $\mu > -\frac{1}{4}$.

(11.18) EXERCISE

*Show that there is a saddlenode bifurcation at* $\mu = -\frac{1}{4}$ *which creates a pair of fixed points* $x_\pm$ *in* $\mu > -\frac{1}{4}$. *Show that* $x_-$ *is always unstable whilst* $x_+$ *is stable if* $-\frac{1}{4} < \mu < \frac{3}{4}$.

When $\mu = \frac{3}{4}$ there is a supercritical period-doubling bifurcation which creates a stable orbit of period two. This then loses stability by another supercritical period-doubling bifurcation giving a stable periodic orbit of period four.

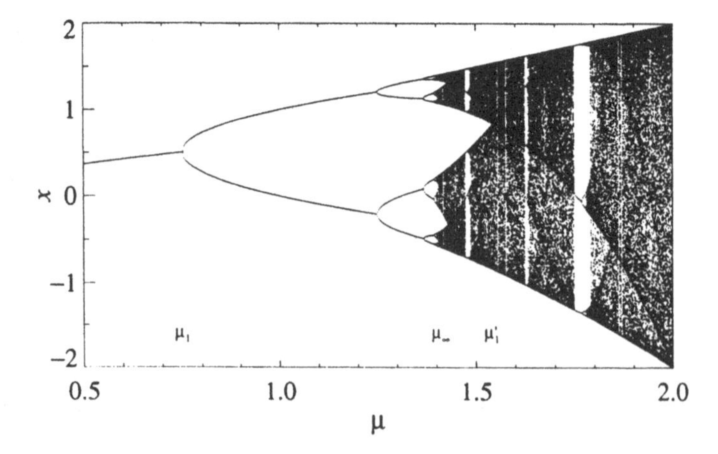


Fig. 11.3 The attractor as a function of $\mu$ for the map $g_\mu(x) = \mu - x^2$.

(11.19) EXERCISE

*By considering $g_\mu^2$ find the value of $\mu$ for which the orbit of period two loses stability by period-doubling.*

This is about as far as we can go by hand. All we have so far is that two fixed points are created in a saddlenode bifurcation at $\mu = -1/4$, one of these has a supercritical periodic-doubling bifurcation when $\mu = 3/4$ and the system is chaotic when $\mu = 2$.

This may not appear to be a great achievement but, together with a little thought, it will enable us to understand some of the gross features of Figure 11.3 described above. To use these results consider the graph of $g_\mu$ for $\mu = \mu_0 = -\frac{1}{4}$, $\mu = \mu_1 = \frac{3}{4}$ and $\mu = \mu_0' = 2$ as shown in Figure 11.4. In each case the box $B_0 = [-q, q]$ is mapped into itself by $g_\mu$, where $q$ is the left hand fixed point of the map, $q = -\frac{1}{2}(1+\sqrt{1+4\mu})$. Let $P_0 = [\mu_0, \mu_0']$ and consider the graph of $g_\mu^2$ for parameter values between $\mu_1$ and $\mu_0'$ (see Fig. 11.5). For $\mu = \mu_1$ there are two boxes, $B_i$, $i = 1, 2$, which are mapped into themselves under $g_\mu^2$ and to each other under $g_\mu$. Furthermore, $g_{\mu_1}^2|_{B_i}$, $i = 1, 2$ looks like $g_{\mu_0}|_{B_0}$ and by the time $\mu = \mu_0'$ the map has come out of the top and bottom of the relevant boxes. Hence, by the continuity of the family $g_\mu$ with $\mu$ there exists a parameter value $\mu_1' \in (\mu_1, \mu_0')$ such that $g_{\mu_1'}^2|_{B_i}$ looks like $g_{\mu_0'}|_{B_0}$ as shown in Figure 11.5c.

Now consider $g_\mu^2$ for $\mu \in P_1 = (\mu_1, \mu_1')$. Restricted to either of the invariant boxes $B_i$, $i = 1, 2$, this must evolve in the same way as $g_\mu$ evolves for $\mu \in P_0$. In particular, there is a parameter value $\mu_2$ where the fixed points of $g_\mu^2$ inside each of the invariant boxes (points of period two for $g_\mu$) period-double producing points of period $2^2$ for $g_\mu$. This is the value calculated in Exercise 11.19. Also, there is a parameter value $\mu_2'$ at

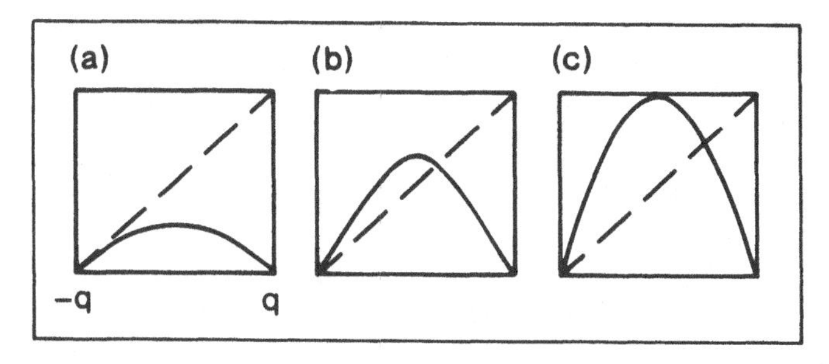


Fig. 11.4 (a) $\mu = \mu_0$; (b) $\mu = \mu_1$; (c) $\mu = \mu_0'$.

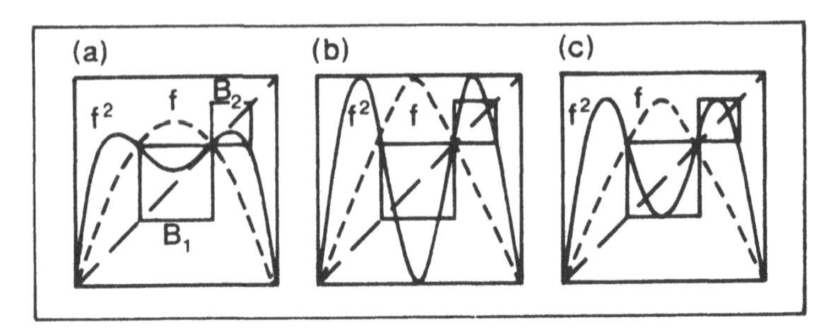


Fig. 11.5 (a) $\mu = \mu_1$; (b) $\mu = \mu_0'$; (c) $\mu = \mu_1'$.

which $g_\mu^{2^2} = (g_\mu^2)^2$ breaks its own box within box invariance. Hence there is a parameter interval $P_2 = (\mu_2, \mu_2')$ on which $g_\mu^{2^2}$ evolves (restricted to appropriate invariant boxes) exactly as $g_\mu^2$ evolves for $\mu \in P_1$ which is exactly as $g_\mu$ evolves for $\mu \in P_0$. Continuing in this way we construct a sequence of nested open intervals $P_0 \supset P_1 \supset P_2 \supset \ldots$ such that for $\mu \in P_n$, $g_\mu^{2^n}$ evolves (restricted to any one of $2^n$ invariant boxes) as $g_\mu$ evolves for $\mu \in P_0$. This nested sequence of intervals converges to a point, $\mu_\infty = \lim_{n\to\infty} \mu_n = \lim_{n\to\infty} \mu_n'$, which is called the boundary of chaos: for $\mu < \mu_\infty$ the map is not chaotic and for $\mu > \mu_\infty$ the map is chaotic (although there may be attracting periodic cycles).

We can now explain some of the features of Figure 11.3 noted earlier: at $\mu_n$ there is a period-doubling bifurcation in which an orbit of period $2^{n-1}$ loses stability, creating an orbit of period $2^n$. This sequence of bifurcations accumulates on $\mu_\infty$. Furthermore, for $\mu \in P_n \backslash P_{n+1}$ there are $2^n$ attracting boxes for $g_\mu^{2^n}$ giving rise to the band structure described above. To understand the parameter regimes with similar structure consider (for example) $g_\mu^3$. There is some value $\nu_0$ such that when $\mu = \nu_0$ a pair of orbits of period three is created in a saddlenode bifurcation (see Fig. 11.6) and when $\mu = \nu_0'$, $g_\mu^3$ (restricted to one of three boxes) is similar to $g_{\mu_0'}$. Hence for $\mu \in M_3 = (\nu_0, \nu_0')$, $g_\mu^3$ evolves in the same way as $g_\mu$ for $\mu \in P_0$. Thus the entire picture is repeated (on a smaller scale and in three invariant boxes), giving cascades of period-doubling bifurcations creating orbits of period $3.2^n$, boxes within boxes and so on.

This is not, by any means, a rigorous argument, but it can be made rigorous. In the next few sections we shall expand on some of the phenomena observed, providing proofs of results which support this general picture.

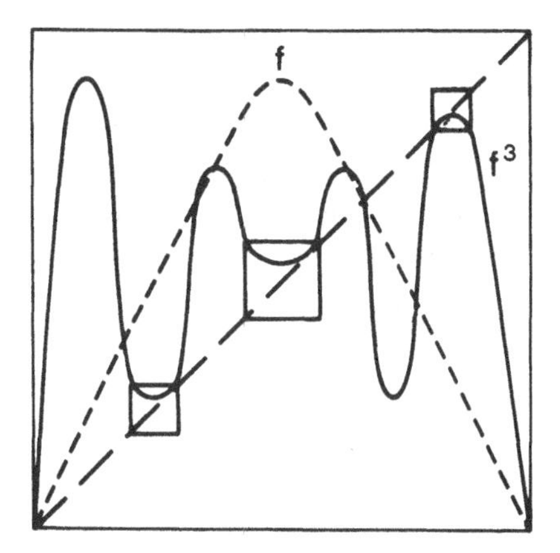


Fig. 11.6 $\mu = \nu_0$.

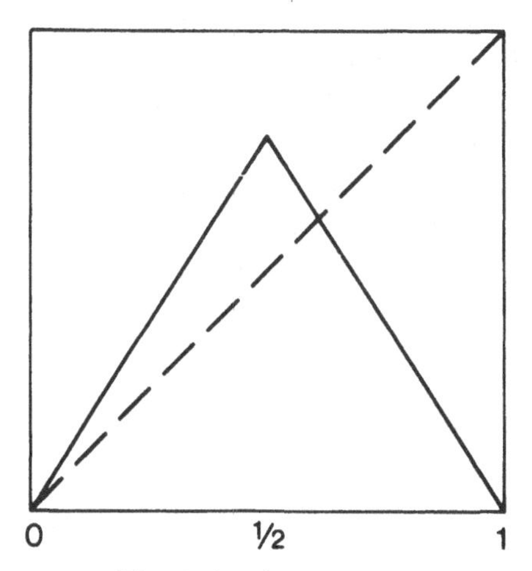


Fig. 11.7 A tent map.

## 11.4 Tent maps

There is a class of particularly simple unimodal maps for which it is possible to prove results about chaos and other topological properties analytically. They are piecewise linear unimodal maps with constant slope and are called tent maps for obvious reasons (see Fig. 11.7). Specifically, a tent map is a map $T_s : [0,1] \to [0,1]$ defined by

$$T_s(x) = \begin{cases} sx, & \text{if } 0 \le x < \frac{1}{2} \\ s(1-x), & \text{if } \frac{1}{2} \le x \le 1 \end{cases} \tag{11.3}$$

where the parameter $s$ is usually restricted to the interval $(1, 2]$.

If $0 < s \leq 1$ then the only recurrent dynamics consists of fixed points (which are boring) and if $s > 2$ the interval $[0,1]$ is not mapped into itself. We shall prove a number of results for tent maps: they are chaotic for $s > 1$, periodic orbits are dense in a union of subintervals of $[0,1]$ and there is a band structure similar to that described in the previous section. Figure 11.8 shows a parameter sweep for tent maps where the band structure is particularly noticeable.

(11.20) LEMMA

*If $\sqrt{2} < s \leq 2$ then $T_s$ is chaotic.*

*Proof*: The proof of this result is almost exactly the same as the proof that period three implies chaos. First note that there is always a fixed point $z \in (\frac{1}{2}, 1)$ for $T_s$ where $z = s/(s+1)$. There is also a point $y \in (0, \frac{1}{2})$ such that $T_s(y) = z$. A little calculation shows that this point is $y = 1/(s+1)$. We will now show that for $\sqrt{2} < s \leq 2$ the second iterate of $T_s$ has a horseshoe and so $T_s$ is chaotic.

By construction, $T_s^2(z) = T_s^2(y) = z$. Furthermore, $T_s(\frac{1}{2}) = \frac{1}{2}s$ and so $T_s^2(\frac{1}{2}) = s(1 - \frac{1}{2}s) = \frac{1}{2}s(2 - s)$. Hence $T_s^2(\frac{1}{2}) < y$ if

$$\tfrac{1}{2}s(2-s) < \frac{1}{s+1}. \tag{11.4}$$

Rearranging this inequality gives

$$s^2 > 2 \tag{11.5}$$

and so provided $s^2 > 2$ we have

$$T_s^2(\tfrac{1}{2}) < y < \tfrac{1}{2} < z = T_s^2(y) = T_s^2(z). \tag{11.6}$$

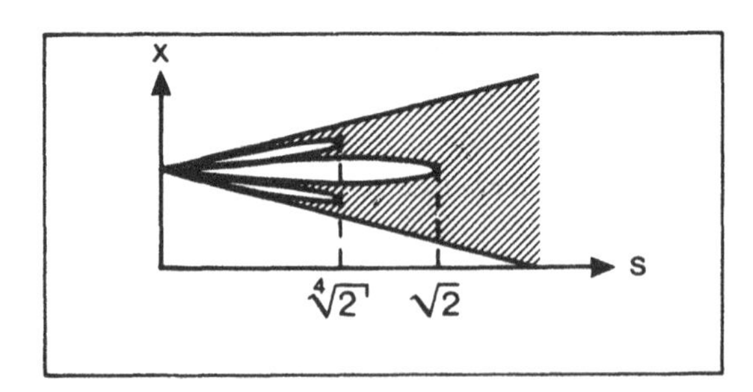


Fig. 11.8 The attractor as a function of $s$ for tent maps.

Now we are finished, since these are precisely the conditions needed to show that $T_s^2$ has a horseshoe (cf. Theorem 11.13). We can, in fact, find the intervals $K_1$ and $K_2$ explicitly (such that $T_s^2(K_i) = [y, z]$). These are given in the next exercise.

(11.21) EXERCISE

*Show that* $K_1 = [y, y + (z - y)/s]$ *and* $K_2 = [z - (z - y)/s, z]$.

To deal with tent maps with $1 < s \leq \sqrt{2}$ we will make use of the idea of renormalization described in the previous section. The most important remark, illustrated in Figure 11.6, is that the second iterate of $T_s$ restricted to appropriate intervals is a tent map with slope $s^2$. Hence if $2^{\frac{1}{4}} < s \leq 2^{\frac{1}{2}}$ the second iterate of $T_s$ looks like $T_{s^2}$ and since $T_{s^2}$ is chaotic (by Lemma 11.20), $T_s$ is chaotic. An inductive argument then leads to the conclusion that $T_s$ is chaotic for all $s \in (1, 2]$. The first step is to prove the important remark. This is done by calculating $T^2$ explicitly.

(11.22) LEMMA

*Suppose* $1 < s \leq \sqrt{2}$, *then there exist intervals* $J_1$ *and* $J_2$ *such that* $T_s(J_1) \subseteq J_2$, $T_s(J_2) \subseteq J_1$ *and* $T_s^2|J_i$ *is (after a linear rescaling)* $T_{s^2}$.

*Proof*: Let $y$ and $z$ be as in the proof of Lemma 11.20 and let $w$ be the point greater than $z$ with $T_s(w) = y$ (see Fig. 11.8). So

$$s(1 - w) = y = 1/(s + 1)$$

i.e.

$$w = \frac{s^2 + s - 1}{s(s + 1)}. \tag{11.7}$$

Now set $J_1 = [y, z]$ and $J_2 = [z, w]$. A little manipulation (which is left as an exercise for the unconvinced) or a glance at Figure 11.9 completes the proof.

This lemma makes the following proposition almost trivial.

(11.23) PROPOSITION

*Let* $T_s$ *be a tent map with* $1 < s \leq 2$. *Then* $T_s$ *is chaotic.*

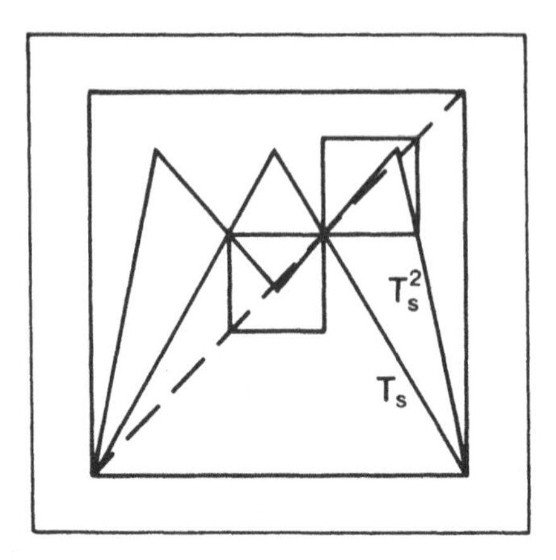


Fig. 11.9 $T_s$ and $T_s^2$ for $1 < s < \sqrt{2}$.

*Proof*: If $2^{\frac{1}{2}} < s \leq 2$ then $T_s^2$ has a horseshoe and $T_s$ is chaotic (Lemma 11.20). If $1 < s \leq 2^{\frac{1}{2}}$ then by Lemma 11.22 there exist intervals $J_i$, $i = 1, 2$, such that $T_s^2|J_i$ is a tent map with slope $s^2$. Hence if $2^{\frac{1}{4}} < s \leq 2^{\frac{1}{2}}$ then $2^{\frac{1}{2}} < s^2 \leq 2$ and so the induced map $T_s^2|J_i$ has a horseshoe and $T_s$ is chaotic.

If $1 < s^2 \leq 2^{\frac{1}{4}}$ consider $T_s^2|J_i$. After rescaling this is a tent map with slope $s^2$ and $1 < s^2 \leq 2^{\frac{1}{2}}$. So provided $2^{\frac{1}{4}} < s^2 \leq 2^{\frac{1}{2}}$ (i.e. $2^{\frac{1}{8}} < s \leq 2^{\frac{1}{4}}$) $T_s^2|J_i$ is chaotic and hence $T_s$ is chaotic. The induction argument should be clear by now: suppose that it has been shown that $T_s$ is chaotic for $2^{1/2^n} < s \leq 2$. Consider $s$ with $2^{1/2^{n+1}} < s \leq 2^{1/2^n}$. By Lemma 11.22 there exist intervals $J_i$, $i = 1, 2$, such that $T_s^2|J_i$ is a tent map with slope $s^2$. Hence $T_s^2|J_i$ is chaotic (since it has slope $s^2$ and $2^{1/2^n} < s^2 \leq 2^{1/2^{n-1}}$) and hence $T_s$ is chaotic. This completes the proof.

It is actually possible to prove rather more than this. Figure 11.10 shows that given a tent map all points (except 0 and 1) tend to the interval $[T_s^2(\frac{1}{2}), T_s(\frac{1}{2})] = [\frac{1}{2}s(2-s), \frac{1}{2}s]$. Call this interval $A$ (for attracting).

(11.24) LEMMA

*Let $T_s$ be a tent map with $2^{\frac{1}{2}} < s \leq 2$, then given any interval $J$ in $A$ there exists $n$ such that $T_s^n(J) = A$.*

Thus any interval eventually expands to the whole of $A$. This has the simple corollary that periodic orbits are dense in $A$ for $2^{\frac{1}{2}} < s \leq 2$,

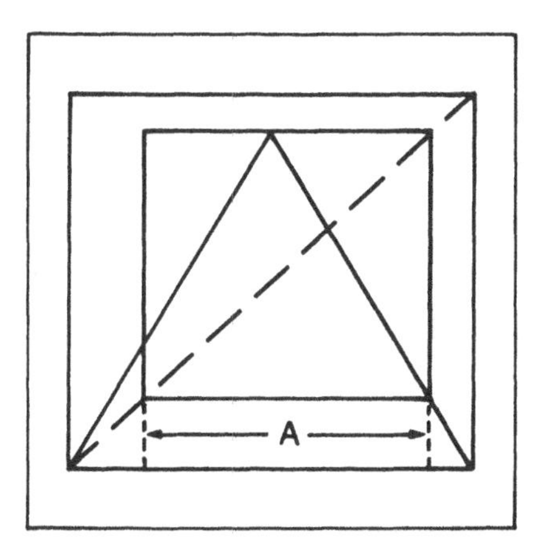


Fig. 11.10 A tent map showing the attracting interval $A$.

since given any neighbourhood of a point $x \in A$ there exists $n$ such that $T_s^n(N) = A$. Hence, using the fixed point lemma of Section 11.2 (Lemma 11.12), there is a fixed point of $T_s^n$ in $N$, which is a periodic point for $T_s$. The proof of the lemma is surprisingly simple.

*Proof of Lemma 11.24:* Take any interval $J$ and let $|J|$ denote the length of $J$. Since the slope of $T_s$ is $s$, if $J$ does not contain the turning point of the map, $x = \frac{1}{2}$, then $|T_s(J)| = s|J|$. However, if $J$ does contain the point $x = \frac{1}{2}$ $J$ is split into two intervals, $U$ and $V$ by $x = \frac{1}{2}$ and $|T_s(J)| = max\{|T_s(U)|, |T_s(V)|\}$ as shown in Figure 11.11. Now, $|U| + |V| = |J|$, so there exists $a \in (0, 1)$ such that

$$|U| = a|J| \quad \text{and} \quad |V| = (1-a)|J|.$$

But $x = \frac{1}{2}$ does not lie inside either $U$ or $V$ so

$$|T_s(U)| = sa|J| \quad \text{and} \quad |T_s(V)| = s(1-a)|J|.$$

Now, the smallest possible value for $|T_s(J)| = max\{|T_s(U)|, |T_s(V)|\}$ is clearly obtained when $|T_s(U)| = |T_s(V)|$, i.e. when $a = \frac{1}{2}$. Hence

$$|T_s(J)| \geq \tfrac{1}{2}s|J|.$$

That was the hardest part of the proof. We now want to show that for any $J$ there exists $n$ such that $\frac{1}{2} \in T_s^n(J)$ *and* $\frac{1}{2} \in T_s^{n+1}(J)$. If this is proved we are finished since if $\frac{1}{2} \in T_s^n(J)$ then $T_s(\frac{1}{2}) \in T_s^{n+1}(J)$, but $\frac{1}{2}$ is also in $T_s^{n+1}(J)$, so $T_s^{n+2}(J) = A$.

To prove that for all $J$ in $A$ there exists $n$ such that $\frac{1}{2} \in T_s^n(J)$ and $\frac{1}{2} \in T_s^{n+1}(J)$ we need to go back to the expansion results. Take any

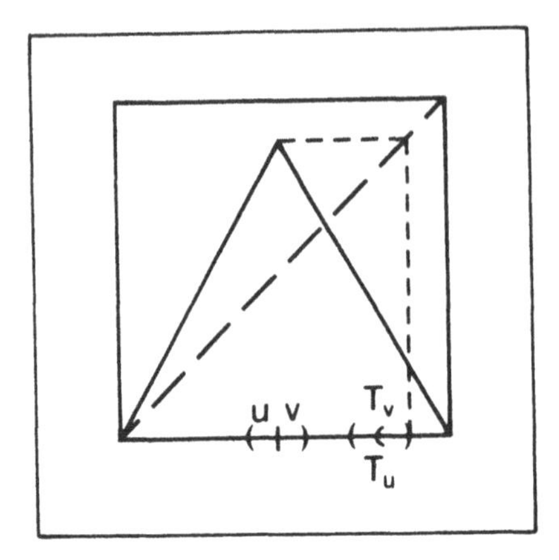


Fig. 11.11

interval in $A$. Then the sequence $|J|$, $|T_s(J)|$, $|T_s^2(J)|$, ... is increasing unless $\frac{1}{2} \in T_s^k(J)$ for some $k$. Since the intervals expand by a factor of $s$ on each iteration and must remain in $A$ (which is finite) such a $k$ must exist. Now suppose that $\frac{1}{2} \in T_s^k(J)$ but $\frac{1}{2} \notin T_s^{k+1}(J)$ then $|T_s^{k+1}(J)| \geq \frac{1}{2}s|T_s^k(J)|$ and $|T_s^{k+2}(J)| \geq \frac{1}{2}s^2|T_s^{k+1}(J)|$. But since $s > 2^{\frac{1}{2}}$, $s^2/2 > 1$ and so the interval continues to expand. Hence, to avoid the length of the interval blowing up, there must exist $n$ such that $\frac{1}{2} \in T_s^n(J)$ and $\frac{1}{2} \in T_s^{n+1}(J)$. This completes the proof.

After so many lemmas it is probably time for a theorem.

(11.25) THEOREM

*Let $T_s$ be a tent map with $2^{1/2^{n+1}} < s \leq 2^{1/2^n}$, $n \geq 0$. Then $T_s$ has periodic orbits of periods $2^k$, $k = 0, 1, ..., n-1$ if $n > 0$ and an attracting set which consists of $2^n$ intervals which are permuted by $T_s$ on which periodic points are dense.*

We leave the full proof of this theorem as an exercise. The proof is very similar to the proof of Proposition 11.23, based on induction. If $2^{\frac{1}{2}} < s \leq 2$ the theorem is true by Lemmas 11.20 and 11.24. Suppose that it is true for $n = m$ and consider the statement for $n = m + 1$. By Lemma 11.22, $T_s^2$ restricted to either one of two intervals is a tent map with slope $s^2$. Hence the induction hypothesis is valid for these maps and the result follows after a little thought.

(11.26) EXERCISE

*Write out a complete proof of Theorem 11.25.*

## 11.5 Unimodal maps II: the non-chaotic case

In this section the structure of non-chaotic unimodal maps of the interval will be described. To do this we must first formalize precisely what is meant by a unimodal map, a task that has been avoided hitherto.

(11.27) DEFINITION

*A unimodal map of the interval is a continuous map* $f : [0,1] \to [0,1]$ *such that*

i) *there exists* $c \in (0,1)$ *such that* $f|(0,c)$ *is strictly increasing and* $f|(c,1)$ *is strictly decreasing;*
ii) $f(0) = f(1) = 0$.

Property (i) is really the only important property, since any continuous function which satisfies (i) can be extended (or restricted) to some other interval in such a way that property (ii) is satisfied. The choice of $[0,1]$ as the interval is equally arbitrary, since any interval can be rescaled to $[0,1]$ by a linear transformation. We will often abuse this definition by referring to maps as unimodal regardless of the interval on which they are defined. So we shall speak of unimodal maps of an arbitrary interval $J$ without further comment.

The main result of this section can be stated in terms of chaotic maps in the following way.

(11.28) THEOREM

*Let* $f$ *be a unimodal map. If* $f$ *has a periodic orbit of period* $m$ *which is not a power of* $2$ *then* $f$ *is chaotic.*

This surprising result is just a corollary of the detailed study of non-chaotic unimodal maps which we are about to undertake. It is worth noting that this result remains true if $f$ is any continuous map of the interval. Rather than prove the theorem as stated we shall prove the following theorem.

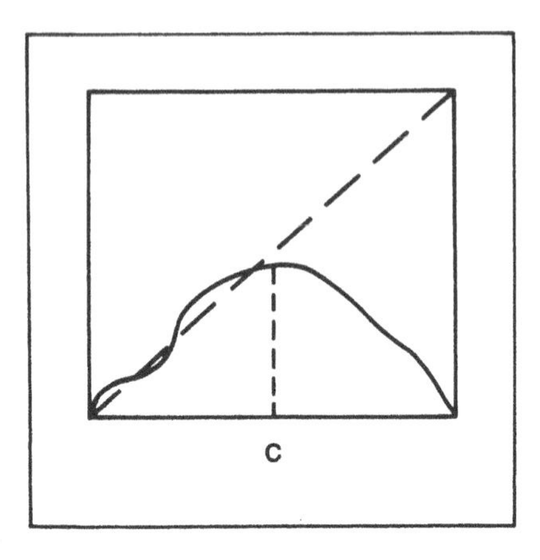


Fig. 11.12 $f(c) < c$.

(11.29) THEOREM

*Let $f$ be a non-chaotic unimodal map. Then there exists $M$, $0 \leq M \leq \infty$ such that every periodic point of $f$ has period $2^n$ for some $n \in \{0, 1, 2, \ldots, M\}$.*

If $M$ is infinite then $f$ can have complicated attracting behaviour which is not periodic but is not chaotic. We shall come to this a bit later. The proof of this theorem requires three ingredients. The first is the idea of renormalization introduced in Section 11.2 and used in Section 11.3. The second is a small bit of analysis for the most simple unimodal maps (it is really little more than a remark).

(11.30) LEMMA

*Let $f$ be a unimodal map with critical point $c$. Then if $f(c) \leq c$ all points are attracted to fixed points of the map.*

This is illustrated in Figure 11.12. The explanation of this lemma (which cannot really be dignified by the term proof) is that since $f(x) < f(c)$ for all $x \in (c, 1)$, $f((c, 1)) \subseteq (0, c)$. For points in $(0, c)$ either $f(x) < x$ or $f(x) > x$ or $f(x) = x$. In the latter case $x$ is a fixed point and the lemma is done so assume that $f(x) > x$. Since $f$ is increasing on $(0, c)$ and $f(x) < f(c) \leq c$ for all $x$ in $(0, c)$ the sequence $(x, f(x), f^2(x), \ldots)$ is increasing and bounded above by $f(c)$ hence it tends to a limit, $y$, and by the continuity of $f$, $f(y) = y$.

(11.31) EXERCISE

*Complete the proof for the case $f(x) < x$.*

The last ingredient involves a definition.

(11.32) DEFINITION

*A unimodal map has an orientation reversing fixed point (ORFP) if it has a fixed point $y$ with $f$ strictly decreasing at $y$.*

Thus an ORFP is a fixed point in $(c, 1)$. Mathematics is full of deep, meaningful statements. The proof of the non-chaotic unimodal map theorem is based about the following statement:

**either f has an ORFP or it does not.**

The strategy is: either $f$ has an ORFP or it does not. If it does not then $f(c) \leq c$ (since if $f(c) > c$ the intermediate value theorem together with $f(1) = 0 < 1$ implies that there is a fixed point in $(c, 1)$) and so by Lemma 11.30 all points tend to fixed points of the map and we stop. If $f$ has an ORFP then we renormalize, looking at $f^2$ restricted to a suitable pair of intervals. We can then ask whether the renormalized map has an ORFP and continue...

We need two further remarks before beginning the proof of the theorem. First note (see Fig. 11.13) that if $f$ is not chaotic and $f(c) > c$ then the interval $[f^2(c), f(c)]$ is mapped into itself by $f$ and that all points which do not tend to fixed points of $f$ outside this interval map eventually into the interval. This follows from Figure 11.13b; if $f$ does not map the interval $[f^2(c), f(c)]$ into itself then $f$ has a horseshoe. The second remark is that if $f$ has an ORFP, $z$, then there exist points $y < c$ and $w > z$ such that $f(y) = z$ and $f(w) = y$ so $f^2$ has the form shown in Figure 11.14.

*Proof of Theorem 11.29:* Either $f$ has an ORFP or it does not. If $f$ has no ORFP then all points tend to fixed points of the map, so $f$ has orbits of period $2^0$ and no other periods and the theorem is proved. If $f$ has an ORFP, $z > c$, then consider $f^2$. By the remark above the statement of the theorem and Figure 11.14, there exist points $y < c$ and $w > c$ such that $f(w) = y$, and $f(y) = z$. Consider $f^2|[y, z]$. The critical point $c$ lies in $[y, z]$ and $f^2$ is decreasing on $(y, c)$ and increasing on $(c, z)$. If $f^2(c) < y$ then (see Fig. 11.14b) $f^2$ has a horseshoe, contradicting the hypothesis that $f$ is not chaotic. Hence $f^2(c) > y$ and $f^2$ is a unimodal

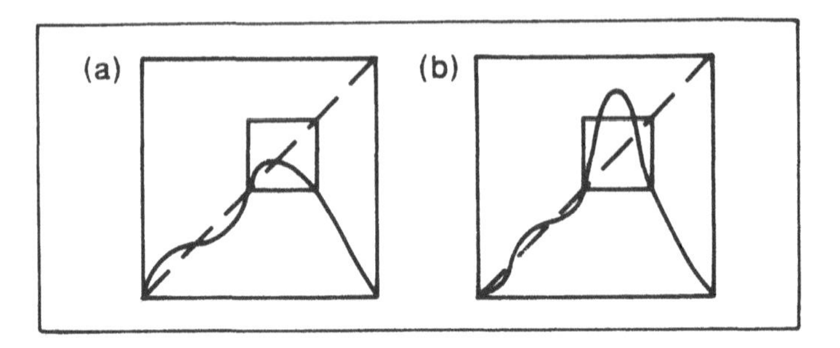


Fig. 11.13 (a) No horseshoe; (b) a horseshoe.

map of the interval $[y, z]$. Similarly, $f^2|[z, w]$ is a unimodal map with maximum value $f(c) < w$. These statements lead to two conclusions. First, the attracting interval $[f^2(c), f(c)]$ is contained in $[y, w]$ and second that the dynamics of $f$ in $[y, w]$ can be deduced from the dynamics of either $f^2|[y, z]$ or $f^2|[z, w]$, since for all $x \in [y, z]$, $f(x) \in [z, w]$ and vice versa.

So, if $f$ has an ORFP then $f$ has fixed points (points of period $2^0$) outside the interval $[y, w]$ and all other points tend under iteration to a set contained in $[y, w]$. To find the dynamics of $f$ in $[y, w]$ consider the induced unimodal map $f^2|[y, z]$. **Either it has an ORFP or it does not.** If it does not then all points in $[y, z]$ tend to fixed points of $f^2$ under iteration (by $f^2$). Hence $f$ has an orbit of period $2^0$ in $[y, w]$, which is the orbit of the ORFP, $z$, and possibly other orbits of period $2^1$. Again the process stops here. On the other hand, if $f^2|[y, z]$ has an ORFP then this is an orbit of period $2^1$ for $f$ and we can look at $f^4|[y, z]$ and $f^2|[z, w]$. We are now in precisely the same situation as before but using $f^2$ instead of $f$ and the proof is completed by induction. There are

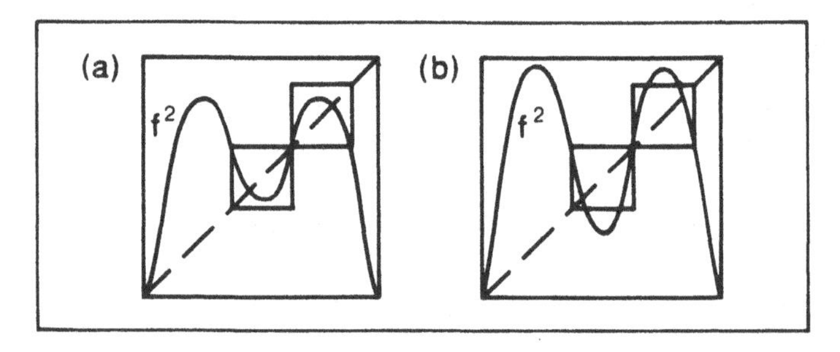


Fig. 11.14 (a) $f^2$ does not have a horseshoe; (b) $f^2$ has a horseshoe.

two possibilities: **either this process ends or it does not.** (Another deep mathematical statement!) If it ends, then $f$ has orbits of period $2^k$ for $k = 0, 1, 2, ..., M$, whilst if it does not end then $f$ has orbits of period $2^n$ for all $n \geq 0$.

Although this completes the theorem as stated it is not quite the end of the story. If the process does not end then the attractor, the set of points to which most points tends, is not a periodic orbit. This shows that it is possible to have non-chaotic but aperiodic behaviour. This 'extra' limit set can be described by taking the analysis of this section a little further. In the construction above suppose that $f$ has an ORFP. Then to a first approximation the limit set is contained in the two intervals $(y, z)$ and $(z, w)$. Since $f^2$ restricted to either of these intervals is a unimodal map with an ORFP there exist four intervals, two inside each of these intervals, joined at the ORFPs of $f^2$. Since $f^4$ restricted to any of these four intervals is unimodal and has an ORFP each of these four intervals contains two subintervals, joined at the ORFPs. On each of these intervals $f^8$ acts as a unimodal map with an ORFP... Continuing in this way and introducing an inductive labelling of the successive approximations of this limit set it is possible to give a complete description of the dynamics on this set. This is not easy and the interested reader is referred to Devaney (1989) for details.

## 11.6 Quantitative universality and scaling

In the mid-1970s Coullet and Tresser and Feigenbaum noticed a remarkable property of the period-doubling cascades described in 11.3. If the parameter values of successive period-doubling bifurcations are labelled $(\mu_n)$ then the quantity

$$\lim_{n\to\infty} \frac{\mu_n - \mu_{n-1}}{\mu_{n+1} - \mu_n} \tag{11.8}$$

converges to 4.667... *independent* of the family of maps being considered (in fact, this quantity does depend upon the nature of the critical point, the number 4.667... is for maps with quadratic maxima). We shall not be able to give a complete explanation of this phenomenon, but can at least indicate why it is true. Indeed, at the time of writing, formal proofs are only just becoming available. The proofs use a formidable arsenal of mathematical techniques and the interested reader should refer to de Melo and van Strien (1993).

The key comes from the proof of the non-chaotic unimodal map theorem. At the accumulation of period-doubling cascades the map has periodic orbits of period $2^n$ for all $n \geq 0$ and hence it can be renormalized infinitely often (i.e. the inductive process in the proof never stops). This implies that at the accumulation value the dynamics of $f^2$ restricted to a suitable subinterval is essentially the same as the dynamics of $f$; both can be renormalized infinitely often, both have periodic orbits of period $2^n$, $n \geq 0$, and, of course, both are unimodal maps. If the two maps, $f$ and $f^2$, were exactly the same then $f$ would satisfy an equation of the form

$$A(f(x)) = f^2(A(x)) \tag{11.9}$$

where $A$ represents a suitable scaling factor and shift of the origin. This is easier to see if we take a slightly different convention for unimodal maps and restrict attention to symmetric maps (this is not essential, but makes the calculations more tractable).

(11.33) DEFINITION

*A unimodal map* $f : [-1,1] \to [-1,1]$ *is in class* $D_\epsilon$ *if there exists* $\epsilon > 0$ *and a homeomorphism* $g : [0,1] \to [0,1]$ *such that* $g(0) = 1$ *and* $f(x) = g(|x|^{1+\epsilon})$.

Note that symmetric unimodal maps with quadratic maxima are in class $D_1$ and that the critical point of the map, $c$, is at $x = 0$. Maps in this class are shown in Figure 11.15 (after Collet and Eckmann, 1980). Now, $f^2(0) = f(1) = -\alpha^{-1}$, say, and we can now consider $f^2$ restricted to $[-\alpha^{-1}, \alpha^{-1}]$. This is again a symmetric unimodal map only upside down and with the direction of $x$ reversed. In order to obtain a map in class $D_\epsilon$ we must change variables so that the map is the right way up and the interval $[-\alpha^{-1}, \alpha^{-1}]$ maps to $[-1,1]$. This can be done by setting $y = -\alpha x$. With this change of variable, $-\alpha f^2(-\alpha^{-1}y)$ is a map in $D_\epsilon$. The statement that $f$ and $f^2$ are equivalent is the same as saying that for all $y \in [-1,1]$

$$f(y) = -\alpha f^2(-\alpha^{-1}y). \tag{11.10}$$

Another way of saying this is that the map $Tf(y) = -\alpha^{-1}f^2(-\alpha y)$ has a fixed point with $Tf = f$.

*Assumption 1:* For each $\epsilon > 0$ there exists $\alpha_\epsilon > 1$ such that there is a map $F_\epsilon \in D_\epsilon$ with $TF_\epsilon = F_\epsilon$.

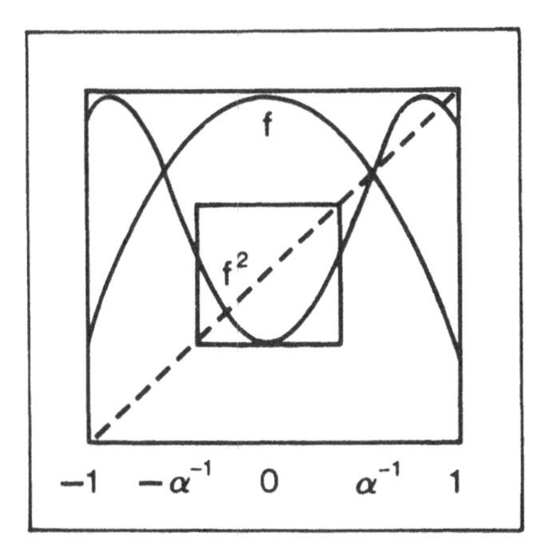


Fig. 11.15

We can now think of $T$ as a map of $D_\epsilon$ in function space and $F_\epsilon$ as a fixed point. As always, the first step in analyzing a fixed point of a map is to determine the spectrum of the linear map (or operator) evaluated at the fixed point.

*Assumption 2:* The derivative of $T$ evaluated at $F_\epsilon$ has an eigenvalue $\delta_\epsilon > 1$. All the other eigenvalues lie inside the unit disc.

This assumption implies that $T$ has a one-dimensional unstable manifold and a codimension one stable manifold, see Figure 11.16. Strictly speaking this assumption is false, but the eigenvalues on the unit circle represent symmetries which are preserved in the class $D_\epsilon$ and do not alter the structure of the argument.

Maps on the stable manifold of $T$ are precisely those maps which are infinitely renormalizable in the sense of the previous section, that is, they have periodic orbits of period $2^n$ and an invariant attracting Cantor set. We will now go through a similar argument to the one used in Section 11.4, but this time we shall work in terms of the map $T$ and use the properties of $T$ assumed above.

We want to understand the structure of maps near $F_\epsilon$, the fixed point of $T$. To do this we will identify some important features of these maps and use $T$ to deduce further properties.

*Assumption 3:* The surface

$$\Sigma_0 = \{f \in D_\epsilon | \text{ there exists } z \text{ with } f(z) = z \text{ and } f'(z) = -1\}$$

is non-empty and intersects the unstable manifold of $F_\epsilon$ transversely.

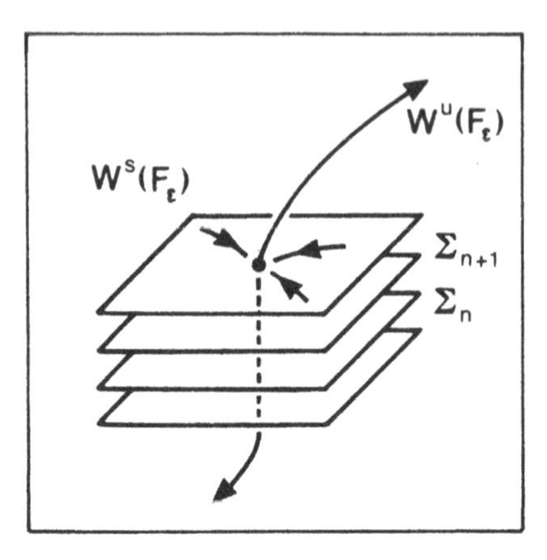


Fig. 11.16 Local structure near the fixed point $F_\epsilon$ (after Collet, Eckmann and Lanford).

This assumption simply states that the geometry of a neighbourhood of $F_\epsilon$ is as shown in Figure 11.16. $\Sigma_0$ is the set of maps in $D_\epsilon$ which have a non-hyperbolic fixed point which is about to undergo a period-doubling bifurcation. Now let $\Sigma_1 = T^{-1}\Sigma_0$. Then maps in $\Sigma_1$ are those which, after renormalization, have a fixed point which is about to period-double. But this fixed point must be an orbit of period 2 for the map (as it is an orientation reversing fixed point of $Tf$) and so maps on $\Sigma_1$ have a periodic orbit of period two which is about to period-double. Furthermore, since $Tf$ is in $\Sigma_0$, $f$ must lie closer to the fixed point, $F_\epsilon$, of $T$ by (approximately) a factor $\delta$ (by Assumption 2) and we have the picture in Figure 11.16. We can now continue. Defining $\Sigma_n = T^{-1}\Sigma_{n-1}$ and going through the same argument, we find that maps on $\Sigma_n$ have a periodic orbit of period $2^n$ which is about to lose stability by period-doubling. Moreover, the surfaces $\Sigma_n$ accumulate on the stable manifold of $F_\epsilon$ at the rate $\delta$. Now consider any family of maps $f_\mu$ in $D_\epsilon$ which pass through the stable manifold of the fixed point of $T$. Such a family can be represented as a curve in $D_\epsilon$ which is parametrized by $\mu$. Assuming that this curve intersects the stable manifold transversely it must pass through an infinite sequence of the manifolds $\Sigma_n$, giving an infinite sequence of period-doubling bifurcations. We can actually do even better. Since the manifolds $\Sigma_n$ accumulate on the unstable manifold at the rate $\delta$, given by the unstable eigenvalue of the linearized map $T$ at $F_\epsilon$, this property is transferred to the parametrization of the family, $\mu$. Thus, if $\mu_\infty$ is the parameter value at which the curve $f_\mu$

intersects $\Sigma$, and the curve intersects the manifolds $\Sigma_n$ at parameter values $\mu_n$ then, asymptotically,

$$\mu_n - \mu_\infty = \delta(\mu_{n+1} - \mu_\infty) \tag{11.11}$$

or

$$\lim_{n\to\infty} \frac{\mu_n - \mu_{n-1}}{\mu_{n+1} - \mu_n} = \delta. \tag{11.12}$$

The astonishing feature of this result is that it is independent of the family of maps being considered: it depends only upon the order of the critical point (i.e. the value of $\epsilon$) and the transversality of the family to $\Sigma$ at $\mu_\infty$. This property is often referred to as universality. As the exercises below show, there are other sequences of bifurcations or behaviour which can be treated in similar ways. In particular, there is a sequence of parameter values $\nu_n$ which accumulate on $\mu_\infty$ at the same rate as the sequence $(\mu_n)$ but from the other side of the stable manifold of $F_\epsilon$ at which $h = T^n f$ satisfies $h^2(0) = h^3(0)$, and so $h$ has a horseshoe (cf. $2 - x^2$).

This universal behaviour refers to universality in parameter space. Similar arguments can be applied to show that there is also universality in the spatial structure. Consider a map on the stable manifold, $\Sigma$, of $F_\epsilon$. In the previous section we saw that infinitely renormalizable maps leave a Cantor set invariant. One way of thinking about the construction of this Cantor set is to concentrate on a neighbourhood of the critical point $x = 0$. To a first approximation the Cantor set lies inside the interval $[-1, 1]$. By renormalization, the Cantor set lies in two intervals, one of which is $[f^2(0), -f^2(0)]$. After rescaling, $f^2$ restricted to this interval is again a map in $D_\epsilon$, $r = Tf$, with the same basic structure. So part of the Cantor set at the next step in the induction procedure lies in the interval $[-r^2(0), r^2(0)]$. Now as $n \to \infty$, $T^n f \to F_\epsilon$ and so the ratio of successive bits in the construction of the Cantor set near $x = 0$ scale like $2F_\epsilon^2(0)$. But this is just the constant $2\alpha^{-1}$ which depends only upon the universality class of $f$, and so we obtain (asymptotic) universal scaling properties for the spatial structure of the dynamics.

Proofs of the statements given above are extremely lengthy and it is not easy to obtain good analytic expressions for either $\alpha$ or $\delta$. However, it is possible to get a very crude estimate of both $\delta$ and $\alpha$ when $\epsilon = 1$ (the 'typical' quadratic case) by expanding power series solutions to the fixed point equation and retaining only the leading order terms.

We begin by looking for the fixed point. By assumption this will be a function of $x^2$ (since $\epsilon = 1$) and so we pose the solution

$$f(x) = 1 - ax^2 + \dots \tag{11.13}$$

with

$$f^2(x) = 1 - a + 2a^2x^2 + \dots \tag{11.14}$$

where the dots denote terms in $x^{2n}$, $n \geq 2$. Hence

$$Tf(x) = -\alpha f^2(-\alpha^{-1}x) = -\alpha(1 - a + 2a^2\alpha^{-2}x^2 + \dots). \tag{11.15}$$

For $f$ to be a fixed point of $T$ we need

$$1 - ax^2 = -\alpha(1 - a + 2a^2\alpha^{-2}x^2 + \dots) \tag{11.16}$$

and so, equating coefficients of $x^n$, $n = 0, 2$, we find

$$\alpha(a-1) = 1 \text{ and } 2a\alpha^{-1} = 1. \tag{11.17}$$

Solving for $a$ gives $2a^2 - 2a - 1 = 0$, or

$$a = \tfrac{1}{2}(1 + \sqrt{3}) \tag{11.18}$$

and since $\alpha = 2a$,

$$\alpha = 1 + \sqrt{3} \tag{11.19}$$

(but remember that this is only a first approximation). This gives a value of $\alpha$ as approximately 2.7 whereas the true value is just greater than 2.5. We can see this another way. Suppose that $f(x) = 1 - b_n x^2$ is a unimodal map with some particular property (for example, there is a period-doubling bifurcation from period $2^{n-1}$ to period $2^n$), then the map

$$-\alpha_n f^2(-\alpha_n^{-1}x) = -\alpha_n(1 - b_n + 2b_n^2\alpha_n^{-2}x + \dots) \tag{11.20}$$

is approximately the unimodal map

$$1 - b_{n-1}x^2 \tag{11.21}$$

provided

$$\alpha_n(b_n - 1) = 1 \text{ and } b_{n-1} = 2b_n^2\alpha_n^{-1}. \tag{11.22}$$

The equations above for $a$ and $\alpha$, (11.17) and (11.18), are obtained by looking for fixed points of this approximate renormalization scheme ($b_{n-1} = b_n$). However, if the map $1 - b_n x^2$ has a period-doubling bifurcation from period $2^{n-1}$ to period $2^n$ then the map $1 - b_{n-1}x^2$ has a period-doubling bifurcation from period $2^{n-2}$ to $2^{n-1}$. In other words (from the relationship between $\delta$ and $\mu_n$ above)

$$\delta \sim \frac{b_{n-1} - b_\infty}{b_n - b_\infty} \tag{11.23}$$

where $b_\infty = a$, the value of $b$ for the fixed point of the renormalization. Substituting for $b_{n-1}$ gives

$$\delta \sim \frac{2b_n^2(b_n - 1) - a}{b_n - a}. \tag{11.24}$$

Now let $b_n = a - \epsilon$ for some small $\epsilon$, then

$$\delta \sim \frac{2(a-\epsilon)^2(a-\epsilon-1) - a}{-\epsilon} \tag{11.25}$$

or

$$-\epsilon\delta \sim 2(a-\epsilon)^2(a-\epsilon-1) - a. \tag{11.26}$$

Looking at terms of order $\epsilon^0$ gives

$$0 = 2a^2 - 2a - 1, \tag{11.27}$$

which we know to be correct, whilst terms of order $\epsilon$ give

$$\delta \sim -2[-2a(a-1) - a^2] = 2[3a^2 - 2a]. \tag{11.28}$$

Evaluating this expression we find $\delta \sim 4 + \sqrt{3} \approx 5.7$. This should be compared to the true value of around 4.667; clearly we have not obtained a particularly good approximation, but it is the right order of magnitude, and taking the calculation to higher order gives a much better result. It is possible to do a similar exercise for functions $f(|x|^{1+\epsilon})$ when $\epsilon$ is very small; we leave this as an exercise.

These results can be seen in yet another way, which brings out the relevance of $\delta$ as an eigenvalue associated with the derivative of the operator $T$. Using the first equation of (11.22) to eliminate $\alpha_n$ from the second gives $b_{n-1} = 2b_n^2(b_n - 1)$. Thus $T$ induces a map on the coefficient of $x^2$ in our approximation of the form

$$b \to 2b^2(b-1).$$

A fixed point of this map, $a$, satisfies $a = 2a^2(a-1)$ from which we obtain $a = \frac{1}{2}(1+\sqrt{3})$ as in (11.18). The derivative of this map is $6b^2 - 4b$, so the stability of the fixed point is determined by $\delta = 6a^2 - 4a = 4 + \sqrt{3}$, cf. (11.28). The interpretation of this eigenvalue is essentially the same as the original interpretation for the eigenvalues of the derivative of $T$ above.

It is also possible to obtain a universal scaling law for the growth of measurements of chaos in $\mu > \mu_\infty$. We shall concentrate on the growth of topological entropy (Definition 11.6). Recall that the topological entropy is the growth rate of the number of $(n, \epsilon)$-separated sets for the map $f$, $C(f, \epsilon, n)$. Now,

$$C(f^m, \epsilon, n) = C(f, \epsilon, nm) \tag{11.29}$$

so

$$\begin{aligned} h(f^m, \epsilon) &= \limsup_{n\to\infty} \frac{1}{n} \log C(f^m, \epsilon, n) \\ &= \limsup_{n\to\infty} \frac{1}{n} \log C(f, \epsilon, mn) \sim mh(f, \epsilon). \end{aligned} \tag{11.30}$$

This suggests that

$$h(f^m) \sim mh(f) \tag{11.31}$$

where we have used the symbol '$\sim$' because we have not been careful about the way limits have been taken. Furthermore, the arguments of the previous section show that when restricted to a suitable subinterval and rescaled the map $f^2_{\mu-\mu_\infty}$ has the same dynamics as $f_{\delta(\mu-\mu_\infty)}$ and so

$$h(f_{\mu-\mu_\infty}) \sim \tfrac{1}{2} h(f^2_{\mu-\mu_\infty}) \sim \tfrac{1}{2} h(f_{\delta(\mu-\mu_\infty)}). \tag{11.32}$$

If we now pose the solution

$$h(f_{\mu-\mu_\infty}) \sim (\mu - \mu_\infty)^\beta$$

equation (11.32) gives

$$(\mu - \mu_\infty)^\beta \sim \tfrac{1}{2}\delta^\beta (\mu - \mu_\infty)^\beta \tag{11.33}$$

and taking logs gives

$$\beta = \frac{\log 2}{\log \delta}. \tag{11.34}$$

Hence the topological entropy in $\mu > \mu_\infty$ grows roughly according to the power law

$$h(f_{\mu-\mu_\infty}) \sim (\mu - \mu_\infty)^{\frac{\log 2}{\log \delta}}. \tag{11.35}$$

## 11.7 Intermittency

Numerical simulations of many maps show that there are parameter regions in which the dynamics of the map is as shown in Figure 11.17.

The interesting feature of these experiments is that the system appears to spend a long time close to a periodic orbit (the so-called laminar region), then oscillate in some complicated manner and then come close to the same periodic orbit again.

To understand this type of behaviour we stay with one-dimensional maps and look at the behaviour of maps near a saddlenode bifurcation. In a neighbourhood of a saddlenode bifurcation the map (or an iterate of the map) varies as shown in Figure 11.18: for $\mu < 0$ trajectories get caught near the impending tangency and spend a long time close to the ghost of the periodic point, at $\mu = 0$ there is a tangency between the graph of the map and the diagonal, creating a non-hyperbolic fixed point which is stable from one side and unstable from the other, and in $\mu > 0$ there are two fixed points locally, one stable and the other unstable. Intermittent trajectories are observed for $\mu < 0$ provided there is some global reinjection mechanism which allows orbits which pass close to the point where the periodic point will be formed to come back close to that point (in fact, the existence of this global reinjection implies, for continuous maps of the interval, that the map is chaotic, so intermittency is not a route to chaos in the same sense as period-doubling). Assuming that the tangency at $\mu = 0$ is quadratic, it is possible to use arguments similar to those of the previous section to show that there is a characteristic time spent in the laminar corridor of Figure 11.18a, which increases to infinity as $\mu$ tends to zero from below.

Let us suppose that for values of $\mu$ near zero and for $x$ close to the point at which the non-hyperbolic periodic point will appear when $\mu = 0$ the map takes the form

$$f_{\mu}(x) = \mu + x - x^2. \tag{11.36}$$

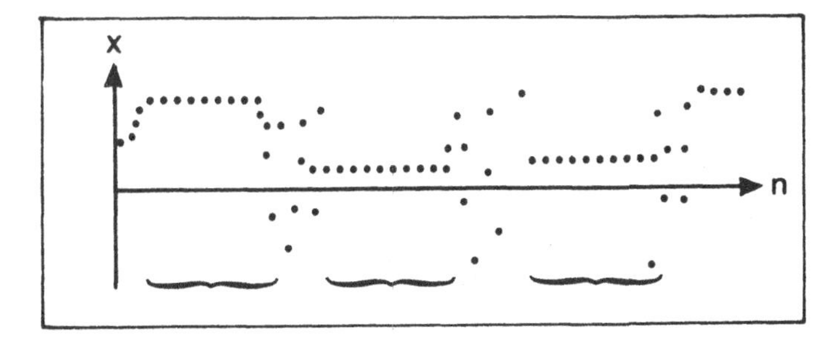


Fig. 11.17 Time sequence for a map showing intermittent laminar regions.

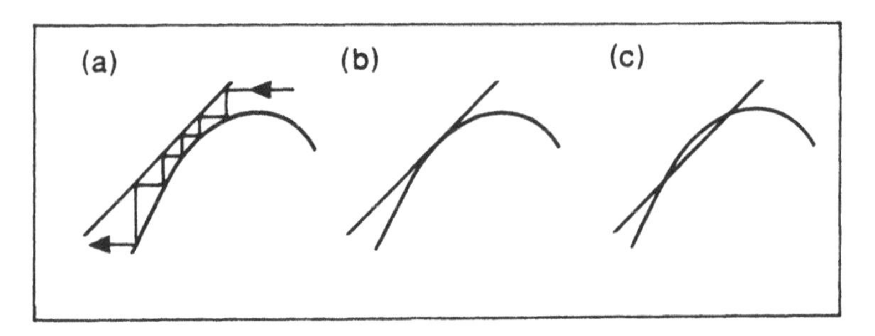


Fig. 11.18 (a) $\mu < 0$; (b) $\mu = 0$; (c) $\mu > 0$.

Now, the second iterate of $f_\mu$ is

$$\begin{aligned} f_\mu^2(x) &= \mu + (\mu + x - x^2) - (\mu + x - x^2)^2 \\ &\approx 2\mu + x - 2x^2 \end{aligned} \tag{11.37}$$

where we have kept only the leading order terms in the approximation. Now if we set $y = 2x$,

$$y_{n+2} = 2x_{n+2} \approx 4\mu + 2x_n - 4x_n^2 = 4\mu + y_n - y_n^2. \tag{11.38}$$

In other words

$$f_{4\mu}(y) \approx 2f_\mu^2(\tfrac{1}{2}y). \tag{11.39}$$

Hence, if an orbit spends on average $N(\mu)$ iterations in the laminar region this equation implies that

$$N(4\mu) \sim \tfrac{1}{2}N(\mu) \tag{11.40}$$

where the factor of $\frac{1}{2}$ comes from the fact that if the orbit spends on average $N(\mu)$ iterations in the laminar region for $f_\mu$ it will spend $\frac{1}{2}N(\mu)$ iterations in the laminar region under $f_\mu^2$. Now looking for a power law, $N(\mu) \sim c|\mu|^\beta$ gives

$$c4^\beta|\mu|^\beta \sim \tfrac{1}{2}c|\mu|^\beta$$

or

$$4^\beta = \tfrac{1}{2}. \tag{11.41}$$

This equation has the obvious solution $\beta = -\frac{1}{2}$ (take logs!) and so we obtain the scaling law, valid for small $|\mu|$, $\mu < 0$,

$$N(\mu) \sim |\mu|^{-\frac{1}{2}}. \tag{11.42}$$

This scaling relationship has been observed in numerous examples, both in maps and in flows (where a periodic orbit plays the role of the fixed point or periodic point). Other more fundamental and universal types of intermittency have been found, associated with saddlenode bifurcations of maps of the circle (cf. Chapter 9). Guckenheimer and Holmes (1983) contains further details and references for the interested reader.

## 11.8 Partitions, graphs and Sharkovskii's Theorem

In this section we will develop some simple tools which will enable us to give a proof of Sharkovskii's Theorem (Theorem 11.16). The treatment follows the proof by Block, Guckenheimer, Misiurewicz and Young (in Nitecki and Robinson, 1980) with modifications from Nitecki (in Katok, 1982). We begin with three definitions which will form the basis for the rest of the analysis.

(11.34) DEFINITION

*A partition of the interval* $[a, b]$ *is a finite set of points* $(x_i)$, $i = 0, 1, \ldots, n$, *with*

$$a = x_0 < x_1 < \ldots < x_{n-1} < x_n = b.$$

*The* $n$ *closed intervals* $J_i = [x_{i-1}, x_i]$, $i = 1, \ldots, n$ *are called elements of the partition.*

(11.35) DEFINITION

*Suppose that* $f : [a, b] \to \mathbf{R}$ *is a continuous map and let* $J_i$ *be elements of a partition of* $[a, b]$. *Then* $J_i$ *f-covers* $J_k$ $m$ *times if there exist* $m$ *disjoint open subintervals* $K_1, \ldots, K_m$ *of* $J_i$ *such that* $f(cl(K_r)) = J_k$, $r = 1, \ldots, m$.

(11.36) DEFINITION

*Let* $f$ *and* $J_i$ *be as in Definition 11.35. An A-graph of* $f$ *is an oriented generalised graph with vertices* $J_i$ *such that if* $J_i$ *f-covers* $J_k$ $m$ *times then there are* $m$ *arrows from the vertex* $J_i$ *to the vertex* $J_k$.

We have already met an example of this in the definition of the horseshoe (Definition 11.2). If $f$ has a horseshoe then there exist intervals $J_1$ and $J_2$ such that $J_1 \cup J_2 \subseteq f(J_i)$ for $i = 1, 2$. Hence each of these two intervals $f$-covers both itself and the other interval at least once, and the resulting A-graph of $f$ must contain the subgraph shown in Figure 11.19.

An *allowed path* for an A-graph is a sequence $J_{a(1)}J_{a(2)} \cdots J_{a(s)}$ where $a(i) \in \{1, \ldots, n\}$ such that there is an arrow from $J_{a(i)}$ to $J_{a(i+1)}$ on the A-graph for $i = 1, \ldots, s-1$. For the A-graph of the horseshoe (Fig. 11.19) every possible sequence of $J_1$s and $J_2$s is an allowed path.

(11.37) LEMMA

*Let* $J_{a(1)}J_{a(2)} \cdots J_{a(s+1)}$ *be an allowed path and suppose that* $a(1) = a(s+1)$. *Then there is a point* $x \in J_{a(1)}$ *such that* $f^s(x) = x$ *and* $f^i(x) \in J_{a(i)}$, $i = 2, \ldots, s$.

*Proof*: Working backwards, using the definition of an $f$-cover (Definition 11.35), we see that there are subintervals $K_i \subseteq J_{a(i)}$ which satisfy $f(K_i) = K_{i+1}$, $i = 1, \ldots, s$, and $K_{s+1} = J_{a(s+1)} = J_{a(1)}$. Hence $f^s(K_1) = J_{a(1)}$ and (by definition) $K_1 \subseteq J_{a(1)}$. Applying the Fixed Point Lemma (Lemma 11.13) to $K_1$ we see that there exists $x \in K_1$ with $f^s(x) = x$. Since $f(K_i) = K_{i+1} \subseteq J_{a(i+1)}$ the last part of the lemma follows immediately.

This lemma is the main ingredient of the proof of Sharkovskii's Theorem. It needs to be applied with a little care: although it can be used to prove the existence of a fixed point for $f^s$ in some element of the partition, this point may not have minimal period $s$. As an example, consider the path $J_1J_2J_1J_2J_1$ allowed by the horseshoe A-graph of Figure 11.19.

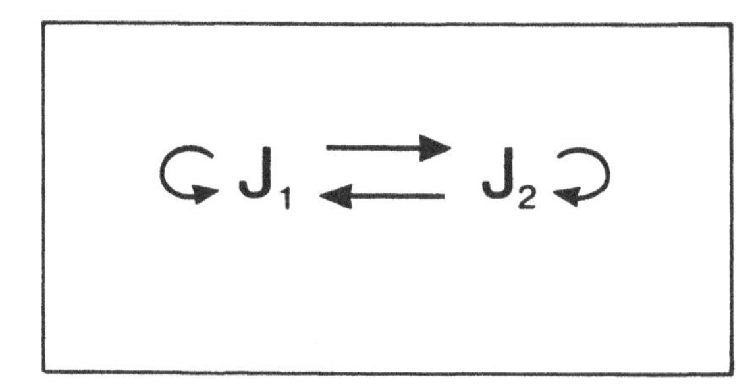


Fig. 11.19 The subgraph for the horseshoe.

Applying Lemma 11.37 we find that there is a fixed point of $f^4$ in $J_1$. However, this closed allowed path (or *loop*) is simply the shorter loop $J_1 J_2 J_1$ repeated twice, and applying the lemma to this loop we find that there is a fixed point of $f^2$ in $J_1$. There is no way of distinguishing the two fixed points using Lemma 11.37, so we cannot deduce the existence of a point with minimal period four from this loop. On the other hand, Figure 11.19 allows the loop $J_1 J_2 J_2 J_2 J_1$, which cannot be a repetition of the same orbit of lower period twice, so there is an orbit of minimal period four associated with this A-graph.

To resolve this ambiguity we say that a loop is *irreducible* if it is not a periodic repetition of a shorter loop. As an example we can give a short proof of Exercise 11.15.

(11.38) LEMMA

*Suppose that $f : [a, b] \to \mathbf{R}$ is a continuous map and has an orbit of (minimal) period three. Then $f$ has orbits of period $n$ for all $n \in \mathbf{N}$.*

*Proof*: Let $p_1 < p_2 < p_3$ be points on the periodic orbit of period three and assume (without loss of generality) that

$$f(p_1) = p_2, \quad f(p_2) = p_3, \quad \text{and} \quad f(p_3) = p_1.$$

Let $J_1 = [p_1, p_2]$ and $J_2 = [p_2, p_3]$. By the Intermediate Value Theorem, $J_2 \subseteq f(J_1)$ and $J_1 \cup J_2 \subseteq f(J_2)$. Hence $f$ has the A-graph given in Figure 11.20. From Figure 11.20 we see immediately that $f$ has a fixed point ($J_2 J_2$ and Lemma 11.37), a point of period two ($J_2 J_1 J_2$) and $f$ has a point of period three by assumption. For any $n > 3$ the loop $J_2 J_1 (J_2)^{n-2} J_2$ is irreducible and allowed. Hence $f$ has an orbit of (minimal) period $n$ for all $N \geq 1$.

This lemma, as described in Section 11.2, is a particular case of Sharkovskii's Theorem (Theorem 11.16). The proof of Sharkovskii's Theorem involves a slightly more sophisticated application of the same principles.

(11.39) LEMMA

*Let $f$ be a continuous map of the interval and suppose that $f$ has an orbit of odd period $m$. Then $f$ has orbits of all even periods less than $m$ and all periods greater than $m$.*

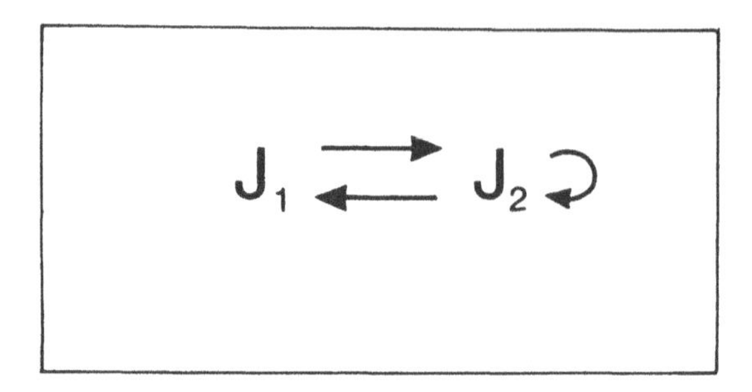


Fig. 11.20 The A-graph for Lemma 11.38.

*Proof*: Without loss of generality we assume that $f$ has no periodic orbits with odd period less than $m$. Let $p_1 < p_2 < \ldots < p_m$ be points on the orbit of (odd) period $m$ and note that $f(p_1) > p_1$ and $f(p_m) < p_m$. This implies that there exists $j < m$ such that $f(p_i) < p_i$ for all $i > j$ and $f(p_j) > p_j$. In particular

$$f(p_j) > p_j \quad \text{and} \quad f(p_{j+1}) < p_{j+1}.$$

We are going to use the periodic points $(p_i)$ as a partition. Let $I_1 = [p_j, p_{j+1}]$ and note that $I_1 \subseteq f(I_1)$, so the A-graph of $f$ has an arrow from $I_1$ to itself and there is a point $z \in I_1$ with $f(z) = z$. Next we want to show that there is an allowed path from $I_1$ to every other element of the partition. Let $V_1 = \{I_1\}$ and let $V_2$ be the set of elements of the partition which are $f$-covered by $I_1$, so $I_1 \in V_2$ but $V_2 \neq \{I_1\}$ as $p_j$ is not periodic of period two. In general, let $V_i$ be the collection of partition elements which are $f$-covered by some element of $V_{i-1}$, so if $I_k \in V_{i+1}$ for some $i > 0$ then there is an allowed path from $I_1$ to $I_k$ of length $i$. It should be clear that $V_i \subseteq V_{i+1}$, so the sets $V_i$ get larger, and since there are only a finite set of partition elements there exists $r$ such that $V_{r+1} = V_r$. If $V_r$ does not contain some element of the partition then we get a contradiction (the point $p_j$ would have period less than $m$) and hence there is a path from $I_1$ to every other partition element.

We now want to show that there is an element of the partition, $I_k$ say, which $f$-covers $I_1$. To do this we use the fact that $m$ is odd. Since $m$ is odd there are more elements of the partition on one side of $z$ than on the other side, and so there must exist $p_r$ such that $f(p_r) > z$ and $f(p_{r+1}) < z$ or vice versa. (If this were not the case then all the points $p_i$ less than $z$ would have to map to points greater than $z$ and vice versa and so the number of points on each side of $z$ would have to be the same, which is impossible if $m$ is odd.) Let $I_k = [p_r, p_{r+1}]$, then $I_1 \subseteq f(I_k)$.

Putting the results of the previous two paragraphs together we obtain the A-graph shown in Figure 11.21: $I_1$ has an arrow to itself, and since there is an allowed path from $I_1$ to every other vertex, there is an allowed path from $I_1$ to $I_k$ which we shall assume to have length $k$ and pass through $I_2$, $I_3$, $\ldots$, $I_{k-1}$. Finally, we have an arrow from $I_k$ to $I_1$.

We now want to investigate $k$ further. Assume that this path is the shortest path from $I_1$ to $I_k$, then each of the intermediate partition elements $I_2, I_3, \ldots, I_{k-1}$ are distinct, otherwise we could find a shortcut contradicting the assumption that we have chosen the shortest path. Hence $k \leq m-1$. However, if $k < m-1$ then one of the allowed loops $I_1 I_2 \ldots I_k I_1$ or $I_1 I_1 I_2 \ldots I_k I_1$ would give rise to an orbit with odd period less than $m$; a contradiction. Hence $k = m-1$ and each element of the partition appears once in the loop from $I_1$ to itself via $I_k$.

The final step of the proof is to show that there exist arrows from $I_k$ (which we will now refer to as $I_{m-1}$) to every odd vertex of the A-graph, giving the A-graph shown in Figure 11.22. The is the only tricky bit of the proof. First note that since we have constructed the shortest non-trivial loop from $I_1$ to itself there can be no arrows from $I_r$ to $I_s$ if $s > r+1$ since if there were such an arrow then we could find a shorter loop back to $I_1$. In particular, $I_1$ $f$-covers $I_2$ but no other element of the partition. There are two ways of achieving this, which turn out to be the same after a change of orientation ($x \to -x$). Either $f(p_j) = p_{j+1}$ and $f(p_{j+1}) = p_{j-1}$, or $f(p_j) = p_{j+2}$ and $f(p_{j+1}) = p_j$. Assume that the former possibility holds. Then $I_2 = [p_{j-1}, p_j]$. Now, $I_2$ cannot $f$-cover $I_1$ (unless $m = 3$, in which case the lemma is already proved by Lemma 11.37) nor can it $f$-cover any of the intervals $I_r$, $r > 3$, or there would be a shortcut in the non-trivial loop from $I_1$ to itself. Hence $I_2$ $f$-covers $I_3$ and no other intervals of the partition. But $I_2 = [p_{j-1}, p_j]$

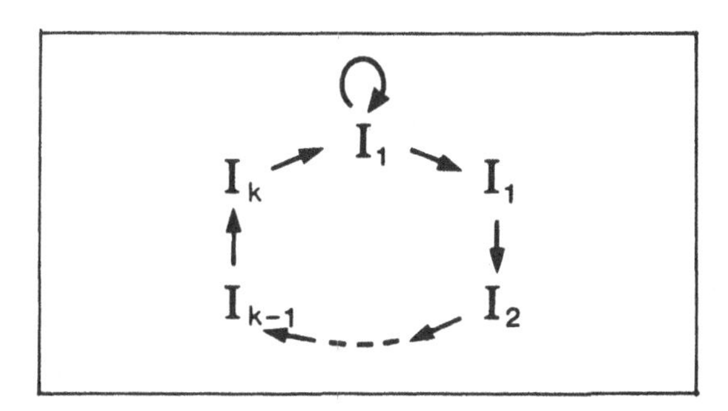


Fig. 11.21 The A-graph halfway through the proof of Lemma 11.39.

and $f(p_j) = p_{j+1}$, so we must have $I_3 = [p_{j+1}, p_{j+2}]$ and $f(p_{j-1}) = p_{j+2}$. Continuing in this way the images of each of the intervals $I_r$ are forced to swap sides and move out from $I_1$ and we find that the order of the intervals on the real line is

$$I_{m-1}I_{m-3}\dots I_2 I_1 I_3 \dots I_{m-2}$$

and that $I_r$ $f$-covers $I_{r+1}$ and no other element of the partition for $2 \leq r \leq m-2$. Now consider $I_{m-1} = [p_1, p_2]$. Since $I_{m-3} = [p_2, p_3]$ $f$-covers $I_{m-2}$ and no other interval we find that $f(p_2) = p_m$. The only point which has not been assigned a preimage by the above argument is $p_j$, and the only point without a prescribed image is $p_1$. Hence $f(p_1) = p_j$. Thus $[p_j, p_m] \subseteq f([p_1, p_2])$ and so $I_{m-1}$ $f$-covers all the elements of the partition to the left of $I_1$, including $I_1$. This gives the arrows from $I_{m-1}$ to every odd vertex as shown in Figure 11.22.

The proof of the lemma is now complete, since periodic orbits of all even periods less than $m$ exist (using the arrows to odd vertices, then the loop back to $I_{m-1}$) and orbits of all periods greater than $m$ exist (going once round the loop and then any number of $I_1$s).

Now we need to deal with the case where $f$ has no orbits of odd period.

(11.40) LEMMA

*Suppose that $f$ has an orbit of period $2^r m$, where $r \geq 1$ and $m$ is odd. Denote the points on the periodic orbit by $(p_i)$ in the usual way. Then $f$ has a fixed point, $z$, in the interval $[p_{2^{r-1}m}, p_{2^{r-1}m+1}]$. Furthermore*

$$f(p_i) \geq p_{2^{r-1}m+1} \quad \text{if} \quad i \leq 2^{r-1}m$$

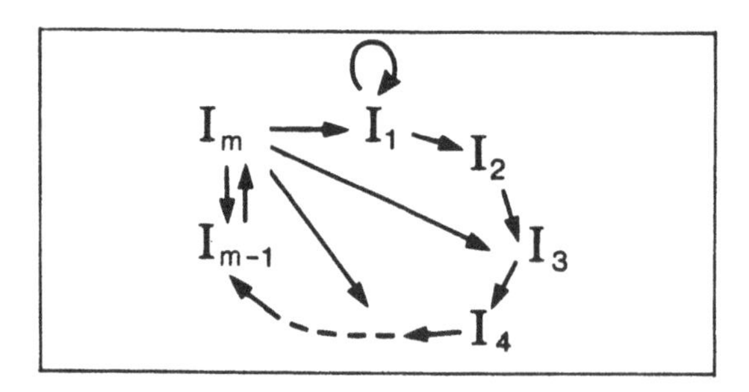


Fig. 11.22 The final A-graph of Lemma 11.39.

*and*

$$f(p_i) \le p_{2^{r-1}m} \quad \text{if} \quad i \ge 2^{r-1}m+1.$$

*Proof*: The only place that the odd property of $m$ was used in the proof of Lemma 11.39 was in showing the existence of an interval $I_k$ which $f$-covers $I_1$. Hence the same argument as in Lemma 11.39 establishes the existence of an element of the partition, $I_1$, which $f$-covers itself, and paths from $I_1$ to every other element of the partition. Suppose that there does exist another element of the partition which $f$-covers $I_1$. Then we obtain the A-graph of Figure 11.20 and so $f$ has periodic orbits with odd period, contradicting the assumptions of the lemma.

So no other element of the partition $f$-covers $I_1$. In order for this to be the case, all the points $(p_i)$ on one side of $I_1$ must either stay on the same side of $I_1$ (a contradiction as the partition is a periodic orbit) or they must map to the other side of $I_1$. Thus $I_1$ must be symmetrically placed in the middle of the partition, i.e. $I_1 = [p_{2^{r-1}m}, p_{2^{r-1}m+1}]$. Since $I_1$ $f$-covers itself $f$ has a fixed point, $z$, in this interval and since the periodic points must map to the other side of $I_1$ we obtain the inequalities of the lemma.

The inequalities of this lemma have an important consequence: when constructing the A-graph of $f$ we see that we are only able to assume that an interval $[p_j, p_{j+1}]$ $f$-covers another interval if they are on opposite sides of the central interval, $I_1$. Indeed, were it to $f$-cover an interval on the same side as itself then it would have to $f$-cover $I_1$, leading to the contradiction used in the proof of the lemma.

This suggests that to gain further information we should look at the *induced A-graph of* $f^2$. The fixed point of $f$ in $I_1$ divides the set of points on the periodic orbit in two. Those on the left move over to the right and vice versa. Hence the points $p_1, \dots, p_{2^{r-1}m}$ (or equivalently the points $p_{2^{r-1}m+1}, \dots, p_{2^r m}$) are periodic with period $2^{r-1}m$ under the second iterate of $f$. The induced A-graph of $f^2$ is the A-graph induced by the action of $f^2$ on this periodic orbit for $f^2$, i.e. there is an arrow from one element of the partition, $I_i$, to another, $I_k$, if and only if $I_i$ $f$-covers some interval $I_j$ and $I_j$ $f$-covers $I_k$. There are two things worth noting about the induced A-graph of $f^2$ when $f$ has no odd periodic orbits: first, if it has an irreducible loop of length $n$, then $f$ has an irreducible loop of length $2n$. Second, the induced A-graph of $f^2$ is slightly different from the A-graph of $f^2$, which may contain more (but not fewer) arrows.

(11.41) EXERCISE

*Prove the two remarks made above.*

We can now prove Sharkovskii's Theorem, which we restate here for convenience.

(11.42) THEOREM (SHARKOVSKII)

*Consider the order on the natural numbers defined by*

$$1 \prec 2 \prec 4 \prec 2^3 \prec \ldots \prec 2^n \prec 2^{n+1} \prec \ldots$$

$$\ldots \prec 2^{n+1}.9 \prec 2^{n+1}.7 \prec 2^{n+1}.5 \prec 2^{n+1}.3 \prec \ldots$$

$$\ldots \prec 2^n.9 \prec 2^n.7 \prec 2^n.5 \prec 2^n.3 \prec \ldots \prec 9 \prec 7 \prec 5 \prec 3.$$

*Let $f : [a,b] \to \mathbf{R}$ be a continuous map of the interval. Suppose $f$ has an orbit of period $n$, then $f$ has an orbit of period $k$ for all $k \prec n$.*

*Proof*: If $n$ is odd then the theorem is proved by Lemma 11.39.

If $n = 2^r m$, where $m$ is odd, then (without loss of generality) $f$ has no orbits with odd period. By Lemma 11.40 $f$ has a fixed point (an orbit of period one) and we look at the induced A-graph of $f^2$. The partition for this A-graph has period $2^{r-1}m$. If $r = 1$ we apply the techniques of Lemma 11.39 to the induced A-graph and the theorem is proved.

If $r \geq 2$ then proceed inductively. Look at the induced graph of the induced graph (the induced graph of $f^4$) restricted to the partition on one side of the fixed point of the induced A-graph of $f^2$. This has a fixed point (a point of period four for $f$) and the partition points have period $2^{r-2}m$. Continuing in this way we obtain, inductively, points of period $2^k$, $k < r$, until the induced A-graph of $f^{2^r}$ has a partition of odd period. Applying Lemma 11.39 again completes the proof if $m \geq 3$, whilst if $m = 1$ we find (looking at the induced A-graph of $f^{2^{r-1}}$) that this has a partition of two points, each of period two, giving the trivial graph with a single vertex and an arrow from that vertex to itself (implying the periodic point of period $2^{r-1}$). In either case the theorem is proved.

## Exercises 11

1. By showing that the map $f_\mu(x) = \mu - x^2$ has a stable point of period three if $\mu^3 - 2\mu^2 + \mu - 1 = 0$, $(\mu > 1)$, prove that a map with a stable periodic orbit can be chaotic.

2. Using methods similar to those of Section 11.4, describe the attracting set for the family of maps $L_s : [0,1] \to [0,1]$, $1 < s \leq 2$, defined by

$$L_s(x) = \begin{cases} 1 - s(x - \frac{1}{2}), & x \in [0, \frac{1}{2}] \\ s(x - \frac{1}{2}), & x \in (\frac{1}{2}, 1] \end{cases}.$$

3. Consider the map

$$H(x) = \begin{cases} 3x, & x \in [0, \frac{1}{2}] \\ -2 + 3x, & x \in (\frac{1}{2}, 1] \end{cases}.$$

Prove that the only points which remain in $[0,1]$ under iteration have the form

$$x = \sum_{n=1}^{\infty} \frac{a_n}{3^n}, \quad a_n \in \{0, 2\}.$$

If $\mathbf{a}(x) = a_1 a_2 a_3 \ldots$, where $a_n \in \{0, 2\}$ as above, show that

$$\mathbf{a}(H(x)) = \sigma \mathbf{a}(x)$$

where $\sigma \mathbf{a} = \mathbf{b}$ if $b_n = a_{n+1}$. For what value of $x$ is $\mathbf{a}(x) = 002002002002\ldots$? Show that this point is periodic of period three for $H$.

4. Repeat Exercise 3 for the map

$$F(x) = \begin{cases} 3x, & \text{if } x \in [0, \frac{1}{2}] \\ 3(1 - x), & \text{if } x \in (\frac{1}{2}, 1] \end{cases}.$$

Show that the set of points which remain in $[0,1]$ under iteration is the same and find $\mathbf{a}(F(x))$ in the cases $a_1 = 0$ and $a_1 = 2$. Consider the point with $\mathbf{a}(x) = 002002002\ldots$ as before. Show that this point is not periodic, but is eventually periodic with period three.

5. Consider the map

$$f(x) = \begin{cases} 2x, & \text{if } x \in [0, \frac{1}{2}] \\ 2(1 - x), & \text{if } x \in (\frac{1}{2}, 1] \end{cases}.$$

Show that $f$ has a horseshoe, and hence that $f$ is chaotic. Let $U$ be any open interval in $[0,1]$. Show that if $\frac{1}{2} \notin U$ then the length of $f(U)$ is twice the length of $U$ and hence deduce that there exists $m$ such that $\frac{1}{2} \in f^m(U)$. Prove that there exists $n$ such that $f^n(U) = [0,1]$. Use this to prove that periodic points are dense in $[0,1]$ and that $f$ has sensitive dependence on initial conditions. Show that given any $\epsilon > 0$ there exist points $x$ and $y$ in $[0,1]$ with $0 < |x-y| < \epsilon$ but $f^n(x) = f^n(y)$ for all $n > 0$.

6. Let $(p_i)$, $0 \le i \le n$, be a partition of the interval into $n$ intervals $I_k = [p_{k-1}, p_k]$. Define the $n \times n$ transition matrix, $A$, by

$$A_{ij} = \begin{cases} 1 & \text{if } I_i \ f\text{-covers } I_j \\ 0 & \text{otherwise} \end{cases}.$$

Prove that $(A^k)_{ij}$ gives the number of independent paths of length $k$ from $I_i$ to $I_j$ in the A-graph of $f$. Deduce further that $\mathrm{Tr} A^k$ gives the number of closed loops of length $k$ and hence that the number of fixed points of $f^k$ is greater than or equal to $\mathrm{Tr} A^k$.

7. Let $(p_i)$, $0 \le i \le 2$, be a partition of the interval by a periodic orbit of period three, with $f(p_0) = p_1$, $f(p_1) = p_2$ and $f(p_2) = p_0$. Show that the associated transition matrix is

$$A = \begin{pmatrix} 0 & 1 \\ 1 & 1 \end{pmatrix}.$$

By calculating $\mathrm{Tr} A^5$ show that $f$ has at least two orbits of period five. Show that the number of fixed points of $f^n$ grows like

$$\left(\tfrac{1}{2}(1+\sqrt{5})\right)^n.$$

8. Classify the different orbits of period four (defined by permutations of the points on the orbits). Which permutations can be realised by unimodal maps? Which permutations imply the existence of orbits of period three?

9. Repeat Exercise 6 for those periodic orbits of period five which can be realised by unimodal maps. Show that the existence of one of these orbits implies the existence of another.

10. By integrating the differential equation $\dot{x} = \mu - x^2$ for $\mu < 0$ show that the amount of time spent in the interval $[-1,1]$ as $\mu \to 0^-$ scales like $(-\mu)^{-\frac{1}{2}}$.

11. Consider a unimodal map which has periodic orbits of period $2^n$ for all $n \geq 0$ and no other periodic orbits (i.e. a map with $\mu = \mu_\infty$). Show that the second iterate of this map, restricted to a suitable subinterval containing the critical point, $c$, has the same property. The orbit of the critical point can be labelled in the following way: let

$$k_0(x) = \begin{cases} L & \text{if } f \text{ is increasing at } x, \\ R & \text{if } f \text{ is decreasing at } x, \\ C & \text{otherwise.} \end{cases}$$

Now let $K(f) = k_0(c)k_0(f(c))k_0(f^2(c))\ldots$. If $F$ denotes the restriction of $f^2$ to the subinterval referred to above, show that $K(f)$ can be obtained from $K(F)$ by the replacement operations

$$C \to CR, \quad R \to LR, \quad L \to RR.$$

Hence show that

$$k(f) = CRLRRRLRLRLRRRLR\ldots.$$

What sequence do the odd terms of $k(f)$ give? What of the even terms?

12. Let $S$ denote the space of differentiable maps of an interval $[a,b]$ which have a hyperbolic fixed point. Show that chaotic maps are dense in $S$ in the $C^0$-topology, but that this is no longer the case in the $C^1$-topology.
*Hint: Distances in the $C^0$-topology are defined by*

$$d_0(f,g) = \sup_{x\in[a,b]} |f(x) - g(x)|$$

*and in the $C^1$-topology by*

$$d_1(f,g) = \sup_{x\in[a,b]} |f(x) - g(x)| + \sup_{x\in[a,b]} |f'(x) - g'(x)|.$$

*Show that given any map $f \in S$ and $\epsilon > 0$ it is possible to construct a map $g \in S$ with $d_0(f,g) < \epsilon$ and $g$ is chaotic by modifying $f$ in a small neighbourhood of a fixed point of $f$ but that this construction is no longer possible if $d_1(f,g) < \epsilon$.*

13. Use the methods of Section 11.8 to prove Lemma 11.4.

# 12

# *Global bifurcation theory*

The treatment of bifurcations in previous chapters has been based on a local analysis of the flow near a stationary point or a periodic orbit. There is another class of bifurcations which is just as fundamental as these local bifurcations but which depends on global properties of the flow. A simple example of such a bifurcation is shown in Figure 12.1. We consider a planar flow depending upon a parameter $\mu$ such that for $\mu > 0$ there is a stable periodic orbit enclosing an unstable focus and a saddle outside the periodic orbit with stable and unstable manifolds as sketched in Figure 12.1a. As $\mu$ decreases to zero the periodic orbit approaches the saddle and when $\mu = 0$ the periodic orbit collides with the saddle to form an orbit called a *homoclinic orbit.* So at $\mu = 0$ one branch of the stable manifold of the saddle coincides with a branch of the unstable manifold of the saddle. By Peixoto's Theorem (Theorem 4.13) this is structurally unstable, and so typical small perturbations of the system will not have a homoclinic orbit, and in $\mu < 0$ (Fig. 12.1c) this connection has been broken and we see that the periodic orbit has disappeared. Hence the homoclinic bifurcation at $\mu = 0$ has destroyed the periodic orbit. We shall see that more complicated homoclinic bifurcations can create chaotic behaviour, but to begin with we shall stick to the two-dimensional situation, starting with an example. Note that homoclinic bifurcations are considerably harder to detect than local bifurcations, since we need to know global properties of the flow, but their effect is just as important. Much of the work on global bifurcations was done by a group in Nizhnii Novgorod, initially under Andronov and then under Shil'nikov, who proved that it is possible to have chaotic behaviour in a neighbourhood of a certain type of homoclinic orbit (Shil'nikov, 1970).

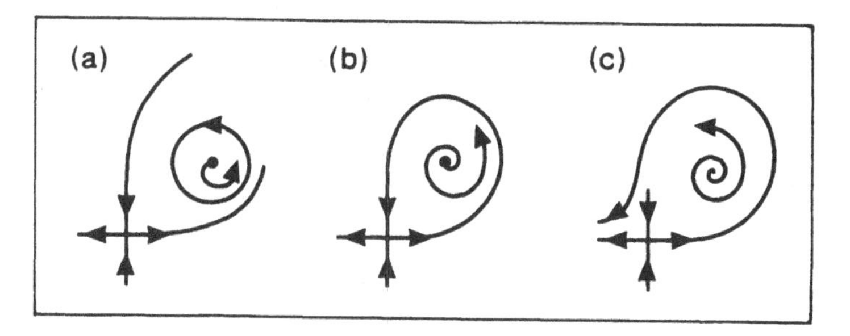


Fig. 12.1 A homoclinic bifurcation: (a) $\mu > 0$; (b) $\mu = 0$; (c) $\mu < 0$.

(12.1) DEFINITION

*A homoclinic orbit is an orbit $\varphi(x,t)$ such that $\varphi(x,t) \to x_0$ as $t \to \infty$ and $t \to -\infty$, where $x_0$ is a stationary point of the flow ($x \neq x_0$).*

## 12.1 An example

As remarked above, it can be very difficult to determine the locus of homoclinic bifurcations analytically. In the rest of this chapter the effect of global bifurcations will be described. This analysis will give an idea of the type of behaviour associated with the bifurcations that might enable one to give strong evidence for the existence of a global bifurcation, in terms of the behaviour of simple periodic orbits and so on. In this section we will go through the analysis of a system for which it is possible to show that there are global bifurcations. Even for this simple prototype, the algebraic manipulations are messy. The theoretical analysis of the later sections is considerably more tractable.

Consider the equation

$$\dot{x} = y, \quad \dot{y} = -a + x^2 + \epsilon(by + xy) \tag{12.1}$$

where $\epsilon > 0$ is a fixed small parameter and $a$ and $b$ are real parameters. This is a simplified version of a more general problem which can be found in Guckenheimer and Holmes (1983). We begin, as ever, by determining the stationary points of solutions and the local bifurcations. Stationary points are solutions of $y = 0$, $x^2 = a$ and so there are two stationary points, $(\pm\sqrt{a}, 0)$ provided $a > 0$. Hence the $b$-axis, $a = 0$, is the locus of a saddlenode bifurcation in parameter space, with a pair of stationary points being created as $a$ increases through zero. Note that if $a < 0$ there are no stationary points and hence (using the Poincaré index) there are no periodic orbits, as any periodic orbit in the plane must enclose at least one stationary point. So all the non-trivial behaviour of this example occurs for $a > 0$. To study this region of parameter space it is convenient to set $a = c^2$, $c > 0$, so the equations become

$$\dot{x} = y, \quad \dot{y} = -c^2 + x^2 + \epsilon(by + xy) \tag{12.2}$$

with stationary points at $(-c, 0)$ and $(c, 0)$. The Jacobian matrix of the flow at $(c, 0)$ is

$$\begin{pmatrix} 0 & 1 \\ 2c & \epsilon(b+c) \end{pmatrix},$$

which has eigenvalues given by the roots of the equation

$$s^2 - \epsilon(b+c)s - 2c = 0. \tag{12.3}$$

Hence the product of the eigenvalues is $-2c < 0$ and the stationary point is a saddle for all $c > 0$. The Jacobian matrix at $(-c, 0)$ is

$$\begin{pmatrix} 0 & 1 \\ -2c & \epsilon(b-c) \end{pmatrix}$$

with eigenvalues given by the roots of $s^2 - \epsilon(b-c)s + 2c = 0$, i.e.

$$s_{\pm} = \frac{\epsilon(b-c) \pm \sqrt{\epsilon^2(b-c)^2 - 8c}}{2}. \tag{12.4}$$

Hence $(-c, 0)$ is stable if $c > b$ and unstable if $b < c$. If $b = c$, the eigenvalues are $s_{\pm} = \pm i\sqrt{2c}$ and so we expect to see a Hopf bifurcation on the curve $b = c$. Note that $c > 0$ so this corresponds to the curve $b = \sqrt{a}$, $b > 0$ in the original $(a, b)$ parameter plane.

(12.2) EXERCISE

*If $b = c > 0$, the Jacobian matrix of the flow at $(-c, 0)$ is*

$$\begin{pmatrix} 0 & 1 \\ 2c & 0 \end{pmatrix}.$$

*Find a coordinate transformation which brings this matrix into the normal form*

$$\begin{pmatrix} 0 & -\sqrt{2c} \\ \sqrt{2c} & 0 \end{pmatrix}$$

*and hence use the standard theory of the Hopf bifurcation to show that there is a subcritical Hopf bifurcation in the system if $b = c$ which creates an unstable periodic orbit in the region $c > b$ (i.e. $\sqrt{a} > b$).*

Putting these local results together we obtain the picture of the $(a, b)$ parameter plane shown in Figure 12.2. This may look perfectly reasonable at first glance, but consider a path in parameter space from the curve of the Hopf bifurcation into $a < 0$, $b < 0$ as shown on the diagram. By the Hopf bifurcation we know that at point $A$ there are two stationary points and a periodic orbit for the system, and at point $B$, having crossed the $b$-axis, the two stationary points have annihilated one another, and (by the Poincaré index argument above) there are no periodic orbits. But we did not notice any bifurcation on this curve which could have destroyed the periodic orbit. To solve this dilemma we claim that this curve in parameter space must cross a curve of homoclinic bifurcations which removes the periodic orbit from the system (this sort of accountancy argument is very useful in bifurcation theory). This bifurcation would be the same as that sketched in Figure 12.1 but with the direction of time reversed.

To support this claim we will use the techniques of perturbation theory developed in Section 7.6. When $\epsilon = 0$ the defining differential equation becomes

$$\dot{x} = y, \quad \dot{y} = -c^2 + x^2 \tag{12.5}$$

where $c^2 = a$, $c > 0$. This unperturbed system is in fact Hamiltonian with

$$H(x, y) = \frac{1}{2}y^2 + c^2 x - \frac{1}{3}x^3. \tag{12.6}$$

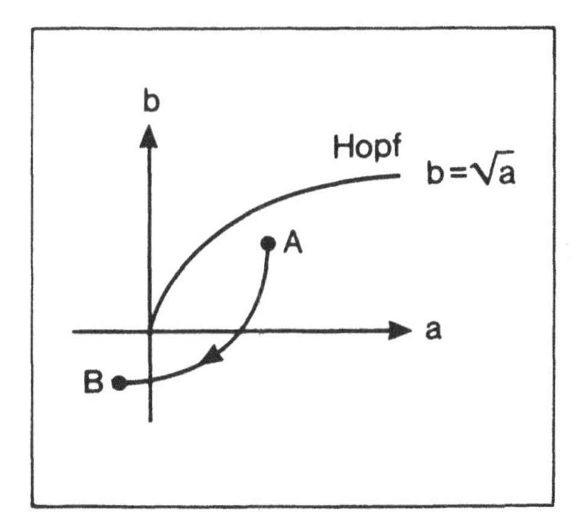


Fig. 12.2 Locus of local bifurcations in the $(a, b)$ plane.

(12.3) EXERCISE

*Verify that* $\dot{x} = \frac{\partial H}{\partial y}$ *and* $\dot{y} = -\frac{\partial H}{\partial x}$.

Trajectories of the unperturbed system therefore lie on curves

$$H(x, y) = E, \qquad (E \text{ constant})$$

as sketched in Figure 12.3. A little manipulation shows that different values of the constant define five possible classes of orbits.

i) If $E < -\frac{2}{3}c^3$ then there are no real solutions to $H(x, y) = E$.
ii) If $E = -\frac{2}{3}c^3$ then the set $H(x, y) = E$ has a single solution, the stationary point $(-c, 0)$.
iii) If $-\frac{2}{3}c^3 < E < \frac{2}{3}c^3$ then the set $H(x, y) = E$ is a closed curve, $\Gamma_E$, which is a periodic orbit of the system.
iv) If $E = \frac{2}{3}c^3$ then the set $H(x, y) = E$ is the union of four trajectories: the stationary point $(c, 0)$, one branch of the unstable manifold of this stationary point which extends to infinity, one branch of the stable manifold of this stationary point which extends to infinity and a homoclinic orbit, $\Gamma_h$, which is bi-asymptotic to $(c, 0)$.
v) If $E > \frac{2}{3}c^3$ then the set $H(x, y) = E$ is the union of two unbounded trajectories.

We now use the perturbation theory of Section 7.6 to determine at which parameter values (if any) the homoclinic orbit is a solution to the

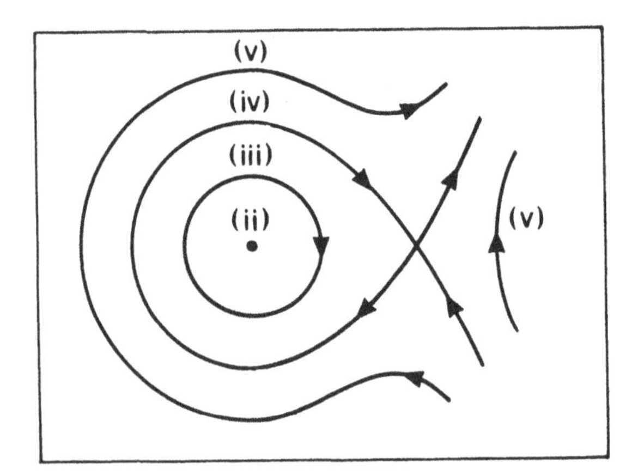


Fig. 12.3 Trajectories of the Hamiltonian system, $H(x, y) = \frac{1}{2}y^2 + c^2x - \frac{1}{3}x^3$.

perturbed equation. Recall that if

$$\dot{x} = f_1(x,y) + \epsilon g_1(x,y), \ \dot{y} = f_2(x,y) + \epsilon g_2(x,y) \tag{12.7}$$

and the unperturbed equation ($\epsilon = 0$) is Hamiltonian, then the perturbed equations have a closed solution $\Gamma_E$ (near the closed solution with $H(x,y) = E$ of the unperturbed system) if

$$\int_0^T (f_1 g_2 - f_2 g_1) dt = 0 \tag{12.8}$$

where the integral is evaluated over a period of the unperturbed periodic solution $(x(t), y(t))$. For the homoclinic orbit $\Gamma_h$, with $E = \frac{2}{3}c^3$ and infinite period this becomes

$$\begin{aligned}\int_{-\infty}^{\infty} (f_1(x_h(t), y_h(t)) g_2(x_h(t), y_h(t)) \\ - f_2(x_h(t), y_h(t)) g_2(x_h(t), y_h(t))) dt = 0\end{aligned} \tag{12.9}$$

where $(x_h(t), y_h(t))$ is the homoclinic orbit. For our example it is easy to see that

$$f_1(x,y) = y, \ g_1(x,y) = 0, \ f_2(x,y) = -c^2 + x^2, \ g_2(x,y) = (b+x)y$$

and so the integral is simply

$$\int_{-\infty}^{\infty} y_h(t)^2 (b + x_h(t)) dt = 0. \tag{12.10}$$

Usually the analysis would end here because of the difficulty of finding the exact solution for the homoclinic orbit $(x_h(t), y_h(t))$ even for the unperturbed system. Fortunately, solutions are known for this particular example.

(12.4) EXERCISE

*Show that $(x_h(t), y_h(t))$ defined by*

$$\left(c - 3c \operatorname{sech}^2\left(t\sqrt{\frac{c}{2}}\right), 3\sqrt{2c^3} \operatorname{sech}^2\left(t\sqrt{\frac{c}{2}}\right) \tanh\left(t\sqrt{\frac{c}{2}}\right)\right)$$

*is a solution of the unperturbed equations $\dot{x} = y$, $\dot{y} = -c^2 + x^2$ with $(x_h(0), y_h(0)) = (-2c, 0)$ and*

$$\lim_{t\to\infty} (x_h(t), y_h(t)) = \lim_{t\to-\infty} (x_h(t), y_h(t)) = (c, 0).$$

Substituting this solution into the integral gives

$$\int_{-\infty}^{\infty} 18bc^3 \text{sech}^4\left(t\sqrt{\frac{c}{2}}\right)\tanh^2\left(t\sqrt{\frac{c}{2}}\right)dt$$
$$+\int_{-\infty}^{\infty} 18c^4\left(1-3\text{sech}^2\left(t\sqrt{\frac{c}{2}}\right)\right)\text{sech}^4\left(t\sqrt{\frac{c}{2}}\right)\tanh^2\left(t\sqrt{\frac{c}{2}}\right)dt$$
$$=0. \qquad (12.11)$$

Setting $\tau = t\sqrt{\frac{c}{2}}$ and dividing out by some miscellaneous constants we obtain

$$\int_{-\infty}^{\infty} b\ \text{sech}^4\tau\ \tanh^2\tau d\tau$$
$$+\int_{-\infty}^{\infty} c(1-3\text{sech}^2\tau)\text{sech}^4\tau\ \tanh^2\tau d\tau = 0. \qquad (12.12)$$

This may look a mess, but using $\text{sech}^2\tau = 1-\tanh^2\tau$ and

$$\int_{-\infty}^{\infty} \text{sech}^2\tau\ \tanh^k\tau d\tau = \left[\frac{\tanh^{k+1}\tau}{k+1}\right]_{-\infty}^{\infty} = \frac{2}{k+1} \qquad (12.13)$$

all the integrals can be evaluated to give

$$b = \tfrac{5}{7}c > 0. \qquad (12.14)$$

Hence there is a homoclinic bifurcation if $b \sim \frac{5}{7}\sqrt{a}$, $b > 0$, and assuming that the bifurcation acts as described in the introduction to this chapter this bifurcation destroys the unstable periodic orbit created in the Hopf bifurcation at $b = \sqrt{a}$, $b > 0$, leaving us with the parameter space shown in Figure 12.4.

## 12.2 Homoclinic orbits and the saddle index

The behaviour of solutions near a homoclinic orbit is dominated by the behaviour near the stationary point, since solutions slow down in the neighbourhood of this point. Hence it is not surprising that the different bifurcations involving homoclinic orbits depend on the type of stationary point that the orbit tends to as $t \to \pm\infty$. Assume that the stationary point is at the origin and that the eigenvalues of the Jacobian matrix evaluated at the origin are distinct and can be divided into two sets, $(\lambda_i)$ with $1 \le i \le m_u$ and $\text{Re}\lambda_i > 0$ and $(\gamma_j)$ with $1 \le j \le m_s$, and $\text{Re}\gamma_j < 0$. We can choose the labelling of these eigenvalues such that $0 < \text{Re}\lambda_i \le \text{Re}\lambda_{i+1}$, $(i < m_u - 1)$, and $0 < -\text{Re}\gamma_j \le -\text{Re}\gamma_{j+1}$,

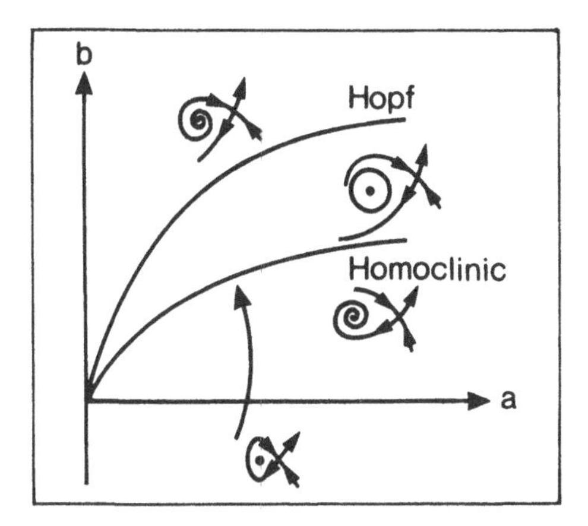


Fig. 12.4 Locus of bifurcations in the $(a, b)$ plane.

$(j < m_s - 1)$. Then typical trajectories approach the origin (as $t \to \infty$) tangential to the eigenspace associated with the eigenvalues with $\mathrm{Re}\gamma_i = \mathrm{Re}\gamma_1$, the eigenvalue whose real part takes the smallest absolute value, and typical trajectories approach the origin as $t \to -\infty$ tangential to the eigenspace associated with the eigenvalues with $\mathrm{Re}\lambda_i = \mathrm{Re}\lambda_1$. By assumption the eigenvalues are distinct, so these tangent eigenspaces are either one-dimensional (if the eigenvalue is real) or two-dimensional (if the eigenvalue is one of a pair of complex conjugate eigenvalues). Thus if we assume that the homoclinic orbit approaches the stationary point tangential to the eigenspace associated with the eigenvalues whose real part has the smallest possible value we find that there are three possibilities: either both of these eigenspaces are one-dimensional, in which case the behaviour is typically two-dimensional (a *saddle*), or one is one-dimensional and the other is two-dimensional (a *saddle-focus*) or both are two-dimensional (a *bi-focus*). These are sketched in Figure 12.5. As we shall see, the stability of solutions depends critically on a quantity called the saddle index, which is defined for each of the three possibilities in the following way.

(12.5) DEFINITION

i) *If the stationary point is a saddle then the saddle index, $\delta$, is the ratio of the leading eigenvalues, $\delta = -\gamma_1/\lambda_1$.*

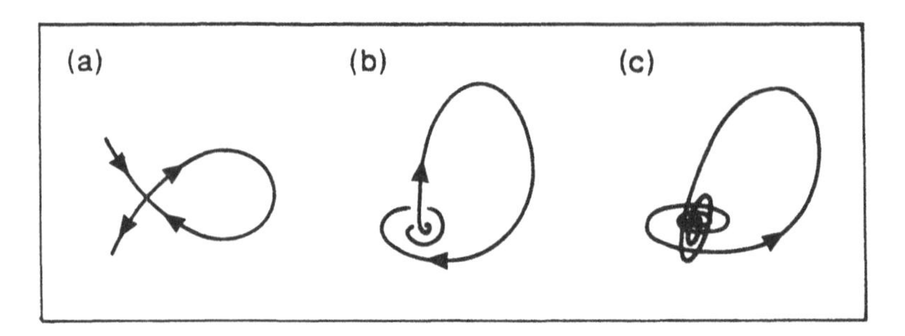


Fig. 12.5 Homoclinic orbits: (a) a two-dimensional saddle; (b) a saddle-focus; (c) a bi-focus.

ii) *If the stationary point is a saddle-focus then it is always possible (after reversing the direction of time if necessary) to make* $\lambda_1$ *real. The saddle index is given by* $\delta = -\text{Re}\gamma_1/\lambda_1$.

iii) *If the stationary point is a bi-focus then* $\delta = -\text{Re}\gamma_1/\text{Re}\lambda_1$.

Note that in cases (*i*) and (*iii*) the direction of time can be chosen such that $\delta \geq 1$ so the saddle index will not play an important role in the qualitative picture of the bifurcations apart from stability properties, but that in case (*ii*) the two possibilities $\delta > 1$ and $\delta < 1$ are completely different. These differences will reappear later in the chapter.

As well as the saddle index and the type of the stationary point there is one further complication which may occur. If the system is symmetric (so the differential equation is invariant under a transformation like $(x, y) \to (-x, -y)$) then the existence of one homoclinic orbit may imply the existence of a second which is the image of the first under the symmetry. More generally one can consider configurations of pairs of homoclinic orbits as drawn in Figure 12.6. Note that Figure 12.6b,c can only be realized for flows in three dimensions or higher. Such configurations can lead to interesting (and chaotic) dynamics in cases such as the saddle, where a single homoclinic orbit only creates a single periodic orbit.

## 12.3 Planar homoclinic bifurcations

In this section we consider differential equations

$$\dot{x} = f_1(x, y, \mu), \quad \dot{y} = f_2(x, y, \mu) \tag{12.15}$$

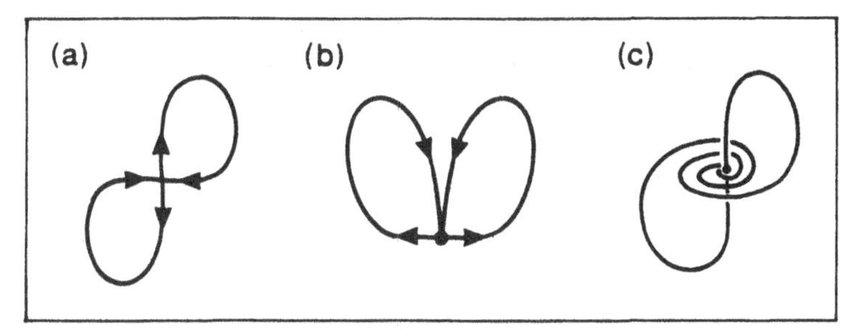


Fig. 12.6 Symmetric pairs of homoclinic orbits: (a) the figure eight; (b) the butterfly; (c) the symmetric saddle-focus.

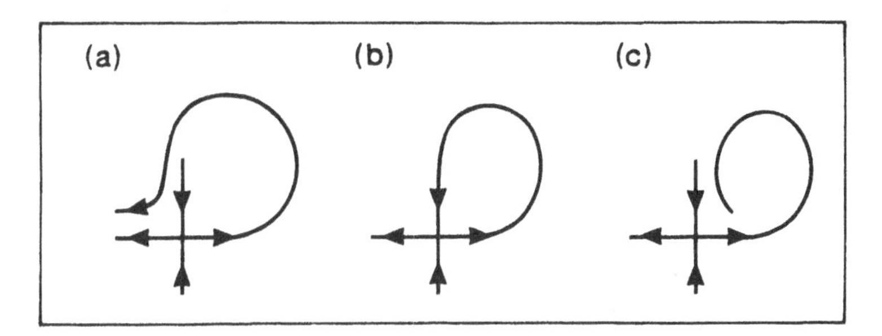


Fig. 12.7 (a) $\mu < 0$; (b) $\mu = 0$; (c) $\mu > 0$.

for which the origin is always a hyperbolic saddle if $\mu$ is in some neighbourhood of $\mu = 0$ and such that if $\mu = 0$ there is a homoclinic orbit biasymptotic to the origin. We assume further that as $\mu$ passes through zero the behaviour of the branches of the stable and unstable manifolds of the origin which form the homoclinic orbit is as shown in Figure 12.7. Note that we know nothing about the other branches of the invariant manifolds. If the eigenvalues of the Jacobian matrix of the flow at the origin are $\lambda(\mu)$ and $\gamma(\mu)$, with $\gamma(\mu) < 0 < \lambda(\mu)$ for $\mu$ in some neighbourhood of $\mu = 0$, then we can use Poincaré's Linearization Theorem to obtain a coordinate system in which the flow is

$$\dot{x} = \lambda(\mu)x, \quad \dot{y} = \gamma(\mu)y \tag{12.16}$$

in some neighbourhood $U$ of the origin. This linear differential equation is easy to solve and so the flow in $U$ is given by

$$\varphi_U(x, y, t) = (xe^{\lambda(\mu)t}, ye^{\gamma(\mu)t}). \tag{12.17}$$

We now want to use this local flow and the geometric information of Figure 12.7 to derive a return map on a suitably chosen local transversal in $U$.

Let $\Sigma = \{(x, y) \in U | y = h\}$ and $\Sigma' = \{(x, y) \in U | x = h\}$, where $h$ is some positive constant sufficiently small so that $\Sigma$ and $\Sigma'$ are non-empty (Fig. 12.8). The local flow $\varphi_U$ induces a map $T_0 : \Sigma \to \Sigma'$ given by $T_0(x, h) = (h, y')$. To find $y'$ we note that a trajectory starting at $(x, h)$ strikes $\Sigma'$ after a time $T$ given by

$$h = xe^{\lambda(\mu)T},$$

i.e.

$$T = \frac{1}{\lambda(\mu)} \log\left(\frac{h}{x}\right) \tag{12.18}$$

and at this value of $t$, the $y$ coordinate of a point which started at $y = h$ will have reached a value $y'$ where

$$y' = h\, e^{\gamma T} = h \left(\frac{h}{x}\right)^{\gamma/\lambda}. \tag{12.19}$$

Hence

$$T_0(x, h) = \left(h,\ h\left(\frac{x}{h}\right)^{-\frac{\gamma}{\lambda}}\right). \tag{12.20}$$

For sufficiently small $|\mu|$ trajectories which leave $\Sigma'$ with $|y|$ small return to $\Sigma$ and do not pass close to any other stationary point. Hence

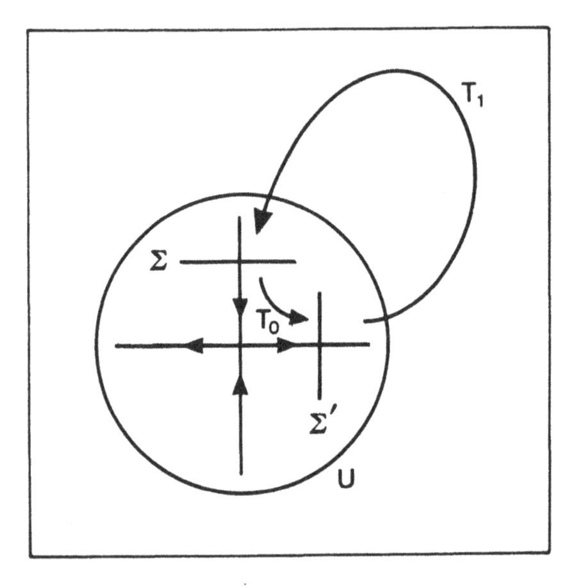


Fig. 12.8 Geometry of the return map.

we obtain a map $T_1 : \Sigma' \to \Sigma$ induced by the global flow outside $U$. When $\mu = 0$ we know that $T_1(h, 0) = (0, h)$, since this is the condition that ensures that the trajectory which leaves $\Sigma'$ on the local unstable manifold of the origin returns to $\Sigma$ on the local stable manifold of the origin to form the homoclinic orbit. By the geometric assumptions of the behaviour of the stable and unstable manifolds of the origin as $\mu$ passes through zero we see that the $x$ coordinate of $T_1(h, 0)$ is greater than zero if $\mu > 0$ and less than zero if $\mu < 0$. Now, since trajectories are bounded away from any other stationary point as they move from $\Sigma'$ to $\Sigma$ the map $T_1$ is as smooth as the original flow and so we can expand it as a Taylor series in $x$ and $\mu$,

$$T_1(h, y) = (a(\mu)\mu + b(\mu)y + O(y^2), h). \tag{12.21}$$

The assumption about the behaviour of the stable and unstable manifolds as $\mu$ passes through zero implies that $a(0) > 0$, and the fact that the flow is planar implies that $b(0) > 0$ (otherwise trajectories must cross). Hence, putting the two maps $T_1$ and $T_0$ together we find a map $T : \Sigma \to \Sigma$ where $T(x, h) = T_1 \circ T_0(x, h)$, valid for sufficiently small $|\mu|$ and $x > 0$. (Remember that if $x < 0$ trajectories move off to the left, close to the branch of the unstable manifold about which we have no information and so are lost from the analysis.) Computing $T$ explicitly by composing the maps $T_0$ and $T_1$ we find

$$T(x, h) = \left( a(\mu)\mu + b(\mu)h \left(\frac{x}{h}\right)^{-\gamma/\lambda} + O(x^{-2\gamma/\lambda}), h \right). \tag{12.22}$$

For sufficiently small $x > 0$ and $|\mu|$ this can be rewritten to leading order as

$$T(x, h) = (a(0)\mu + Bx^\delta + \ldots, h), \quad x > 0 \tag{12.23}$$

where $B > 0$ and $\delta = -\gamma(0)/\lambda(0)$ is the saddle index.

This map can now be analysed like any other map (see Fig. 12.9). First consider the case if $\delta > 1$. If $\mu > 0$ there is a fixed point in $x > 0$ with $x = a\mu + O(\mu^\delta)$. As $\mu$ tends to zero from above this fixed point tends to $x = 0$ and if $\mu < 0$ there are no fixed points (at least for $x > 0$ sufficiently small). A fixed point of the return map corresponds to a periodic orbit of the flow and so we deduce that there is a periodic orbit in $\mu > 0$ which tends to the homoclinic orbit (i.e. $x = 0$) as $\mu$ tends to zero. The stability of the fixed point is obtained from the slope of the map at $x \sim a\mu$, i.e. $\delta Bx^{\delta-1}$. For sufficiently small $\mu$, $\delta > 1$ implies that the slope of the map is less than one and so the fixed point is stable.

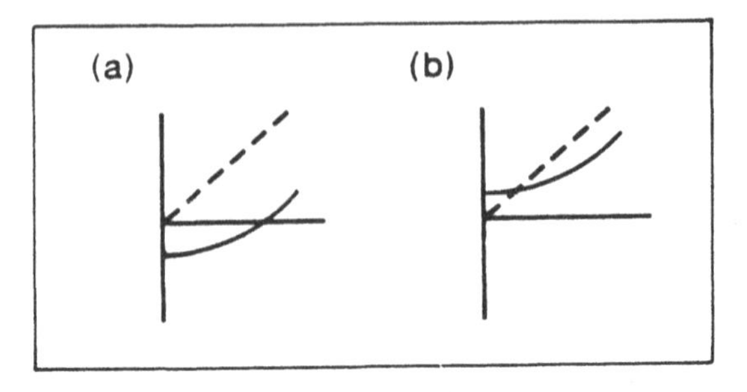


Fig. 12.9 (a) $\mu < 0$; (b) $\mu > 0$.

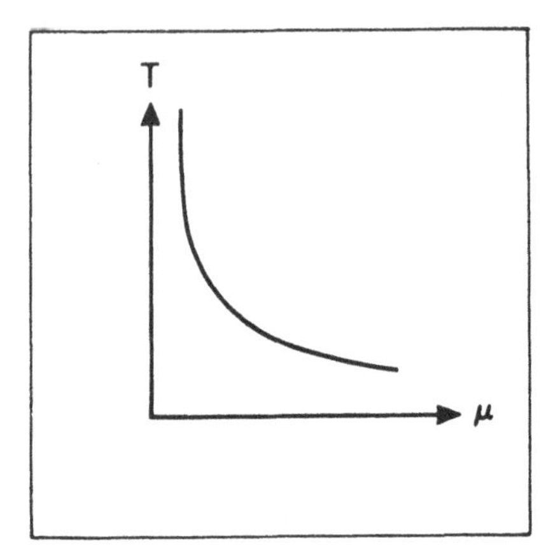


Fig. 12.10 Period against parameter for the periodic orbit.

We can also deduce information about the period of the orbit. The time taken from $(a\mu, h) \in \Sigma$ to its image in $\Sigma'$ is $T$ where, from the linear analysis above,

$$T = -\frac{1}{\lambda}\log\left(\frac{a\mu}{h}\right) \tag{12.24}$$

and the time taken to return to $\Sigma$ from $\Sigma'$ is of order 1. Hence the total period of the periodic orbit is, to leading order,

$$-\frac{1}{\lambda}\log\mu. \tag{12.25}$$

This information is conveniently encoded in a graph of the period of the orbit against the parameter value as shown in Figure 12.10. The period increases exponentially with $\mu$ and there is a sense in which we can view the homoclinic orbit at $\mu = 0$ as a periodic orbit of infinite period. The case $\delta < 1$ is similar except that the periodic orbit exists for $\mu < 0$ and is unstable. We leave this analysis as an exercise.

## 12.4 Homoclinic bifurcations to a saddle-focus

In this section we shall consider the bifurcations associated with a homoclinic orbit which is biasymptotic to a saddle-focus. Recall that the leading eigenvalues for a saddle-focus are $\lambda > 0$ and the complex conjugate pair $\gamma$ and $\gamma^*$, where $\text{Re}\gamma < 0$. Hence we can take $\gamma = -\rho + i\omega$ with $\rho, \omega > 0$ and, working with flows in three dimensions, choose coordinates such that in a neighbourhood, $U$, of the origin the flow is given by

$$\dot{x} = -\rho x - \omega y \tag{12.26a}$$

$$\dot{y} = \omega x - \rho y \tag{12.26b}$$

$$\dot{z} = \lambda z \tag{12.26c}$$

which can be solved to give the local flow

$$\varphi_U(x, y, z, t) = \left(e^{-\rho t}[x \cos \omega t + y \sin \omega t], e^{-\rho t}[-x \cos \omega t + y \sin \omega t], z e^{\lambda t}\right) \tag{12.27}$$

As in the previous section we shall derive a return map on a local transversal in $U$ by composing two maps, one induced by the local flow and the second by the flow near the homoclinic orbit outside $U$. Note that the local stable manifold in $U$ is $\{(x, y, z) \in U | z = 0\}$ and the local unstable manifold is $\{(x, y, z) \in U | x = y = 0\}$. We shall assume that if $\mu = 0$ the upper branch of the unstable manifold forms a homoclinic orbit as shown in Figure 12.11b, and that for $\mu < 0$ the behaviour is as in Figure 12.11a, and for $\mu > 0$ the behaviour is as in Figure 12.11c.

Let $\Sigma = \left\{(x, y, z) \in U | y = 0, he^{-2\pi\rho/\omega} < x < h\right\}$ for some small positive constant $h$. Then trajectories strike $\Sigma$ once as they spiral in towards

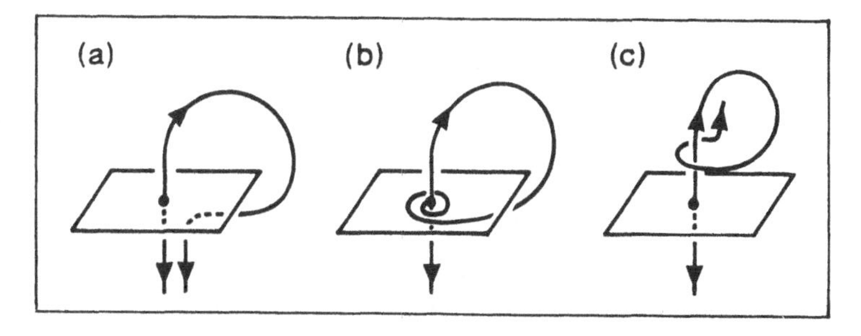


Fig. 12.11 (a) $\mu < 0$; (b) $\mu = 0$; (c) $\mu > 0$.

the origin and $\Sigma$ is a local transversal. Now, if $\Sigma' = \{(x,y,z) \in U | z = h\}$ the local flow induces a map $T_0 : \Sigma \to \Sigma'$ for sufficiently small $z > 0$ with $T_0(x,0,z) = (X,Y,h)$. To find $X$ and $Y$ as a function of $x$ and $z$ note that a trajectory which starts at $(x,0,z)$ will strike $z = h$ after a time $T$ where

$$h = ze^{\lambda T},$$

i.e.

$$T = -\frac{1}{\lambda}\log\left(\frac{z}{h}\right) \tag{12.28}$$

and so, using the local flow $\varphi_U$ again

$$X = xe^{-\rho T}\cos\omega T, \quad Y = -xe^{-\rho T}\sin\omega T. \tag{12.29}$$

Thus, for sufficiently small $z > 0$, using (12.28) and (12.29), we find that $T_0(x,0,z) = (X,Y,h)$ where

$$X = x\left(\frac{z}{h}\right)^{\rho/\lambda}\cos\left\{-\frac{\omega}{\lambda}\log\left(\frac{z}{h}\right)\right\} \tag{12.30a}$$

$$Y = -x\left(\frac{z}{h}\right)^{\rho/\lambda}\sin\left\{-\frac{\omega}{\lambda}\log\left(\frac{z}{h}\right)\right\}. \tag{12.30b}$$

The flow near the unstable manifold of the origin now defines a second map $T_1 : \Sigma' \to \Sigma$ which takes points $(X,Y,h)$ to $(x',0,z')$ for sufficiently small $X^2 + Y^2$. The existence of the homoclinic orbit if $\mu = 0$ implies that $T_1(0,0,h) = (r,0,0)$ if $\mu = 0$ for some $r \in (he^{-2\pi/\omega}, h)$ and since the homoclinic orbit is bounded away from other stationary points of the flow the map $T_1$ is smooth, hence we can expand it as a Taylor series to obtain

$$\begin{pmatrix} x' \\ z' \end{pmatrix} = \begin{pmatrix} r + a(\mu)\mu \\ b(\mu)\mu \end{pmatrix} + A\begin{pmatrix} X \\ Y \end{pmatrix} + \ldots \tag{12.31}$$

where $b(0) > 0$ so that the geometric conditions in Figure 12.9 are satisfied and $\det A \neq 0$. Composing the two maps $T_0$ and $T_1$ we obtain the return map $T : \Sigma \to \Sigma$ with $T(x,0,z) = (x',0,z')$ where

$$x' = r + a(\mu)\mu + \alpha xz^{\delta}\cos\left\{-\frac{\omega}{\lambda}\log z + \Phi_1\right\} + \ldots \tag{12.32a}$$

$$z' = b(\mu)\mu + \beta xz^{\delta}\cos\left\{-\frac{\omega}{\lambda}\log z + \Phi_2\right\} + \ldots \tag{12.32b}$$

where $\alpha$, $\beta$, $\Phi_1$ and $\Phi_2$ are constants. This map is only valid for sufficiently small $|\mu|$ and $z > 0$; if $z < 0$ then trajectories arrive below the stable manifold of the origin and are lost from the analysis. We can now treat this map in the usual way, looking for fixed points and determining their stability. Note that fixed points of $T$ correspond to periodic orbits of the flow which pass through $U$ once each period, and that if $T$ has a periodic orbit of period $k$ then this corresponds to a periodic orbit of the flow which passes $k$ time through $U$ per period. As suggested in Section 12.2 the cases $\delta > 1$ and $\delta < 1$ are significantly different and so we will treat them separately.

**Case (i):** $\delta > 1$.

The equation for a fixed point of the map, (12.32), is

$$x = r + O(\mu, xz^{\delta}) \tag{12.33a}$$

$$z = b\mu + O(\mu^2, xz^{\delta}) \tag{12.33b}$$

and since $\delta > 1$ and $z > 0$ is small we find a fixed point at approximately $(x, z) = (r, b\mu)$ provided $\mu > 0$ and no fixed points if $\mu < 0$. Furthermore, the period of the periodic orbit associated with the fixed point can be obtained from the amount of time spent in $U$, i.e.

$$T \sim -\frac{1}{\lambda}\log\mu \tag{12.34}$$

and so the situation is almost precisely the same as in the previous section. The stability of this fixed point (and hence the orbital stability of the associated periodic orbit for the flow) is determined by the eigenvalues of the Jacobian matrix of the map: the fixed point is stable if both eigenvalues lie inside the unit circle. The Jacobian matrix can be found by differentiating the map to be

$$\begin{pmatrix} A & B \\ C & D \end{pmatrix}$$

where

$$A = \alpha z^{\delta}\cos\left\{-\frac{\omega}{\lambda}\log z + \Phi_1\right\},$$

$$B = \alpha x z^{\delta-1}\left[\delta\cos\left\{-\frac{\omega}{\lambda}\log z + \Phi_1\right\} + \frac{\omega}{\lambda}\sin\left\{-\frac{\omega}{\lambda}\log z + \Phi_1\right\}\right],$$

$$C = \beta z^{\delta}\cos\left\{-\frac{\omega}{\lambda}\log z + \Phi_2\right\},$$

$$D = \beta x z^{\delta-1}\left[\delta\cos\left\{-\frac{\omega}{\lambda}\log z + \Phi_2\right\} + \frac{\omega}{\lambda}\sin\left\{-\frac{\omega}{\lambda}\log z + \Phi_2\right\}\right].$$

Since $z$ is small at the fixed point, the trace of the Jacobian is $O(z^{\delta-1})$, which is small, and the determinant is $O(z^{2\delta-1})$, which is also small.

Hence the sum of the eigenvalues and the product of the eigenvalues is small and so the eigenvalues are small. This implies that the fixed point is stable. Hence, in the case $\delta > 1$ the bifurcation is almost exactly the same as in the planar case: a single stable periodic orbit is destroyed as $\mu$ decreases through zero (see Fig. 12.12).

**Case (ii)** $\delta < 1$.
The case $\delta < 1$ is more subtle as might be guessed from the fact that $\delta - 1 < 0$ and so we cannot ignore terms of order $z^\delta$ compared to terms of order $z$ because $z^\delta / z$ tends to infinity as $z$ tends to zero. Going back to the equation of the return map for $(x', z')$ (equation (12.32)) we see that the leading order terms of the equation for a fixed point when $\delta < 1$ are

$$x = r + O(\mu, z^\delta) \tag{12.35a}$$

$$z = b\mu + \beta r z^\delta \cos\left\{-\frac{\omega}{\lambda}\log z + \Phi_2\right\} + O(z, z^{2\delta}). \tag{12.35b}$$

As the sketch in Figure 12.13 shows, there is a finite number of fixed points of this map if $\mu \neq 0$ and if $\mu = 0$ there is an infinite set of stationary points at (approximately) $(r, z_n)$ where $z_n > 0$ and

$$\cos\left\{-\frac{\omega}{\lambda}\log z_n + \Phi_2\right\} \sim 0. \tag{12.36}$$

Hence

$$-\frac{\omega}{\lambda}\log z_n + \Phi_2 \sim \frac{\pi}{2} + n\pi \tag{12.37}$$

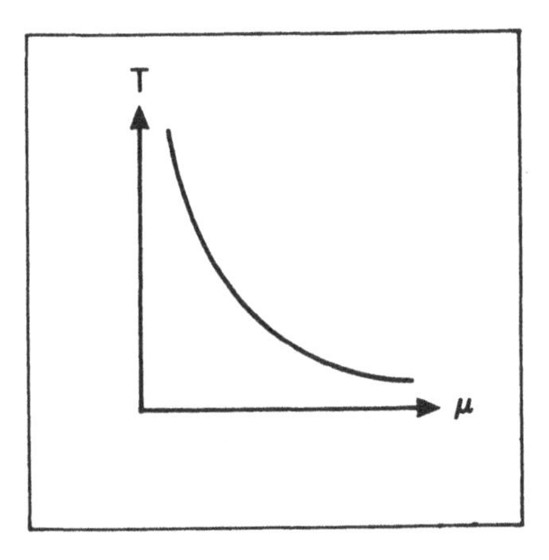


Fig. 12.12 Period against parameter for the simple periodic orbit if $\delta > 1$.

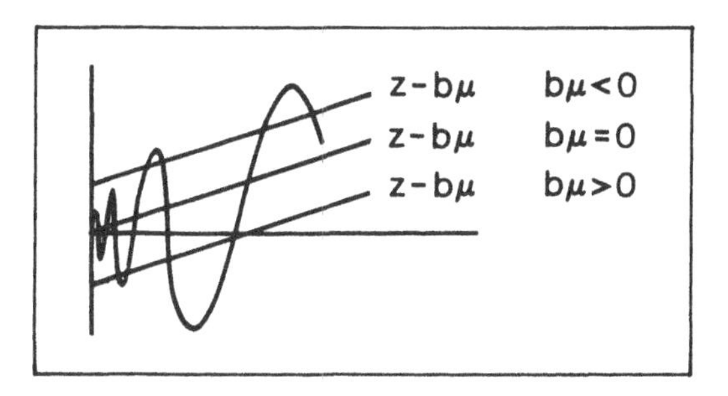


Fig. 12.13 Fixed points of equation (12.35b).

or

$$\log z_n \sim -\frac{\lambda}{\omega}\left(\frac{\pi}{2} + n\pi - \Phi_2\right). \tag{12.38}$$

Now, estimating the period of the associated periodic orbit of the flow by the amount of time spent in $U$ we find that

$$T_n \sim -\frac{1}{\lambda}\log z_n \tag{12.39}$$

and so

$$T_{n+1} - T_n \sim -\frac{1}{\lambda}\log\left(\frac{z_{n+1}}{z_n}\right) = \frac{\pi}{\omega}. \tag{12.40}$$

This shows that if $\mu = 0$ there is an infinite number of periodic orbits close to the homoclinic orbit, and that the periods of these orbits satisfy a simple scaling law. Each of these periodic orbits is a saddle, since the Jacobian matrix evaluated at $(r, z_n)$ is approximately

$$\begin{pmatrix} A & B \\ C & D \end{pmatrix}$$

with $A, B, C$ and $D$ as defined earlier. But for $n$ large enough $C \sim 0$ (as $\cos(-\frac{\omega}{\lambda}\log z_n + \Phi_2) \sim 0$) and so the eigenvalues of the Jacobian matrix are $A$ and $B$ which are (respectively) of order $z_n^\delta$ and $z_n^{\delta-1}$. Since $z_n$ is small and $\delta < 1$ one of these eigenvalues lies inside the unit circle and the other outside.

Putting this information together we find that if $\mu = 0$ and $\delta < 1$ there is an infinite set of periodic orbits which accumulate on the homoclinic orbit and whose periods $(T_n)$ satisfy $T_{n+1} - T_n = \frac{\pi}{\omega}$. None of these orbits is stable (in either positive or negative time). With a little more mathematical machinery it is possible to prove that the flow is chaotic if $\mu = 0$, but we will be satisfied with the existence of infinitely many periodic orbits here.

To describe the fixed points of the map if $\mu \neq 0$ it is more convenient to work with the period of the orbit given by $T \sim -\frac{1}{\lambda}\log z$. The equation for a fixed point of the return map is

$$0 \sim b\mu + \beta r z^{\delta} \cos\left\{-\frac{\omega}{\lambda}\log z + \Phi_2\right\} \tag{12.41}$$

and so substituting for $z$ using $T \sim -\frac{1}{\lambda}\log z$ we obtain

$$0 \sim b\mu + \beta r e^{-\rho T}\cos(\omega T + \Phi_2). \tag{12.42}$$

This curve (see Fig. 12.14) gives the locus of the fixed points of the return map in the $(\mu, T)$ plane. The wiggles of the curve are clearly saddlenode bifurcations which create a pair of fixed points at parameter values $(\mu_n)$ and $(\mu_n')$ with $\mu_n, -\mu_n' > 0$ and $\lim_{n\to\infty}\mu_n = \lim_{n\to\infty}\mu_n' = 0$. By inspection of the mathematical expression of this wiggly curve, (12.42), it is obvious that the parameter values $\mu_n$ occur when the periodic orbit has period $T_n$ with

$$\sin(\omega T_n + \Phi_2) \sim 0, \text{ and } \cos(\omega T_n + \Phi_2) \sim -1$$

if $\beta > 0$. Hence

$$\omega T_n + \Phi_2 \sim (2n+1)\pi$$

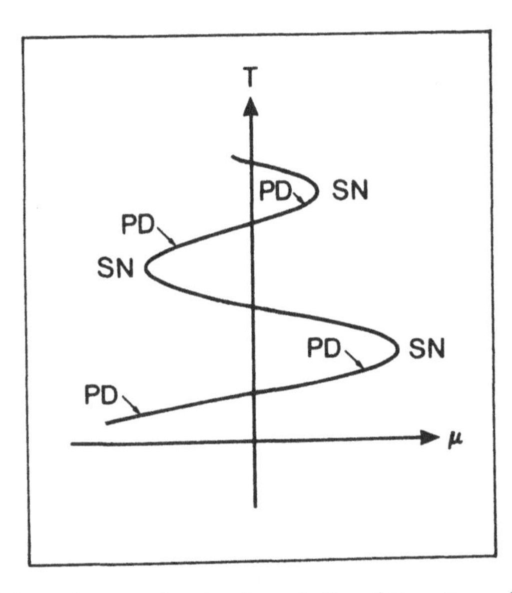


Fig. 12.14 Bifurcations on the simple periodic orbit as its period approaches infinity (after Glendinning and Sparrow, 1984).

or

$$T_n \sim \frac{1}{\omega}[(2n+1)\pi - \Phi_2]. \tag{12.43}$$

Substituting this value of $T_n$ into the equation defining the wiggly curve, (12.42), gives

$$b\mu_n \sim \beta r e^{-\rho T_n} \sim \beta r e^{-\frac{\rho}{\omega}((2n+1)\pi - \Phi_2)}$$

and hence

$$\frac{\mu_{n+1}}{\mu_n} \sim e^{-2\pi\rho/\omega}$$

or

$$\lim_{n\to\infty} \frac{\mu_{n+1} - \mu_n}{\mu_n - \mu_{n-1}} = e^{-2\pi\rho/\omega}. \tag{12.44}$$

It only remains now to determine the stability of these periodic orbits. A saddlenode bifurcation creates either a stable fixed point and a saddle or an unstable fixed point and a saddle (where we take unstable here to mean stable in negative time). Looking at the Jacobian matrix at the fixed point we see that the determinant (and hence the product of the eigenvalues) is $O(z^{2\delta-1})$ as $z \to 0$ and so the product of the eigenvalues is small if $\delta > \frac{1}{2}$ and large if $\delta < \frac{1}{2}$. Thus the only possible products of a saddlenode bifurcation are a stable (respectively unstable) fixed point and a saddle if $\delta > \frac{1}{2}$ (respectively $\delta < \frac{1}{2}$). However, if $\mu = 0$ we know that all the fixed points are saddles, so there must be a further bifurcation in which the stable (or unstable) fixed point becomes a saddle *before* $\mu = 0$. A careful examination of the Jacobian matrix (which we leave as an exercise) shows that this happens via a period-doubling bifurcation as shown in Figure 12.14. A similar analysis in $\mu < 0$ near the bifurcation points $\mu'_n$ completes the picture.

Figure 12.14 is already complicated enough, but it is only the tip of the iceberg so far as the complete description of the orbit structure near the homoclinic orbit is concerned. There is one further feature that we can study without serious mathematical problems: the existence of further homoclinic bifurcations in $\mu > 0$. If $\mu > 0$ the unstable manifold of the stationary point at the origin leaves $\Sigma'$ at $(0, 0, h)$ and so, using the formula for $T_1$ derived above, strikes $\Sigma$ for the first time at the point $(r + O(\mu), 0, b\mu)$. Now, since $b\mu > 0$ the trajectory moves through $U$ and around the homoclinic loop to return to $\Sigma$ with a $z$-coordinate given by

$$z \sim b\mu + \beta r (b\mu)^\delta \cos(-\frac{\omega}{\lambda}\log(b\mu) + \Phi_2). \tag{12.45}$$

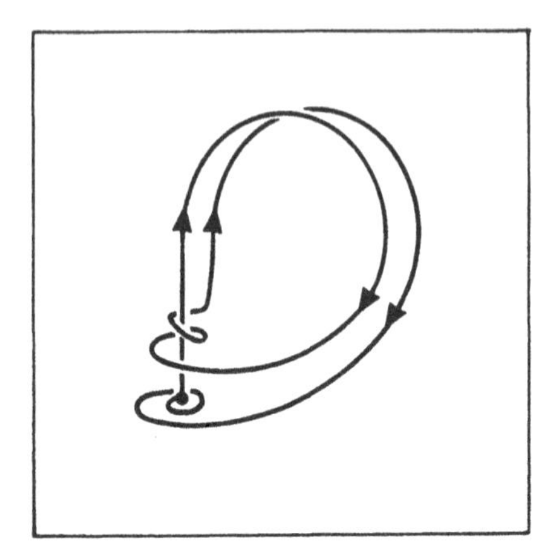

Fig. 12.15 A double-pulse homoclinic orbit.

If this $z$-coordinate is zero, then the trajectory strikes $\Sigma$ on the local stable manifold of the origin and spirals into the stationary point, so if $z = 0$ we have another homoclinic orbit which passes through $U$ twice (see Fig. 12.15). These homoclinic orbits are called double-pulse homoclinic orbits, and they exist for values of $\mu$ such that

$$-b\mu \sim \beta r(b\mu)^{\delta} \cos\left(-\frac{\omega}{\lambda}\log(b\mu) + \Phi_2\right). \qquad (12.46)$$

Hence there is an infinite sequence of such $\mu$ values, $\mu = \nu_n$, with $\nu_n \to 0$ as $n \to \infty$, $\nu_n > 0$ which are given asymptotically by solutions of

$$\cos\left(-\frac{\omega}{\lambda}\log(b\nu_n) + \Phi_2\right) = 0. \qquad (12.47)$$

Hence

$$b\nu_n \sim e^{-\frac{\lambda}{\omega}((n+\frac{1}{2})\pi - \Phi_2)}$$

and so

$$\lim_{n\to\infty} \frac{\nu_{n+1} - \nu_n}{\nu_n - \nu_{n-1}} = e^{-\lambda\pi/\omega}. \qquad (12.48)$$

Each of these homoclinic orbits undergoes a homoclinic bifurcation as $\mu$ passes through $\nu_n$ and so can be treated in the same way as the homoclinic bifurcation we are discussing (after reducing the size of $U$ appropriately). There is a sequence of saddlenode bifurcations accumulating on $\nu_n$ for each $n$ large enough, and so on! We could continue this discussion for some time, looking for more and more complicated homoclinic bifurcations and periodic orbits, but life is short and the analysis has to stop somewhere; here will do nicely.

## 12.5 Lorenz-like equations

We have already mentioned the Lorenz equations as an example of a differential equation with complicated asymptotic behaviour, or at least behaviour which appears complicated when simulated on a computer, but we have not given any indication of how this complexity might arise. Recall that the Lorenz equations are

$$\dot{x} = \sigma(y - x) \tag{12.49a}$$

$$\dot{y} = rx - y - xz \tag{12.49b}$$

$$\dot{z} = -bz + xy \tag{12.49c}$$

where the parameters $\sigma$, $b$ and $r$ are real and positive. The values chosen by Lorenz are $\sigma = 10$, $b = \frac{8}{3}$ and $r = 28$, but in the context of bifurcation theory it is usual to treat $\sigma$ and $b$ as fixed and allow $r$ to vary.

The Lorenz equations are symmetric under the operation $(x, y, z) \to (-x, -y, z)$, a fact that will be useful later and has stationary points at the origin $(0, 0, 0)$ and at solutions of $x = y$ (from $\dot{x} = 0$) and $bz = x^2$ (from $\dot{z} = 0$) and so

$$b(r - 1)x - x^3 = 0$$

(from $\dot{y} = 0$). Hence there are two other stationary points,

$$C_\pm = (\pm\sqrt{b(r-1)}, \pm\sqrt{b(r-1)}, r - 1) \tag{12.50}$$

provided $r > 1$. A little linear analysis shows that the origin is stable if $0 < r < 1$ and loses stability in a pitchfork bifurcation at $r = 1$, creating the two non-trivial stationary points which are (initially) stable. To determine the stability of these stationary points we look at the Jacobian matrix

$$\begin{pmatrix} -\sigma & \sigma & 0 \\ r - z & -1 & -x \\ y & x & -b \end{pmatrix}$$

which has eigenvalues given by the roots of

$$s^3 + (\sigma + b + 1)s^2 + b(\sigma + r)s + 2\sigma b(r - 1) = 0 \tag{12.51}$$

when evaluated at either $C_+$ or $C_-$. Note that since $C_-$ is the image of $C_+$ under the symmetry the stability properties of the two stationary points must be the same. Now we look for bifurcations of the stationary points, i.e. values of the parameters for which either $s = 0$ or $s = i\omega$ are solutions of the eigenvalue equation. Setting $s = 0$ we find that $r = 1$, giving the pitchfork bifurcation which we already know about. Setting $s = i\omega$ and equating real and imaginary parts of the equation we find

$$-(\sigma + b + 1)\omega^2 + 2\sigma b(r - 1) = 0, \quad -\omega^3 + b(r + \sigma)\omega = 0$$

i.e.

$$\omega^2 = \frac{2\sigma b(r-1)}{\sigma + b + 1} = b(r + \sigma). \tag{12.52}$$

Rearranging terms a little, this implies that there is a Hopf bifurcation at

$$r_H = \frac{\sigma(\sigma + b + 3)}{\sigma - b - 1} \tag{12.53}$$

provided $\omega^2 > 0$ at this value of $r$, i.e. provided

$$\omega^2 = \frac{2b\sigma(\sigma + 1)}{\sigma - b - 1} > 0, \tag{12.54}$$

which will always be satisfied provided $\sigma > b + 1$. For the standard choice of parameter values, $\sigma = 10$ and $b = \frac{8}{3}$ this is clearly satisfied and evaluating $r_H$ on a pocket calculator gives $r_H \sim 22.74$. With a little more work, similar to the manipulations of the three-dimensional Lotka-Volterra model in Section 9.4, it is possible to show that the Hopf bifurcation is subcritical, creating an unstable periodic orbit in $r < r_H$. For $r > r_H$ both $C_+$ and $C_-$ are saddles and so there are no stable periodic orbits that we know about and no stable stationary points. This presents us with two problems: it is easy to show that for $r < 1$ the origin is globally asymptotically stable (Exercise 12.5) and so there are no periodic orbits if $r < 1$. But the Hopf bifurcation implies that a pair of periodic orbits which are images of each other under the symmetry $(x, y, z) \to (-x, -y, z)$ bifurcate into $r < r_H$, so they must be destroyed in some bifurcation before $r = 1$. Also, a bounding function can be constructed for all $r > 0$ (Exercise 12.6) and so for $r > r_H$ we know that solutions remain bounded and yet we know of no stable or attracting objects. These two observations will be resolved by looking at the effect of a homoclinic bifurcation at the origin for some value $r_h$ in $(1, r_H)$.

Numerical evidence for the existence of a homoclinic bifurcation is obtained by computing the unstable manifold of the origin. Figure 12.16 shows the positive branch of the unstable manifold at two parameter values, $r = 13.8$ and $r = 14$. At $r = 13.8$ the manifold spirals into the stationary point $C_+$ (which is stable) without crossing into $x < 0$ whilst at $r = 14$ the unstable manifold spirals into $C_-$ after crossing into the region with $x < 0$. We deduce (by continuity of solutions with parameters) that for some $r$-value between 13.8 and 14, $r = r_h$, there is a homoclinic orbit biasymptotic to the origin and further that due to the symmetry of the equations there is a second homoclinic orbit involving the negative branch of the unstable manifold of the origin at the same parameter value, giving the configuration shown in Figure 12.16c.

The Jacobian matrix at the origin is

$$\begin{pmatrix} -\sigma & \sigma & 0 \\ r & -1 & 0 \\ 0 & 0 & -b \end{pmatrix}$$

and so there is one eigenvalue of $-b$ with the $z$-axis as its associated eigenvector and the other eigenvalues $s_{\pm}$ are the solutions of $s^2 + (\sigma + 1)s - \sigma(r-1) = 0$, i.e.

$$s_{\pm} = \frac{-(\sigma+1) \pm \sqrt{(\sigma+1)^2 + 4\sigma(r-1)}}{2}, \qquad (12.55)$$

which gives $s_- < 0$ and $s_+ > 0$. Indeed, for the parameter values considered here

$$s_- < -b < 0 < s_+ \qquad (12.56)$$

and so the dominant eigenvalues are $-b$ and $s_+$, so the homoclinic orbit approaches the origin tangential to the $z$-axis as $t \to \infty$. In this case numerical experiments suggest that both homoclinic orbits approach the origin tangential to the $z$-axis from above. Finally we note that the saddle index, $\delta = b/s_+$, is less than 1. This information is enough for us to start investigating the effect of a homoclinic bifurcation to the origin involving two homoclinic orbits.

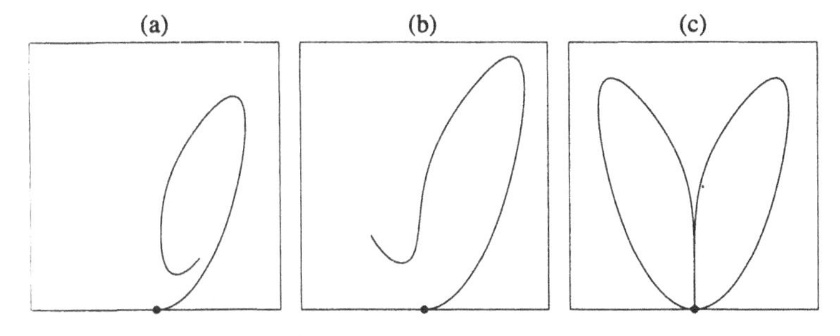


Fig. 12.16 One branch of the unstable manifold for (a) $r < r_h$; (b) $r > r_h$; and (c) both branches of the unstable manifold for $r = r_h$.

We shall now step back from the Lorenz equations to consider general systems in $\mathbf{R}^3$ which have a pair of homoclinic orbits biasymptotic to the origin if $\mu = 0$ (for some real parameter $\mu$) and such that

i) there is a neighbourhood $U$ of the origin such that for $\mu$ in a neighbourhood of 0 coordinates can be found such that in $U$ the differential equation is

$$\dot{x} = \lambda_1 x$$
$$\dot{y} = \lambda_2 y$$
$$\dot{z} = \lambda_3 z$$

with $\lambda_2 < \lambda_3 < 0 < \lambda_1$.

ii) The system is invariant under the symmetry $(x, y, z) \to (-x, -y, z)$.

iii) If $\mu = 0$ there is a pair of homoclinic orbits biasymptotic to the origin which are the images of each other under the symmetry (ii). Both orbits tend to the origin as $t \to \infty$ tangential to the positive $z$-axis.

With these three assumptions we can proceed to obtain a return map defined on a suitably chosen transversal, $\Sigma$, in $U$ by following the same steps as in the two previous sections. The linear flow in $U$ is easy to write down:

$$\varphi_U(x, y, z, t) = (xe^{\lambda_1 t}, ye^{\lambda_2 t}, ze^{\lambda_3 t}). \tag{12.57}$$

Now let $\Sigma = \{(x, y, z) \in U | z = h\}$ for some small $h > 0$ and $\Sigma^{\pm} = \{(x, y, z) \in U | x = \pm h\}$ (see Fig. 12.17). The linear flow $\varphi_U$ induces maps $R_0^{\pm} : \Sigma \to \Sigma^{\pm}$ in the usual way.

Let $(x, y, h) \in \Sigma$ with $x > 0$, then the trajectory through $(x, y, h)$ will strike $\Sigma^+$ at $(h, y', z')$ after a time $T$ where

$$h = xe^{\lambda_1 T}, \quad \text{i.e. } T = -\frac{1}{\lambda_1} \log\left(\frac{x}{h}\right). \tag{12.58}$$

Using this value of $T$ we find that

$$y' = y\left(\frac{x}{h}\right)^{-\lambda_2/\lambda_1}, \text{ and } z' = h\left(\frac{x}{h}\right)^{-\lambda_3/\lambda_1} \tag{12.59}$$

and so we have $R_0^+(x, y, h) = (h, y', z')$ for $x > 0$. Similarly, if $x < 0$ we obtain a map $R_0^- : \Sigma \to \Sigma^-$ with $R_0^-(x, y, h) = (-h, y'', z'')$ where

$$y'' = y\left(\frac{|x|}{h}\right)^{-\lambda_2/\lambda_1}, \text{ and } z'' = h\left(\frac{|x|}{h}\right)^{-\lambda_3/\lambda_1}. \tag{12.60}$$

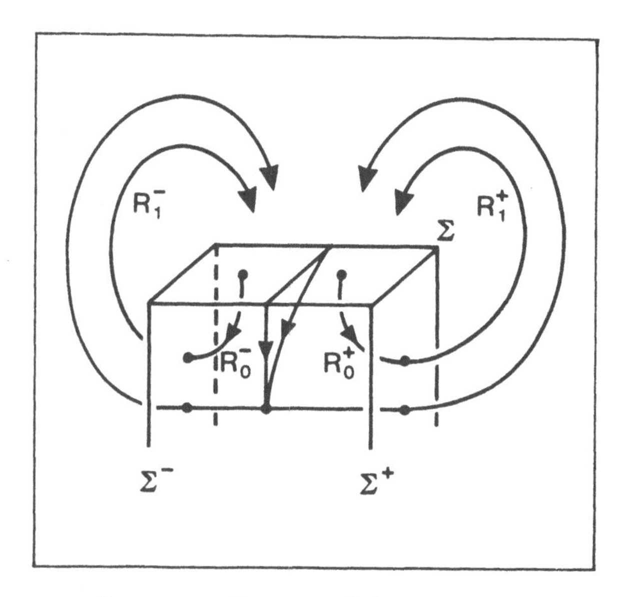


Fig. 12.17 Geometry of the return map.

Note that the map is undefined if $x = 0$ since $x = 0$ is the local stable manifold of the origin in $U$ and hence points on $\Sigma$ with $x = 0$ will tend to the origin and will not reappear on $\Sigma^{\pm}$. We now define two maps $R_1^{\pm} : \Sigma^{\pm} \to \Sigma$ by following the two branches of the unstable manifold of the origin back to $\Sigma$. The condition for a homoclinic orbit to exist must be that $R_1^+(h, 0, 0) = (0, r, h)$ for some small constant $r \neq 0$, in which case (using the symmetry) $R_1^-(-h, 0, 0) = (0, -r, h)$ and so the existence of one homoclinic orbit implies the existence of the second (note that $(\pm h, 0, 0) = \Sigma^{\pm} \cap W^u_{loc}(0)$). Hence, expanding the flow as a Taylor series about this point in both the parameter and phase space gives $R_1^+(h, Y, Z) = (x, y, h)$ where

$$\begin{pmatrix} x \\ y \end{pmatrix} = \begin{pmatrix} a(\mu)\mu \\ r + b(\mu)\mu \end{pmatrix} + A\begin{pmatrix} Y \\ Z \end{pmatrix} + O(Y^2 + Z^2) \qquad (12.61)$$

where we choose $a(0) < 0$ so that the positive branch of the unstable manifold strikes $\Sigma$ with $x > 0$ if $\mu < 0$ and $x < 0$ if $\mu > 0$ as suggested by the numerical experiment on the Lorenz equations (Fig. 12.16). The matrix $A$ is invertible, so $\det A \neq 0$. The map $R_1^- : \Sigma^- \to \Sigma$ is defined by symmetry, so $R_1^-(-h, Y, Z) = (-x, -y, h)$ where $x$ and $y$ are as above.

Putting these results together we obtain a map $R : \Sigma \to \Sigma$ where

$$R(x,y,h) = \begin{cases} R_1^+ \circ R_0^+(x,y,h) & \text{if } x > 0 \\ \text{undefined} & \text{if } x = 0 \\ R_1^- \circ R_0^-(x,y,h) & \text{if } x < 0 \end{cases}. \tag{12.62}$$

More explicitly, $R(x,y,h) = (x', y', h)$ where

$$x' = \begin{cases} a(\mu)\mu + A_{11}y\left(\frac{x}{h}\right)^{-\lambda_2/\lambda_1} + A_{12}h\left(\frac{x}{h}\right)^{\delta} + \dots & \text{if } x > 0 \\ -\pi_x R(|x|, y, h) & \text{if } x < 0 \end{cases}$$

and

$$y' = \begin{cases} r + b(\mu)\mu + A_{21}y\left(\frac{x}{h}\right)^{-\lambda_2/\lambda_1} + A_{22}h\left(\frac{x}{h}\right)^{\delta} + \dots & \text{if } x > 0 \\ -\pi_y R(|x|, y, h) & \text{if } x < 0 \end{cases}$$

where $\delta = -\lambda_3/\lambda_1$ is the saddle index and $\pi_k$ is the usual projection operator onto the $k$-axis, $k = x, y$, so $\pi_x R(x,y,h) = x'$ and so on. Since $\delta < -\lambda_2/\lambda_1$ the $x'$ equation is decoupled from $y$ to lowest order and we find the one-dimensional map

$$x' = \begin{cases} a\mu + cx^{\delta} & \text{if } x > 0 \\ -a\mu - c|x|^{\delta} & \text{if } x < 0 \end{cases} \tag{12.63}$$

as an approximate model of the flow near the homoclinic orbits (i.e. with $|x|$ and $|\mu|$ sufficiently small). It is this map that we will concentrate upon, but note that it is only an approximation of the full two-dimensional map. The evolution of this map as $\mu$ passes through zero is sketched in Figure 12. 18 for the two cases $c > 0$ and $c < 0$ (recall that $a < 0$). Since the derivative of the map is $\delta c|x|^{\delta-1}$ and $|x|$ is small the derivative is large and tends to infinity as $x$ tends to zero. This implies immediately that there are no stable periodic orbits for the map with $|x|$ small. Note that the map has a horseshoe (and is therefore chaotic) if $\mu > 0$ and $c > 0$ or $\mu < 0$ and $c < 0$ but that most orbits eventually leave the neighbourhood of $x = 0$ in which the map is a valid approximation to the flow for all $\mu$ near zero.

To apply this map to the Lorenz equations we need to determine whether we are in the case $c < 0$ or $c > 0$. To do this note that we know from Figure 12.16 that for $r < r_h$ the positive branch of the unstable manifold spirals into $C_+$ without passing into $x < 0$ and for $r > r_h$ it spirals into $C_-$ after crossing into $x < 0$ on the first circuit through a neighbourhood of the origin. In the map, the behaviour of

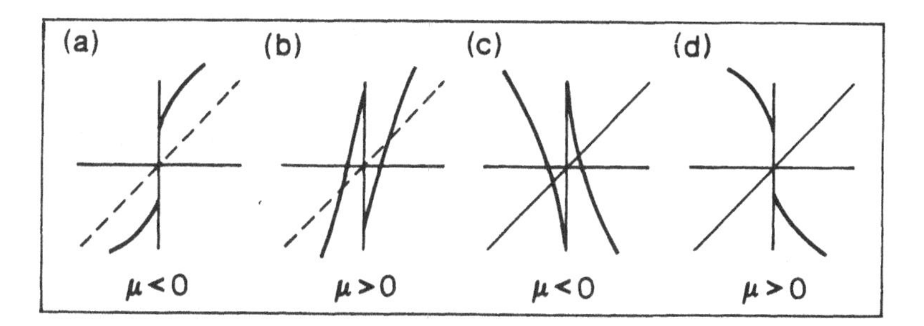


Fig. 12.18 The one-dimensional return map: (a,b) $c > 0$; (c,d) $c < 0$.

the positive branch of the unstable manifold of the origin is modelled by the behaviour of $x = 0^+$, i.e. the limit of $x$ as $x$ tends to 0 from above. We can describe this symbolically by labelling motion through a neighbourhood of the homoclinic orbit in $x > 0$ by the symbol $R$ and motion through a neighbourhood of the homoclinic orbit in $x < 0$ by $L$ (for 'right' and 'left'), so the change just described for the Lorenz equations as $r$ passes through $r_h$ is from $RRRRR\ldots$ to $RLLLLL\ldots$. Now, from the form of the maps in Figure 12.18 we see that if $c < 0$ the change as $\mu$ passes through zero of the orbit of $0^+$ is from

$$RRLLLLL\ldots \text{ in } \mu < 0 \text{ to } RLRLRLRL\ldots \text{ in } \mu > 0$$

and the map is chaotic if $\mu < 0$ whilst if $c > 0$ the change is from

$$RRRRRRR\ldots \text{ in } \mu < 0 \text{ to } RLLLLLLL\ldots \text{ in } \mu > 0$$

and the map is chaotic if $\mu > 0$. Hence the homoclinic bifurcation for the Lorenz equations at $r = r_h$ has $c > 0$ and the direction of $r$ is the same as the direction of $\mu$. In particular, two fixed points of the map, one in $x < 0$ and one in $x > 0$, bifurcate into $\mu > 0$. These correspond to a pair of periodic orbits for the flow and hence account for the two unstable periodic orbits created into $r < r_H$ in the Hopf bifurcation described earlier.

The chaotic set which is created into $\mu > 0$ (i.e. $r > r_h$ for the Lorenz equations) is unstable; almost all trajectories near the set eventually leave the small neighbourhood of $x = 0$ where the return map has been derived. However, *if we assume that* the one-dimensional return map remains a reasonable approximation to the dynamics of the differential equations at parameter values outside a neighbourhood of $\mu = 0$ we can ask whether this set can become attracting. For this to happen we need to find a trapping region for the map, i.e. an interval which is mapped into itself. This can only happen (assuming the form of the map remains unchanged) if there is a parameter value $\mu_c$ at which the map takes the form shown in Figure 12. 19a, and so for $\mu > \mu_c$ it evolves as shown in Figure 12.19b. Thus the chaotic set can become attracting if the positive branch of the unstable manifold of the origin tends directly to the simple periodic orbit in $x < 0$ (which is represented by the fixed point in $x < 0$ in the map) and the negative branch of the unstable manifold tends to the simple periodic orbit in $x > 0$. For the Lorenz equations numerical simulations suggest that this happens if $r \sim 22.06$ and so for $r$ a little greater than this critical value there is a stable chaotic set (a strange attractor) and a pair of stable stationary points, $C_\pm$. Note that the stability of the strange attractor is not related to the occurrence of the Hopf bifurcation at $r = r_H$. For $r > r_H$ we are left only with the strange attractor, which appears (again on the basis of numerical experiments) to persist up to $r \sim 31$ where some stable periodic motion starts to be observable at some parameter values. It is worth emphasising that these remarks are not theorems – there is still no proof that the Lorenz equations have stable chaotic motion – but the combination of numerical simulations and the theoretical arguments described here strongly suggest that some such object exists.

Finally, note that as $\mu$ increases from $\mu_c$ there must be many more homoclinic bifurcations of more complicated geometric forms, since if $\mu = \mu_c$ there are periodic orbits with arbitrarily many successive turns in $x < 0$ or $x > 0$, whilst if $\mu > \mu_c$ (Fig. 12.19b) the maximum number of successive turns on one side or the other is limited by the behaviour of the left and right end points of the invariant box. For more details of the Lorenz equations (and there are a great many more details!) the

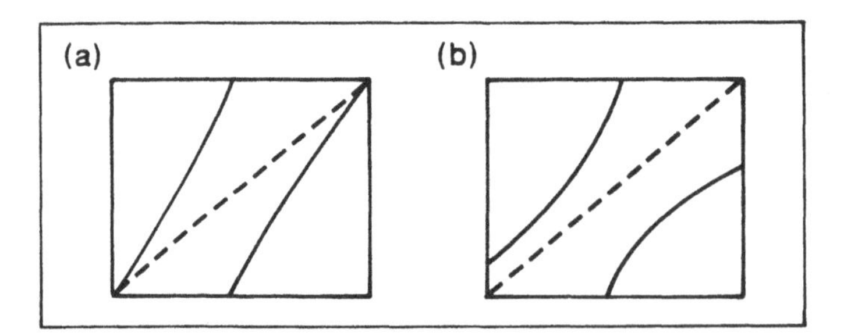


Fig. 12.19 (a) The map at $\mu = \mu_c$; (b) $\mu > \mu_c$, the number of successive iterates on either side of the discontinuity is bounded.

book by Sparrow (1981) is worth looking at, being both readable and informative.

## 12.6 Cascades of homoclinic bifurcations

The derivation of the return map in the previous section is valid for both $\delta < 1$ and $\delta > 1$, but because we have been interested in the Lorenz equations we have focused on the case $\delta < 1$. The other case, $\delta > 1$, is just as fascinating, although one has to assume again that the map remains a good model of the flow for parameter values outside a small neighbourhood of $\mu = 0$ in order to find interesting behaviour. This case was first analysed by Arnéodo, Coullet and Tresser (1981), and we follow their treatment below. Rescaling $\mu$ we can fix $a = -1$ so the map can be written as

$$x' = f(x,\mu) = \begin{cases} -\mu + cx^\delta & \text{if } x > 0 \\ \mu - c|x|^\delta & \text{if } x < 0 \end{cases} \tag{12.64}$$

where we assume now that $\delta > 1$. As in the previous section the two cases $c > 0$ and $c < 0$ give different results and we shall concentrate on the case $c > 0$ (see Exercise 12.10 for the case $c < 0$). We shall begin by analysing the map for small $|\mu|$ and $|x|$, which is the region where the map is strictly valid, and then extend the analysis to investigate the behaviour for larger $|\mu|$ and $|x|$, where the description of unimodal maps in the previous chapter will enable us to deduce the existence of complicated sequences of homoclinic bifurcations. Throughout the rest of this section we assume that $c > 0$.

If $\mu < 0$ and $|\mu|$ is small then $f(x,\mu) > 0$ for all $x > 0$ and the slope of $f(x,\mu)$ is $c\delta x^{\delta-1}$, which is small and positive for all sufficiently small $x$. Hence there is a stable fixed point in $x > 0$ at $x = -\mu + O(\mu^\delta)$ which attracts all orbits with sufficiently small $x > 0$. By the symmetry of the equations $x \to -x$ a similar fixed point exists in $x < 0$.

If $\mu > 0$ and $|\mu|$ is small then $f(x,\mu)$ maps the interval $[-\mu,\mu]$ into itself since $f(0^+,\mu) = -\mu$, $f(\mu,\mu) = -\mu + c\mu^\delta < 0$ for $\mu$ sufficiently small and similar results hold in $x < 0$ (see Fig. 12.20). This implies that $f((0^+,\mu),\mu) \subset (-\mu, 0)$ and $f((-\mu,0^-),\mu) \subset (0,\mu)$. Hence all points are attracted to orbits of period two.

The previous two paragraphs show that as $\mu$ passes through zero two fixed points come together at the origin and for $\mu > 0$ there is a stable orbit of period two. This is sometimes called a gluing bifurcation: in

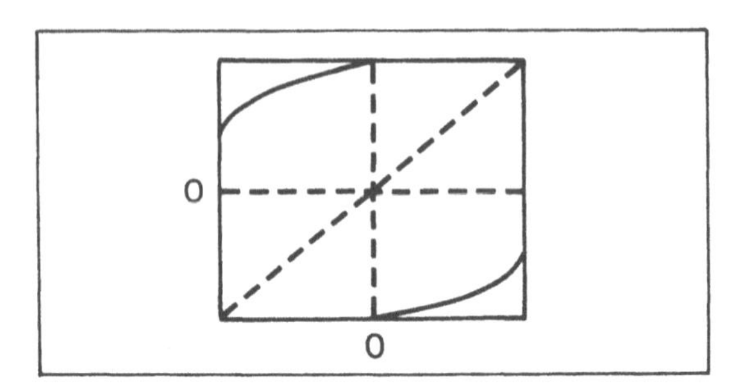


Fig. 12.20

terms of the flow two periodic orbits become homoclinic to the origin and form a single periodic orbit which can be thought of as an orbit obtained by gluing the original two orbits together.

None of this is very exciting, but if we assume that this map is a valid approximation of the corresponding flow *outside* a neighbourhood of $\mu = 0$ and $x = 0$ we start to see much more interesting phenomena. To understand the complications that arise we want to relate the map under consideration to a standard family of unimodal maps (Fig. 12.21). Let

$$g(x,\mu) = \begin{cases} -f(x,\mu) & \text{if } x > 0, \\ f(x,\mu) & \text{if } x < 0 \end{cases} \tag{12.65}$$

then, apart from the ambiguity about what happens at $x = 0$, $g(x, \mu)$ is the standard unimodal family

$$g(x,\mu) = \mu - c|x|^{\delta} \tag{12.66}$$

with $\delta > 1$. Hence (cf. Section 11.5) as $\mu$ increases there is a sequence of period-doubling bifurcations and all the complicated bifurcation structure associated with families of unimodal maps. Note that both maps have a simple symmetry property:

$$f(x,\mu) = -f(-x,\mu) \text{ and } g(x,\mu) = g(-x,\mu). \tag{12.67}$$

(12.6) EXERCISE

*An important consequence of these symmetry properties is that*

$$f^n(x,\mu) = -f^n(-x,\mu) \text{ and } g^n(x,\mu) = g^n(-x,\mu)$$

*for all $n > 1$. These relations will be used in the proofs of the results below. Using induction, or otherwise, prove these two identities.*

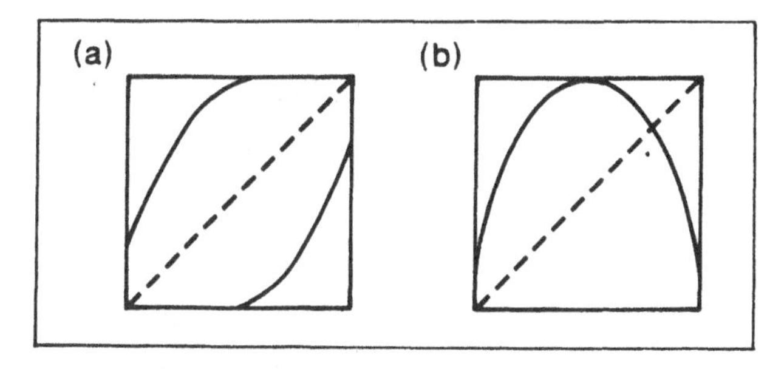


Fig. 12.21 (a) $f$; (b) $g$.

Our aim is to reinterpret results for the unimodal family $g(x,\mu)$ to obtain information about the behaviour of $f(x,\mu)$ via the following result.

(12.7) LEMMA

*Let $f(x,\mu)$ and $g(x,\mu)$ be as defined above. Then*

$$f^n(x,\mu) = sign\left(\frac{\partial g^n}{\partial x}(x,\mu)\right).g^n(x,\mu)$$

*and*

$$\left(\frac{\partial f^n}{\partial x}(x,\mu)\right) = \left|\left(\frac{\partial g^n}{\partial x}(x,\mu)\right)\right|.$$

*Proof*: The proof of the first statement follows by induction. Take $n = 1$, then

$$f(x,\mu) = \begin{cases} -g(x,\mu) & \text{if } x > 0 \\ g(x,\mu) & \text{if } x < 0 \end{cases}$$

and

$$sign\left(\frac{\partial g^n}{\partial x}(x,\mu)\right) = \begin{cases} -1 & \text{if } x > 0 \\ 1 & \text{if } x < 0 \end{cases} \tag{12.68}$$

and so the result is clearly true for $n = 1$. Suppose the result is true for $n = N$, i.e.

$$f^N(x,\mu) = sign\left(\frac{\partial g^N}{\partial x}(x,\mu)\right) g^N(x,\mu). \tag{12.69}$$

Suppose $g^N(x,\mu) < 0$. Then

$$g^{N+1}(x,\mu) = g(g^N(x,\mu),\mu) = f(g^N(x,\mu),\mu). \tag{12.70}$$

There are two cases.

i) If $sign\left(\frac{\partial g^N}{\partial x}(x,\mu)\right) = 1$ then $g^N(x,\mu) = f^N(x,\mu)$ and so $g^{N+1}(x,\mu) = f^{N+1}(x,\mu)$. But by the chain rule

$$\frac{\partial g^{N+1}}{\partial x}(x,\mu) = \left(\frac{\partial g}{\partial x}(g^N(x,\mu),\mu)\right)\cdot\left(\frac{\partial g^N}{\partial x}(x,\mu)\right) \tag{12.71}$$

and since both terms on the right-hand side of this expression are positive (the first since $g^N(x,\mu) < 0$ and the second by assumption) we have that

$$sign\left(\frac{\partial g^{N+1}}{\partial x}(x,\mu)\right) = 1 \tag{12.72}$$

and the result is true.

ii) If $sign\left(\frac{\partial g^N}{\partial x}(x,\mu)\right) = -1$ then $g^N(x,\mu) = -f^N(x,\mu)$ and so

$$g^{N+1}(x,\mu) = f(-f^n(x,\mu),\mu) = -f^{N+1}(x,\mu) \tag{12.73}$$

using the symmetry of the equation $f$. But, using the chain rule as in case (i),

$$sign\left(\frac{\partial g^{N+1}}{\partial x}(x,\mu)\right) = -1 \tag{12.74}$$

and so, once again, the result is true.

It now remains to check that the same argument works if $g^N(x,\mu) > 0$. It does, although we leave the full proof as an exercise. The second part of the lemma is easy and is left as an exercise as well.

Lemma 12.7 allows us to reinterpret the behaviour of $g$, which we know about, in terms of $f$, which we would like to know about.

(12.8) LEMMA

*Suppose that $g$ has an orbit of least period $p$, then*

i) *$f$ has a symmetric pair of periodic orbits of period $p$ if*

$$sign\left(\frac{\partial g^p}{\partial x}(x,\mu)\right) > 0;$$

ii) *$f$ has a symmetric periodic orbit of period $2p$ if*

$$sign\left(\frac{\partial g^p}{\partial x}(x,\mu)\right) < 0;$$

iii) *f has a symmetric pair of orbits which pass through the point* $x = 0$ *if*

$$sign\left(\frac{\partial g^p}{\partial x}(x,\mu)\right) = 0.$$

*Proof*: Suppose $g^p(x) = x$ and $\frac{\partial g^p}{\partial x}(x) > 0$. Then by Lemma 12.7 $f^p(x) = x$ and $f^p(-x) = -f^p(x) = -x$ so the points $x$ and $-x$ are both periodic of period p. We want to show that these orbits are distinct. Suppose that $f^q(x) = -x$ for some $q < p$. By Lemma 12.7

$$f^q(x) = \pm g^q(x) \tag{12.75}$$

and if we take the $-$ sign we find $g^q(x) = x$, contradicting the minimality of $p$, whilst if we take the $+$ sign, $g^q(x) = -x$. In this case, since $g(x) = g(-x)$, $g^{q+1}(x) = g(x)$ and so $g(x)$ is periodic of period $q$. But $x$ and $g(x)$ are, by assumption, on the same periodic orbit with minimal period $p$, so $g(x)$ cannot be periodic with period $q < p$. Hence the orbits of $x$ and $-x$ are distinct under $f$. These two orbits are symmetric in the sense that $f^k(x) = -f^k(-x)$, so the orbits map to each other under the transformation $x \to -x$.

Now suppose that $g^p(x) = x$ and $\frac{\partial g^p}{\partial x}(x) < 0$. Then $f^p(x) = -x$ and so (by the symmetry) $f^p(-x) = x$. Hence there is a symmetric periodic orbit of period $2p$ for $f$.

Finally, suppose that $g^p(x) = x$ and $\frac{\partial g^p}{\partial x}(x) = 0$. Then $x = 0$ must be on the periodic orbit through $x$, i.e. $g^p(0) = 0$. This is a slight problem, since we have not really defined the map $f$ at $x = 0$ (for the flow, trajectories which strike the return plane at $x = 0$ lie on the stable manifold of the origin and so they tend to the origin without leaving the neighbourhood $U$ of the origin). However, it is not hard to see, using techniques similar to those already described above, that

$$f^{p-1}(\mu) = f^{p-1}(-\mu) = 0. \tag{12.76}$$

This implies that $f^p(0^+) = f^p(0^-) = 0$, giving the required result.

We now need to do two things: first use Lemma 12.8 to reinterpret the standard period-doubling bifurcation sequences of $g(x,\mu)$ for the maps $f(x,\mu)$, and second, translate the bifurcations of $f(x,\mu)$ into a coherent description of the bifurcations in the associated differential equations.

Recall that for the unimodal maps $g$ there is a sequence $(\mu_n)$ of parameter values as $\mu$ increases at which a periodic orbit of period $2^{n-1}$

loses stability, creating a stable periodic orbit of period $2^n$. This sequence accumulates at some parameter $\mu_\infty$ at a rate which depends on the order of the critical point ($\delta$) and for $\mu > \mu_\infty$ the map has positive topological entropy. Let $x(\mu)$ denote a point on the stable periodic orbit of period $2^n$ for $\mu_n < \mu < \mu_{n+1}$. When $\mu = \mu_n$, the point $x(\mu_n)$ lies on the non-hyperbolic periodic orbit of period $2^{n-1}$, which is about to undergo a period-doubling bifurcation, so

$$\frac{\partial g^{2^{n-1}}}{\partial x}(x(\mu_n), \mu_n) = -1 \tag{12.77}$$

and hence

$$\frac{\partial g^{2^n}}{\partial x}(x(\mu_n), \mu_n) = 1. \tag{12.78}$$

Furthermore, if $\mu = \mu_{n+1}$ the point is about to period-double with

$$\frac{\partial g^{2^n}}{\partial x}(x(\mu_{n+1}), \mu_{n+1}) = -1. \tag{12.79}$$

Thus there exists a parameter value $\mu = \nu_n \in (\mu_n, \mu_{n+1})$ such that

$$\frac{\partial g^{2^n}}{\partial x}(x(\nu_n), \nu_n) = 0 \tag{12.80}$$

and the parameter values $\nu_n$ accumulate on $\mu_\infty$ at the same rate as the sequence $\mu_n$. Assuming that $\nu_n$ is defined uniquely let us consider the implications of these results for $f(x, \mu)$.

When $\mu_n < \mu < \nu_n$, $x(\mu)$ is a point of least period $2^n$ for $g$ and so

$$g^{2^n}(x(\mu), \mu) = x(\mu) \text{ with } \frac{\partial g^{2^n}}{\partial x}(x(\mu), \mu) > 0. \tag{12.81}$$

Hence $f$ has a symmetric pair of periodic orbits of period $2^n$ (both of which are stable). As $\mu$ tends to $\nu_n$ from below a point on each of these orbits tends to $x = 0$. For $\nu_n < \mu < \mu_{n+1}$, $x(\mu)$ is a point of least period $2^n$ for $g$ with

$$\frac{\partial g^{2^n}}{\partial x}(x(\mu), \mu) < 0 \tag{12.82}$$

and so $f$ has a single symmetric periodic orbit of period $2^{n+1}$ which is stable. As $\mu$ increases further this sequence is repeated with higher powers of 2, so for $\mu_{n+1} < \mu < \nu_{n+1}$ there is a symmetric pair of orbits with period $2^{n+1}$ which are formed from the symmetric orbit of period $2^{n+1}$ by a *symmetry-breaking* bifurcation (see Exercise 12.9).

To describe the bifurcations of the flows which these maps are supposed to model first note that the symmetry properties of the map are a reflection of the symmetry of the differential equation under the transformation $S : (x, y, z) \to (-x, -y, z)$. Hence if the map has a symmetric pair of periodic orbits of period $p$ the flow has a pair of periodic orbits, $\Gamma_0$ and $\Gamma_1$, which pass through a neighbourhood $U$ of the origin $p$ times per period, and such that $\Gamma_i = S\Gamma_{1-i}$, $i = 0, 1$. If the map has a symmetric periodic orbit of period $p$ then the map has a periodic orbit $\Gamma$ which passes through $U$ $p$ times per period and $\Gamma = S\Gamma$. Finally the orbits which pass through the origin $x = 0$ for the map correspond to homoclinic orbits of the flow. Hence if $\mu \in (\mu_n, \nu_n)$ we expect to see a symmetric pair of periodic orbits which become homoclinic orbits if $\mu = \nu_n$ creating a symmetric orbit by 'gluing' the symmetric pair together. This symmetric orbit is stable for $\mu \in (\nu_n, \mu_{n+1})$ and if $\mu = \mu_{n+1}$ it loses stability in a symmetry-breaking bifurcation creating a stable symmetric pair of periodic orbits. This sequence of bifurcations is repeated, accumulating at $\mu_\infty$ at a rate which depends only on the saddleindex, and for $\mu > \mu_\infty$ there is the possibility of chaotic behaviour and further (more complicated) sequences of homoclinic cascades. This sequence was first explained by Arnéodo, Coullet and Tresser in 1981. Their example, which seems to behave in this way, is

$$\dot{x} = \alpha(x - y) \tag{12.83a}$$

$$\dot{y} = -4\alpha y + xz + \mu x^3 \tag{12.83b}$$

$$\dot{z} = -\delta\alpha z + xy + \beta z^2 \tag{12.83c}$$

with $\alpha = 1.8$, $\beta = -0.07$ and $\delta = 1.5$. The simple pair of homoclinic orbits exists if $\mu \sim 0.076071$, and the homoclinic cascade occurs as $\mu$ decreases (at $\mu = 0.02$ the system appears chaotic). Readers with access to a computer might like to investigate the bifurcations in this system numerically. The first few bifurcations are sketched in Figure 12.22.

## Exercises 12

1. Describe the bifurcations and stability associated with a homoclinic orbit in a planar flow in the cases $\delta < 1$ and $\delta > 1$, where $\delta$ is the saddle index. If $\delta = 1 + \epsilon$, ($\epsilon \ll 1$), describe the bifurcations in the $(\mu, \epsilon)$ parameter plane in the map

$$x_{n+1} = \mu + a x_n^{1+\epsilon}, \quad x_n > 0$$

in the cases $0 < a < 1$ and $a > 1$.

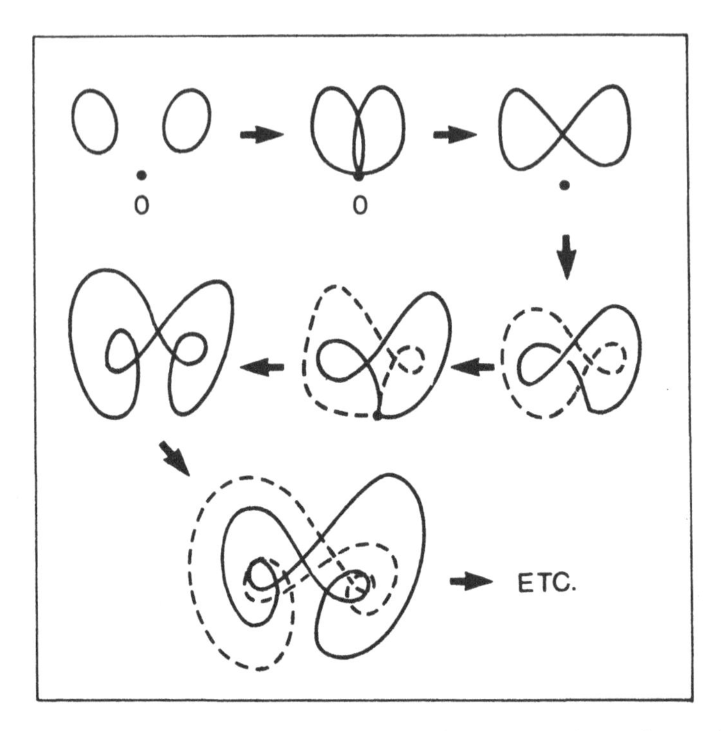


Fig. 12.22 Cascades of homoclinic orbits (after Arnéodo et al., 1981).

2. A family of differential equations in $\mathbf{R}^2$ depends upon two parameters. If both are zero the system has two homoclinic orbits, $\Gamma_0$ and $\Gamma_1$, biasymptotic to the same stationary point. Choosing parameters $\mu$ and $\nu$ so that a homoclinic orbit close to $\Gamma_1$ exists if $\mu = 0$ and a homoclinic orbit close to $\Gamma_0$ exists if $\nu = 0$, describe the bifurcations and periodic orbits in the $(\mu, \nu)$ plane near $(0, 0)$.

3. Consider the equations

$$\begin{aligned}
\dot{x} &= \lambda_1 x + P_1(x, y, z, \mu) \\
\dot{y} &= -\lambda_2 y + P_2(x, y, z, \mu) \\
\dot{z} &= -\lambda_3 z + P_3(x, y, z, \mu)
\end{aligned}$$

where $P_i(0, 0, 0, \mu) = 0$ for all $\mu$, and the first derivatives of $P_i$ vanish at $(x, y, x, \mu) = (0, 0, 0, 0)$. Suppose that if $\mu = 0$ there is a homoclinic orbit, $\Gamma$, biasymptotic to the origin. If $\lambda_2 > \lambda_1$, $\lambda_3 > 0$, derive a return map on a piece of the plane $z = h \ll 1$. Show that to lowest order the

$x$-coordinate of this return map decouples to give the map

$$x \to a\mu + bx^{\lambda_3/\lambda_1}, \quad x > 0.$$

Hence describe the periodic orbits which exist in a tubular neighbourhood of $\Gamma$ for $|\mu|$ small in the cases $b < 0$ and $b > 0$. (As ever, the cases of saddle index greater than one and less than one should also be treated separately).

4. Reconsider Exercise 3 at the end of Chapter 7. Find the locus of a (symmetric) homoclinic bifurcation.

5. By considering the Liapounov function

$$V(x, y, z) = x^2 + \sigma y^2 + \sigma z^2$$

show that in the Lorenz equations, (12.49), the origin is globally asymptotically stable if $0 < r < 1$.

6. By considering the function

$$E(x, y, z) = rx^2 + \sigma y^2 + \sigma(z - 2r)^2$$

$(r, \sigma, b > 0)$ show that all trajectories in the Lorenz equations eventually lie in a bounded ellipsoid.

7. Discuss the occurrence of double-pulse homoclinic orbits in the systems

$$\begin{aligned}
\dot{x} &= -\rho x + \omega y + P_1(x, y, z, \mu) \\
\dot{y} &= -\omega x - \rho y + P_2(x, y, z, \mu) \\
\dot{z} &= \lambda z + P_3(x, y, z, \mu)
\end{aligned}$$

where $\lambda > \rho > 0$, $\omega > 0$, the functions $P_i$ are as in Exercise 4 and the system is invariant under the transformation

$$(x, y, z) \to (-x, -y, -z).$$

To exploit the symmetry, define the return map on two symmetrically placed return planes

$$\Sigma_1 = \left\{(x, 0, z) |\ x_0 < x < x_0 e^{\pi\rho/\omega}\right\}$$

and

$$\Sigma_2 = \left\{(x, 0, z) |\ -x_0 > x > -x_0 e^{\pi\rho/\omega}\right\}$$

where $x_0 > 0$ is constant. Let $\mu$ be the $z$-coordinate of the first intersection of the positive branch of $W^u(0)$ (which leaves 0 in the positive $z$

direction) with $\Sigma_1$, so, by symmetry, $-\mu$ is the $z$-coordinate of the first intersection of the negative branch of $W^u(0)$ with $\Sigma_2$. Now derive the return map on $\Sigma_1 \cup \Sigma_2$ following the same procedure as in Section 12.4.

8. Consider non-symmetric differential equations of the same form as in the previous question, but with $\rho > \lambda > 0$. Take $\mu \in \mathbf{R}^2$, so $\mu = (\mu_1, \mu_2)$. Assume that if $\mu = (0,0)$ then the system has a pair of homoclinic orbits, $\Gamma_1$ and $\Gamma_2$, and that if $\mu_1 = 0$ there is a homoclinic orbit similar to $\Gamma_1$ whilst if $\mu_2 = 0$ there is a homoclinic orbit similar to $\Gamma_2$. Define two return planes, $\Sigma_1$ and $\Sigma_2$, as in the previous question and let $\mu_1$ represent the $z$-coordinate of the first intersection of the positive branch of $W^u(0)$ with $\Sigma_1$ and $-\mu_2$ represent the $z$-coordinate of the first intersection of the negative branch of $W^u(0)$ with $\Sigma_2$. Show that there are double-pulse homoclinic orbits (which pass once through a tubular neighbourhood of each of the homoclinic orbits $\Gamma_i$, $i = 1, 2$) for systems with

$$\mu_1 < 0 \text{ and } -\mu_2 \approx \beta_2 |\mu_1|^\delta \cos\left\{-\tfrac{\omega}{\lambda} \log|\mu_1| + \Phi_2\right\}$$

and

$$\mu_2 < 0 \text{ and } -\mu_1 \approx \beta_1 |\mu_2|^\delta \cos\left\{-\tfrac{\omega}{\lambda} \log|\mu_2| + \Phi_1\right\},$$

where $\delta = \rho/\lambda$. Sketch these curves in the $\mu$-plane and describe the periodic orbits which exist in the various regions of the plane determined by these curves and the axes.

9. Consider the continuous one-dimensional map $x_{n+1} = f(x_n)$. If $f(x) = -f(-x)$ show that the origin is always a fixed point. Suppose that $f^k(y) = -y$, $y \neq 0$. Show that $f^{2k}(y) = y$ and deduce that $(f^{2k})'(y) > 0$. Hence show that a symmetric orbit cannot undergo a period-doubling bifurcation. What bifurcations can occur in this case?

10. By considering the second iterate of the map, or otherwise, describe the cascade of homoclinic bifurcations in the maps

$$f(x,\mu) = \begin{cases} \mu - x^\delta & \text{if } x > 0, \\ -\mu + |x|^\delta & \text{if } x < 0 \end{cases}$$

where $\delta > 1$. How does this differ from the case described in Section 12.6?

# *Notes and further reading*

## Chapter 1

Coddington and Levinson (1955) and Hartman (1964) are two of the classic texts in differential equations. Neither is easy, but both are well worth looking at. Problems of the existence and uniqueness of solutions are described in the first two chapters of Coddington and Levinson (1955), whilst Hartman (1964) addresses the existence of solutions in chapter II and the dependence of solutions on initial conditions and parameters in chapter V. These problems, together with an elegant description of phase space, are also dealt with in Hirsch and Smale (1974). A more detailed account of invariant sets and definitions associated with ideas of recurrence can be found in Bhatia and Szëgo (1967). Percival and Richards (1982) give an introductory account of Hamiltonian systems such as the examples of Section 1.1. A simple non-technical introduction to chaos can be found in Stewart (1989).

## Chapter 2

Most books on the qualitative behaviour of differential equations will include a chapter on stability and Liapounov functions. The main weakness of Chapter 2 is that it does not really deal with stability for non-autonomous differential equations, or time-dependent Liapounov functions. Verhulst (1990) considers these problems in a very readable way.

## Chapter 3

A full description of the algebra used in this chapter can be found in most undergraduate texts on algebra and matrices; see, for example, Birkhoff and Maclane (1966). Hirsch and Smale (1974) contains a much more detailed account of normal forms for linear differential equations than Section 3.2. Arnol'd (1973) covers much the same material. The section on Floquet Theory (3.5) is largely taken from Iooss and Joseph (1980, Section VII.6), and more of the technical detail of these ideas can be found there.

## Chapter 4

Arnol'd (1983) contains most of the material presented in Chapter 4. It is not always an easy book to read, but it more than repays any effort put in to it. Chapter 3 deals with some aspects of structural stability, whilst Chapter 5 contains several different linearisation theorems. The proof of Poincaré's Linearisation Theorem in Section 4.1 is taken from there, and the 'proof' of the Stable Manifold Theorem (Section 4.6) is based on the same ideas. The book by Arrowsmith and Place (1990) gives a very readable account of structural stability. Guckenheimer and Holmes (1983) has (deservedly) become the standard graduate introduction to the field of nonlinear dynamics. All the ideas of hyperbolicity, structural stability and invariant manifolds can be found there, together with practical applications.

## Chapter 5

Almost any introduction to the qualitative theory of differential equations will treat the material of Chapter 5 somewhere; see, for example, Arrowsmith and Place (1982), Jordan and Smith (1987) or Grimshaw (1990). The classic books by Coddington and Levinson (1955) and Hartman (1964) give a lot more detail on the effect of nonlinear perturbations to the linear flow (Section 5.1). Hirsch and Smale (1974) has a slightly more algebraic approach and is well worth reading for comparison.

## Chapter 6

The books by Iooss and Joseph (1980) and Verhulst (1990) both cover Floquet Theory in more detail than given here. Guckenheimer and Holmes (1983) and Arrowsmith and Place (1990) give definitions of hyperbolicity for maps. Arnol'd (1983) is, as ever, excellent on linearisation whilst Marsden and McCraken (1976), see below, have a good section on return maps (section 2B). Devaney (1989, chapter 2.6) gives a proof of the Stable Manifold Theorem for maps.

## Chapter 7

Hinch (1991) provides a good introduction to the ideas of perturbation theory. Both Minorsky (1974) and Nayfeh and Mook (1979) give many more examples and different methods applied to differential equations. Arnol'd (1983) and Guckenheimer and Holmes (1983) provide some rigorous results about the use of perturbation methods, but these are not applicable to the method of multiple scales.

## Chapter 8

Andronov et al. (1973) and Minorsky (1974) give a good impression of the early uses of bifurcation theory (both these books were written much earlier than the given publication dates - Andronov died in 1954!). Ruelle (1989b) provides a more mathematical account of the theory. Wiggins (1991) also gives a more mathematical treatment of the results and sketches a similar proof of the Hopf Bifurcation Theorem. More details and applications of this result can be found in Marsden and McCracken (1976) and Hassard, Kazarinoff and Wan (1981). For a comprehensive account of the Centre Manifold Theorem see Carr (1981), although the treatment of this theorem in Section 8.1 follows Guckenheimer and Holmes (1983). A proof of the Implicit Function Theorem can be found in Burkill and Burkill (1970) and Loomis and Sternberg (1968).

## Chapter 9

Another proof of the period-doubling theorem using the Implicit Function Theorem can be found in Devaney (1989). The description of Arnol'd tongues in Section 9.4 owes a lot to Iooss (1979), which also contains many technical results about smoothness conditions. Arrowsmith and Place (1990) also give a very comprehensible account of these results, together with a description of simple circle maps (as described in the exercises at the end of the chapter). For more details about circle maps see Devaney (1989). Arnol'd (1983) contains a good discussion of the strong resonances.

## Chapter 10

More details about singularity theory and catastrophes can be found in Bröcker (1975), Thom (1983) and Zeeman (1977). The little book by Arnol'd (1984) is a riot, and contains a mixture of acute mathematical reasoning and robust historical scepticism. Many other topics could have found a place here: complex iteration (Devaney, 1989, Beardon, 1992), fractals (Mandelbrot, 1977), chaos and Newton's method (Peitgen, 1989), biology of nerves (Cronin, 1987), hyperbolic geometry (Bedford, Keane and Series, 1991), knot theory (Holmes in Bedford and Swift, 1988), ergodic theory (Sinai, 1977), ... . The list is potentially endless.

## Chapter 11

Collet and Eckmann (1980) is still an excellent introduction to unimodal maps. Nitecki (in Katok, 1982) describes the topological dynamics of maps in more detail than has been attempted here. Block and Coppel (1992) gives a more general treatment of one-dimensional dynamics and includes a very thorough discussion of topological entropy, including the proof of Theorem 11.7 (that a continuous map is chaotic if and only if it has positive topological entropy). The article by van Strien in Bedford and Swift (1988) describes the basic theory of unimodal maps and de Melo and van Strien (1993) is bound to become *the* graduate textbook on the subject. It also contains a discussion of recent work by Sullivan on universality. The treatment of intermittency follows Guckenheimer and Holmes (1983).

## Chapter 12

Wiggins (1988) goes into far more detail than has been possible here. Guckenheimer and Holmes (1983) covers most of the material and also treats global bifurcations in periodically forced systems (as does Wiggins, 1988). For a survey of more recent results see Glendinning (in Bedford and Swift, 1988) and Lyubimov et al. (1989). The derivation of the Lorenz equations is described in Lorenz (1963), and Sparrow (1982) builds up a picture of the behaviour of these equations over a range of parameter values.

# *Bibliography*

A.A. Andronov, E.A. Leontovich, I.I. Gordon and A.G. Maier (1973) *Qualitative theory of second-order dynamic systems*, John Wiley & Sons, New York.

A. Arnéodo, P. Coullet and C. Tresser (1981) A possible new mechanism for the onset of turbulence, *Physics Letters A*, **81** 197–201.

V.I. Arnol'd (1973) *Ordinary Differential Equations*, MIT Press, Cambridge, Mass.

V.I. Arnol'd (1983) *Geometrical Methods in the Theory of Ordinary Differnetial Equations*, Springer, New York.

V.I. Arnol'd (1984) *Catastrophe Theory*, Springer-Verlag, Berlin.

D.K. Arrowsmith and C.P. Place (1982) *Ordinary Differential Equations*, Chapman and Hall, London.

D.K. Arrowsmith and C.P. Place (1990) *An Introduction to Dynamical Systems*, Cambridge University Press.

A.F. Beardon (1991) *Iteration of Rational Functions*, Springer-Verlag, New York.

T. Bedford and J. Swift (eds.) (1988) *New Directions in Dynamical Systems*, Cambridge University Press.

T. Bedford, M. Keane and C. Series (eds.) (1991) *Ergodic theory, symbolic dynamics, and hyperbolic spaces*, Oxford University Press.

N.P. Bhatia and G.P. Szegö (1967) *Dynamical Systems: Stability Theory and Applications*, Springer-Verlag, Berlin.

G. Birkhoff and S. MacLane (1966) *A Survey of Modern Algebra*, Macmillan, New York (3rd edition).

L.S. Block and W.A. Coppel (1992) *Dynamics in One Dimension*, Springer-Verlag, Berlin.

L.S. Block, J. Guckenheimer, M. Misiurewicz and L.S. Young (1980) *Periodic Points and Topological Entropy of One-dimensional Maps*, in Nitecki and Robinson (1980).

T. Bröcker (1975) *Differentiable Germs and Catastrophes*, Cambridge University Press.

J.C. Burkill and H. Burkill (1970) *A Second Course in Mathematical Analysis*, Cambridge University Press.

J. Carr (1981) *Applications of Center Manifold Theory*, Springer, New York.

E.A. Coddington and N. Levinson (1955) *Theory of Ordinary Differential Equations*, McGraw-Hill, New York.

P. Collet and J.-P. Eckmann (1980) *Iterated maps on the interval as dynamical systems*, Birkhäuser, Boston.

P. Collet, J.-P. Eckmann and O.E. Lanford (1980) Universal properties of maps on an interval, *Comm. Math. Phys.* **76** 211–254.

J. Coste, J. Peyraud and P. Coullet (1979) Asymptotic behaviours in the dynamics of competing species, *SIAM J. Appl. Math.* **36** 516–543.

P. Coullet and C. Tresser (1978) Iterations d'endomorphismes et groupe de renormalisation, *J. de Physique* **C5** 25–28.

J. Cronin (1987) *Mathematical Aspects of Hodgkin-Huxley Neural Theory*, Cambridge University Press.

W. de Melo and S. van Strien (1993) *One-dimensional Dynamics*, Springer-Verlag, Berlin.

R.L. Devaney (1989) *An Introduction to Chaotic Dynamical Systems*, Addison-Wesley, Redwood City, Calif. (2nd edition).

M.J. Feigenbaum (1978) Quantitative universality for a class of nonlinear transformations, *J. Stat. Phys.* **19** 1–25.

P. Glendinning and C. Sparrow (1984) Local and global behaviour near homoclinic orbits, *J. Stat. Phys.* **35** 645–696.

R. Grimshaw (1990) *Nonlinear Ordinary Differential Equations*, Blackwell Scientific Publications, Oxford.

J. Guckenheimer and P. Holmes (1983) *Nonlinear Oscillations, Dynamical Systems, and Bifurcations of Vector Fields*, Springer, New York.

P. Hartman (1964) *Ordinary Differential Equations*, John Wiley & Sons, New York.

B.D. Hassard and Y.-H. Wan (1978) Bifurcation formulae derived from center manifold theory, *J. Math. Anal. Appl.* **63** 297–312.

B.D. Hassard, N.D. Kazarinoff and Y.-H. Wan (1981) *Theory and Applications of Hopf Bifurcation*, Cambridge University Press.

E.J. Hinch (1991) *Perturbation Methods*, Cambridge University Press.

M.W. Hirsch, C.C. Pugh and M. Shub (1977) *Invariant Manifolds*, Springer, New York.

M.W. Hirsch and S. Smale (1974) *Differential Equations, Dynamical Systems and Linear Algebra*, Academic Press, New York.

G. Iooss (1979) *Bifurcation of Maps and Applications*, North-Holland, Amsterdam.

G. Iooss and D.D. Joseph (1980) *Elementary Stability and Bifurcation Theory*, Springer, New York.

M.C. Irwin (1980) *Smooth Dynamical Systems*, Academic Press, New York.

D.W. Jordan and P. Smith (1987) *Nonlinear Ordinary Differential Equations*, Clarendon, Oxford (2nd Edition).

A. Katok (ed.) (1982) *Ergodic Theory and Dynamical Systems II*, Birkhäuser, Boston.

A. Kelley (1967) The stable, center stable, center, center unstable and unstable manifolds, *J. Diff. Equ.* **3**, 546–570.

T. Li and J.A. Yorke (1975) Period three implies chaos, *Amer. Math. Monthly* **82** 985–992.

L.H. Loomis and S. Sternberg (1968) *Advanced Calculus*, Addison-Wesley, Reading, Mass..

E.N. Lorenz (1963) Deterministic non-periodic flows, *J. Atmos. Sci.* **20** 130–141.

D.V. Lyubimov, A.S. Pikovsky and M.A. Zaks (1989) *Universal Scenarios of Transitions to Chaos via Homoclinic Bifurcations*, Harwood Academic Publishers, Chur.

B.B. Mandelbrot (1977) *Fractals*, W.H. Freeman, San Fransisco.

J.E. Marsden and M. McCracken (1976) *The Hopf Bifurcation and Its Applications*, Springer, New York.

J. Marsden and J. Scheurle (1987) The construction and smoothness of invariant manifolds by the deformation method, *SIAM J. Math. Anal.* **18** 1261–1274.

N. Minorsky (1974) *Nonlinear Oscillations*, Kreiger, New York.

A.H. Nayfeh and D.T. Mook (1979) *Nonlinear Oscillations*, John Wiley & Sons, New York.

Z. Nitecki and C. Robinson (eds.) (1980) *Global Theory of Dynamical Systems*, Springer, New York.

H.-O. Peitgen (ed.) (1989) *Newton's Method and Dynamical Systems*, Kluwer Academic Publishers, Dordrecht.

I. Percival and D. Richards (1982) *Introduction to Dynamics*, Cambridge University Press.

H. Poincaré (1899) *Les Methodes Nouvelles de la Mécanique Celeste*, Gauthier-Villars, Paris.

D. Ruelle (1989a) *Chaotic Evolution and Strange Attractors*, Cambridge University Press.

D. Ruelle (1989b) *Elements of Differentiable Dynamics and Bifurcation Theory*, Academic Press, Boston.

L.P. Shil'nikov (1970) A contribution to the problem of the structure of an extended neighbourhood of a rough equilibrium of saddle-focus type, *Math. USSR Sb.* **10** 91–102.

Ya.G. Sinai (1977) *Introduction to Ergodic Theory*, Princeton University Press.

S. Smale (1967) Differentiable dynamical systems, *Bull. Amer. Math. Soc.* **73** 747–817.

C.T. Sparrow (1982) *The Lorenz Equations: bifurcations, chaos and strange attractors*, Springer, New York.

I. Stewart (1989) *Does God Play Dice?*, Basil Blackwell, Oxford.

R. Thom (1983) *Mathematical Models of Morphogenesis*, Ellis Harwood, Chichester.

S. Wiggins (1988) *Global Bifurcation Theory and Chaos*, Springer, New York.

S. Wiggins (1991) *Introduction to Applied Nonlinear Dynamical Systems and Chaos*, Springer, New York.

E.C. Zeeman (1977) *Catastrophe Theory*, Addison-Wesley, Reading, Mass.

# *Index*